Designing Microwave Circuits by Exact Synthesis

For a complete listing of the *Artech House Microwave Library*,
turn to the back of this book.

Designing Microwave Circuits by Exact Synthesis

Brian J. Minnis

Artech House
Boston • London

Library of Congress Cataloging-in-Publication Data
Minnis, Brian J.
Designing microwave circuits by exact synthesis / Brian J. Minnis.
p. cm.
Includes bibliographical references and index.
ISBN 0-89006-741-4
1. Microwave circuits—Design and construction. 2. Electric network synthesis.
I. Title.
TK7876.M5916 1996
621.381'32—dc20 95-49985
CIP

British Library Cataloguing in Publication Data
Minnis, Brian J.
Designing microwave circuits by exact synthesis
1. Microwave circuits 2. Microwave circuits—Design
I. Title
621.3'8132

ISBN 0-89006-741-4

Cover design by Dutton & Sherman Design.

685 Canton Street
Norwood, MA 02062

International Standard Book Number: 0-89006-741-4
Library of Congress Catalog Card Number: 95-49985

10 9 8 7 6 5 4 3 2 1

To my lovely wife Patricia and all loved ones past, present, and future.

Contents

Preface

This book is the culmination of my 20 years of experience working within the U.K. microwave industry, where I designed microwave circuits and components for many different applications. I have been constantly fascinated by exact network synthesis techniques throughout this period, since their elegance and versatility appear to be almost inexhaustible. While I have been using these techniques for designing all kinds of microwave circuits and components, it has become abundantly clear that they are not widely used by the rest of the microwave community, largely because they are thought to be too complicated and of use only for designing filter networks. It has also become clear that there is no universal design procedure for microwave circuits and components in widespread use among the microwave community, except perhaps for one based on intelligent guesswork followed by substantial computer optimization.

It is in response to these two fundamental observations that I present in this book a universal design procedure for microwave components and integrated circuits. Based on exact network synthesis, the procedure can be applied to almost any type of microwave component, the only proviso being that the passive parts of the component be modeled as a generalized ladder of lumped inductors, capacitors, resistors, or transmission lines. It is intended especially for the design of hybrid microwave integrated circuits (MIC) and monolithic microwave integrated circuits (MMIC) fabricated on gallium arsenide (GaAs). The circuits may be either passive or active. They may employ lumped or distributed elements or a combination of the two and they may also use active devices operating either in linear or nonlinear modes.

The universal design procedure allows completely new circuits to be created from a basic set of circuit elements. With only a few exceptions, any combination of elements may be selected that is consistent with the frequency response and physical realization required. The complexities of the network synthesis are confined to computer software. It is a progressive design procedure, ensuring the creation of optimum networks with a minimum number of elements. The need for computer optimization is held to an absolute minimum.

All aspects of the design procedure will be described, including the rules and guidelines relating to the choice of network topology, the specification of frequency responses, the prototype network synthesis, and the conversion into physical dimensions. Network synthesis theory will be presented, but only those parts relevant to the universal design procedure and mainly for the purpose of providing background knowledge. The book will try to promote the versatility of the synthesis procedures and the simplicity of their application.

In addition to some theoretical examples, an extensive set of practical examples will be given. These comprise microwave filters, bias Ts, baluns, small-signal amplifiers, and power amplifiers fabricated with several different kinds of technology.

It may be worthwhile to emphasize that there is a great deal of material already published in the literature on the theory of exact network synthesis techniques. This book does not intend to add substantially to this theory, but it does intend to add substantially to the relatively little that has been published on the *application* of the synthesis techniques.

Without the support of my wife, Patricia, and my four children, writing this book would not have been possible. I am indebted to them for their patience and tolerance. My sincere gratitude extends to Philips Research Laboratories, who provided the facilities and funds necessary to carry out the technical work, and to my fellow colleagues who helped with the fabrication of the practical demonstrators. I would also like to thank Dr. R. J. Collier of the University of Kent at Canterbury for his support and encouragement over a research period of three years.

CHAPTER 1

Introduction

To the best of the author's knowledge, nothing approaching what might be called a universal design procedure for microwave circuits and components has yet been reported. Design procedures have tended to be developed for particular types of components. Consequently, when a new component has to be designed, most designers will try to obtain from the literature or draw from past experience a design procedure for an existing component with a performance as close as possible to the one required. Difficulties arise when the performance of the new component lies just outside the range of validity of the design equations for the existing component. In these circumstances, some computer optimization might be employed to stretch the performance toward the desired target specification. However, without special understanding of the relationship between the structure of the network and its electrical performance, it may not be clear what changes, if any, can bring about a satisfactory solution. In the worst-case scenario, where a new component has to be designed whose performance is quite unlike anything already in existence, fundamental questions arise, such as what sort of structure would be capable of meeting the specification, what types of circuit elements are required, in what order they should be assembled, and what their initial values should be. Without a universal approach, new components like these need special study, which is both time-consuming and expensive.

The result of several years of research by the author, this book presents a universal design procedure for microwave circuits and components that is based on the use of exact synthesis techniques [1]. Classic works on these techniques have included the early paper by Brune [2] published in 1931 and the book by Bode [3] published in 1945, and since then many others have contributed to the considerable volume of network synthesis theory now to be found in the literature. Unfortunately, the techniques have gathered a reputation among nonspecialists for being rather complicated and applicable only to passive components and filters.

It is true that the *mathematics* of network synthesis are relatively complicated. However, this should not really be of any concern, because the complexities of the mathematics can be handled by computer software, leaving the circuit designer to concentrate on only the application of the techniques. In fact, exact synthesis techniques are extremely simple to apply once certain basic rules and constraints have been properly understood. They can be applied to almost any microwave circuit or component, including those containing active devices. They offer the opportunity to build new circuits from scratch from a selection of basic circuit elements. There is no dependency on existing circuit structures and there are few restrictions on network topology. New circuit designs progress in a step-by-step, logical sequence from the specification of a frequency response at the beginning, through the derivation of a network prototype and the conversion into physical elements, to culminate in the layout of the final circuit at the end. Unlike any process based on numerical optimization, synthesis is *guaranteed* to find a valid network solution for a given target performance specification, whatever the nature of the specification.

In the absence of exact synthesis, the procedure most engineers are believed to follow for the design of microwave circuits is generally referred to as the *classical* or *approximate* design procedure. In fact, there are many and various approximate design procedures, and a comprehensive exposition of these relevant for most passive networks can be found in [4]. Nearly all of them result in a set of component-specific design equations. The details of the procedures vary from component to component, but certain basic principles (and topological restrictions) nearly always apply. In all cases, it is necessary for the prototype network of the relevant component to be a cascade of nearly identical sections and, therefore, be uniform along most of its length. This allows a relatively simple derivation of the distributed elements of each section from a lumped element prototype, the derivation being based on equating image impedances. The prototype is usually a low-pass network whose element values are obtained from published design tables. It is in the conversion of the lumped elements into distributed elements where most of the approximations have to be made, but small errors also arise from the need to terminate the whole network in real, resistive loads rather than complex image impedances. The approximations become worse as frequency bandwidth increases. Approximate procedures are therefore most appropriate for narrowband circuits and, indeed, have been widely successful in the design of narrowband filters. However, they are not well suited to wideband circuits, often necessitating the use of a substantial amount of computer optimization.

The universal design procedure based on exact synthesis imposes none of the topological restrictions of the approximate procedures and allows, at least in principle, any frequency response to be generated. The synthesis creates a two-port generalized ladder network containing *either* lumped or distributed elements. It is an exact process in the sense that, for a given set of frequency parameters, it will create a prototype network with element values that are exactly those required to reproduce a specified frequency response. No conversion of lumped elements into distributed elements or vice versa is involved. Only at the very end of the procedure, when physical elements are involved, may there be a need to introduce any approximations.

In an extension of the exact techniques, noncommensurate line networks and networks containing a mixture of lumped and distributed elements can be accommodated. This is accomplished by using band-pass prototype networks that correspond to distributed networks whose elements are much shorter than a quarter of a wavelength in the vicinity of the passband. With one or two exceptions, element length can then be traded against element impedance, and in the extreme case of physical length tending toward zero, some lumped elements can be used in place of the distributed elements.

Almost any microwave circuit, including one containing active devices, can be designed by the universal design procedure. Active circuits merely require partitioning into separate active and passive parts before the technique is applied. The technique offers one of the most direct routes to an optimum circuit design that requires little or no numerical optimization. It is also worthwhile noting that while the synthesis techniques are concerned specifically with linear networks, the universal design procedure as a whole can be applied to nonlinear active circuits as well. Nonlinear circuits simply require a wider control of matching network behavior at harmonic as well as fundamental frequencies.

With regard to the software tools available for microwave circuit design, most commercial computer-aided design (CAD) packages in use today still concentrate on analysis and optimization. Some companies such as HP-EEsof and Compact Software do offer a synthesis tool as part of their CAD software products, but this tends to interest designers much less than the analysis/optimization tools. This is a pity because despite the undoubted versatility of an analysis tool such as LIBRA [5], it can only verify or adjust the performance of a circuit. It will not tell the designer how to choose the structure of a new network, how to set the initial element values, or how to change the structure of an existing network to achieve some desired performance. This remains very much the task of the circuit designer. However, it is a task that could be made very much easier by adopting the use of the exact synthesis techniques.

The technical content of the book begins with Chapter 2, which first reviews the *classical*, nonsynthesis approach to microwave circuit design. It then gives an initial outline of the proposed universal design procedure, dealing with each of the strategic steps of the procedure only briefly so that the overall concept can be grasped as soon as possible. At the heart of the procedure is the network synthesis process itself, and to help with the understanding and execution of the process, all the relevant theory is presented. This includes definition of important basic terms, the many options that exist for frequency responses, the options and restrictions that apply to network topology, and the way in which input data for the synthesis must be specified to create a particular type of network with a particular type of frequency response. Essential techniques for the manipulation and transformation of networks are also given.

A major section in the latter half of Chapter 2 deals with the conversion of network prototypes into real, physical circuits. It describes the physical form of most of the lumped and distributed elements relevant to microwave integrated circuit design. For distributed elements, the section includes lines, stubs, and coupled-line sections realized in some of the most common planar transmission line media. Formulas are given that relate prototype element values to the basic parameters of the transmission lines, but reference is made to the literature for the calculation of physical dimensions. The penultimate section of Chapter 2 contains a review of the whole design procedure, reinforcing the basic concepts and emphasizing some of the more practical aspects. Working advice is given on how electrical performance and physical constraints affect the choice of circuit structure, prototype structure, and synthesis parameters. Chapter 2 concludes with a discussion on the availability of synthesis software, including a design example based on the use of one of the leading commercial packages, which is called E-Syn [6].

Chapters 3, 4, and 5 demonstrate the application of the generalized synthesis procedure to a wide range of practical microwave circuits and components. Chapter 3 describes the design of a series of wideband printed circuit filters. The first of these is a relatively simple coupled-line filter that is, for the purposes of comparison, designed by both approximate and the exact synthesis procedures. The second type is an enhanced coupled-line filter that has significantly more selectivity and that can only be designed by synthesis. Finally, the chapter describes four classes of printed circuit filters that incorporate a mixture of lumped and distributed elements. These have passbands and stopbands whose widths can be specified independently and whose physical size is substantially smaller than most other filters of comparable performance.

In Chapter 4, the application of the generalized procedure to wideband bias Ts, baluns, and impedance transformers is described. These are specialized passive components that would not normally be classified as filters, but are readily designed by synthesis. The bias Ts have more than a decade bandwidth and were constructed in both microstrip and stripline transmission line media on plastic substrates. The balun circuits have bandwidths of greater than an octave and were constructed in microstrip as monolithic microwave integrated circuits (MMIC) on gallium arsenide (GaAs). In the last section of the chapter, various types of impedance-transforming circuits are described. A set of design formulas are presented for each type, which must be used in conjunction with the synthesis procedure. They are suitable for realization in most of the popular printed circuit transmission line media.

Chapter 5 describes the design of some important active circuits by exact synthesis. As an introduction, the chapter covers some elementary theory governing the limits imposed on frequency bandwidth when matching into complex impedances. This is followed by a section describing the design of distributed amplifiers, which gives details of the design of a 6- to 18-GHz MMIC amplifier with a variable gain slope and the details of an 18- to 30-GHz hybrid design that uses a novel technique to extend the maximum operating frequency. The next section describes a reactively matched power amplifier MMIC that achieves 2.5W of output power over the 7.5- to 10.5-GHz frequency range. When first reported, the amplifier set a new state of the art for a single-chip wideband power amplifier in Europe. Its design was only made possible by the use of exact synthesis techniques.

Section 4 of Chapter 5 introduces a new form of cascadable low-noise MMIC amplifier, which resembles a single field-effect transistor (FET) distributed amplifier on its input side, but a reactively matched amplifier on its output side. It combines the best characteristics of both types of amplifiers, having the good terminal match properties and flat gain of a distributed amplifier and the small size and high efficiency of a reactively matched amplifier. Its good input match is a consequence of developing what has been called the *traveling-wave matching technique*, which is a powerful concept, allowing, perhaps for the first time, very-high-gain multistage amplifiers to be realized on a single chip of GaAs. A working, practical example of the single-stage amplifier will be given. The matching technique can also be used to good effect in the interstage matching network of a two-stage amplifier. This is demonstrated by the design of a 0.75W MMIC amplifier with a 12-dB gain over the 7- to 14-GHz band. As a final demonstration of the capability of the technique, the last section of the chapter describes a five-stage amplifier fabricated as a single MMIC with 1.25W output power and a gain of more than 25 dB. Exact synthesis played a vital role in the design of these circuits.

Chapter 6 brings the technical work to a close with a brief summary and some concluding remarks.

References

[1] Guillemin, E. A., *Synthesis of Passive Networks*, New York: Wiley & Sons, 1957.

[2] Brune, O., "Synthesis of a Two Terminal Network Whose Driving-Point Impedance Is a Prescribed Function of Frequency," *J. Mathematics and Physics*, Vol. 10, No. 3, October 1931, pp. 191–236.

[3] Bode, H. W., *Network Analysis and Feedback Amplifier Design*, New York: D. Van Nostrand, 1945.

[4] Matthaei, G. L., L. Young, and E. M. T. Jones, *Microwave Filters, Impedance-Matching Networks and Coupling Structures*, New York: McGraw-Hill, 1964.

[5] HP-EEsof., *LIBRA—Microwave and RF Circuit Design Software*, Bracknell, Berkshire, UK: Hewlett-Packard Ltd., February 1994.

[6] HP-EEsof., *E-Syn (Series IV) User's Guide—Microwave and RF Design Synthesis*, Bracknell, Berkshire, UK: Hewlett-Packard Ltd., February 1994.

CHAPTER 2

Microwave Circuit Design by Synthesis: A Universal Procedure

2.1 OUTLINE OF THE CLASSICAL, NONSYNTHESIS DESIGN APPROACH

A proposal of a universal design procedure for microwave circuits and components based on exact synthesis will be made in the following section and described in full detail throughout the remainder of the chapter. However, before making the proposal, it may be helpful to first examine the sort of procedure that most microwave circuit designers are believed to follow who do not use exact synthesis techniques. The examination is necessarily a general analysis covering most cases where synthesis is not used. It is based on the personal observations of the author over the course of many years working within the microwave community, and inevitably, therefore, there will be cases for which it is an oversimplification.

A flow diagram of the perceived nonsynthesis approach to microwave circuit design is presented in Figure 2.1. As indicated, the procedure begins with the choosing of a physical structure that is most likely to satisfy a given combination of electrical specifications and physical constraints. It is a choice made from what is effectively a library of circuit structures available to the designer, comprising articles in the scientific literature (i.e., journals, symposium digests, and books) as well as the designer's own working experience. In the happy event that a structure can be found with the potential for satisfying the electrical and physical constraints, the procedure continues with the derivation of an equivalent circuit. This will often be an ideal or pseudoideal representation of the chosen physical structure and will not include parasitic circuit elements. Parasitic elements are included at a later stage.

In many cases, the derivation of the equivalent circuit will be through a set of design equations based on a mixture of experimental and analytical data.

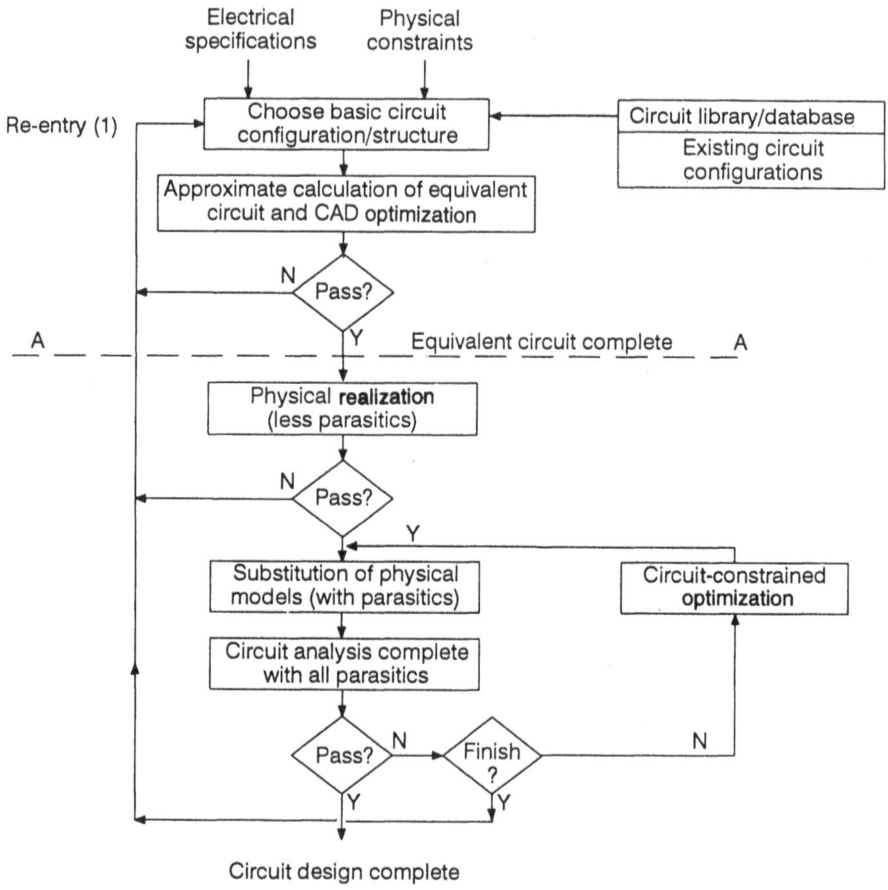

Figure 2.1 The classical, nonsynthesis circuit design procedure.

Consequently, this is an *approximate* (as opposed to an *exact*) process, producing a circuit whose responses are close but not identical to those originally specified. In other cases, the equivalent circuit might be derived with the aid of an alternative tool such as a Smith Chart. This has the advantage of reducing the reliance on a set of equations that are valid for only a single physical structure. However, the usefulness of the Smith Chart is very much confined to narrowband applications where frequency bandwidths are less than 20% and where networks contain fewer than three or four elements.

After deriving the ideal equivalent circuit, its performance would be verified using a circuit simulation tool such as the commercially available package called LIBRA (HP-EEsof software). Because of the approximate nature of the design equations, the circuit may also require some optimization to correct for any significant deviations from the required performance. This is the first of two optimization stages embodied in the overall design procedure. As illustrated in Figure 2.1, if after the optimization the performance is acceptable, the procedure continues to the next stage. If, however, the performance is not successful, then the only option is to return to the start of the procedure and continue the search for another, more suitable physical structure.

Once a satisfactory equivalent circuit has been derived, the procedure moves into its practical phase, during which the physical dimensions of the circuit are calculated. As will be seen, this part of the procedure is common to both the nonsynthesis and synthesis design approaches and is represented by the part of the flow diagram below the dotted line (AA) in Figure 2.1. As a first step, the dimensions of the physical elements are calculated without taking account of the various parasitics in the circuit. Most of the relevant physical models for this are to be found in the literature, but are also available in convenient software packages such as LINECALC, which is supplied by HP-EEsof as part of the LIBRA design suite. LINECALC deals with a comprehensive set of both lumped and distributed circuit elements. When the element dimensions have been calculated, a check is made of their physical realizability. A pass allows progress to the next stage of the procedure, but if one or more of the circuit elements are not realizable, the procedure has to be halted and the search for an alternative circuit started once again.

If the structure passes the dimensional check, the element dimensions are assigned to the appropriate physical models in the circuit simulator so that another analysis of circuit performance can be made, which this time includes the effects of the element parasitics. At this point other, more general circuit parasitics such as those associated with transmission line junctions and end effects might also be added. After analysis, if the electrical performance meets the specification, then the circuit design could be considered complete and the procedure finished. If it does not meet specification, however, a second optimization of the circuit would usually be performed, altering the element dimensions within some sensible range, in an attempt to recover a satisfactory performance. This would continue until either a good result is obtained or until it is clear, after several attempts, that a good result is impossible to obtain. Failure forces another return to the beginning of the design process.

There are at least two obvious weaknesses in this nonsynthesis design approach. The first concerns the choice of physical structure that has to be made. It is critical to the success of the procedure that at the outset a physical structure is identified that is capable of meeting the given performance specifications. If no such structure can be identified, then the design cannot proceed at all without some special study of the problem. The second weakness concerns the lack of knowledge as to how the structure should be modified to bring about the required result. There is nothing to stop changes being made, should these be necessary, but the procedure gives no indications as to what changes, if any, can achieve the desired effect. Furthermore, substantial changes to the structure of the circuit will invariably invalidate the design equations, and if valid design equations are deemed to be an important requirement, more effort will have to be expended on deriving a new set. As far as making changes to the element values is concerned, this may seem relatively straightforward compared to changing the element types, since the changes can be made by computer optimization. However, optimization is a process particularly prone to failure because of its dependence on previous knowledge of the performance capability of the circuit. It will fail if performance targets are set that exceed the capability of the circuit, that are too far from the existing circuit performance, or that are in a region where there are several local minima in the error function. As a general rule, optimization only works well if the answer is already known!

In summary, the nonsynthesis design approach can and does produce satisfactory circuit designs. However, it is a rather hit-and-miss procedure, relying heavily on prior art and is comparatively ineffective for designing entirely new circuits. It is most applicable to narrowband microwave circuits.

2.2 OUTLINE OF THE UNIVERSAL DESIGN PROCEDURE BASED ON EXACT SYNTHESIS

An outline of the proposed universal design procedure will now be given that will highlight the major role played by exact synthesis and the main advantages over the previous, nonsynthesis approach. This section will give only a brief description of each of the essential stages of the procedure so that an overall picture can be acquired by the reader as soon as possible. Subsequent sections will go into the details of each individual step, and Section 2.6 will review the whole procedure in sufficient practical detail to enable real new circuits to be designed. The initial description will concentrate on the design of two-port passive networks operating between resistive terminating loads. The variations required to deal with active circuits will be covered briefly in Section 2.6.1 and then in much more practical detail in Chapter 5.

A flow diagram of the complete design procedure is given in Figure 2.2. As in the nonsynthesis procedure, the input parameters are a set of frequency specifications and physical constraints; the output is a working circuit that is consistent with all of the input parameters. In comparing the flow diagrams of Figures 2.1 and 2.2, the most striking difference is the much greater length of the diagram in Figure 2.2. Below the AA line, both procedures are the same. However, above the AA line, the universal design procedure employs a much greater number of processing steps, giving a considerable increase in design flexibility. Instead of only one reentry point into the procedure at the top of the diagram, there are now at least three. Should any of the various checks on electrical performance or physical realizability now fail, there are at least two more reentry points into the upper stages of the procedure that can be tried before there is a need to return to the

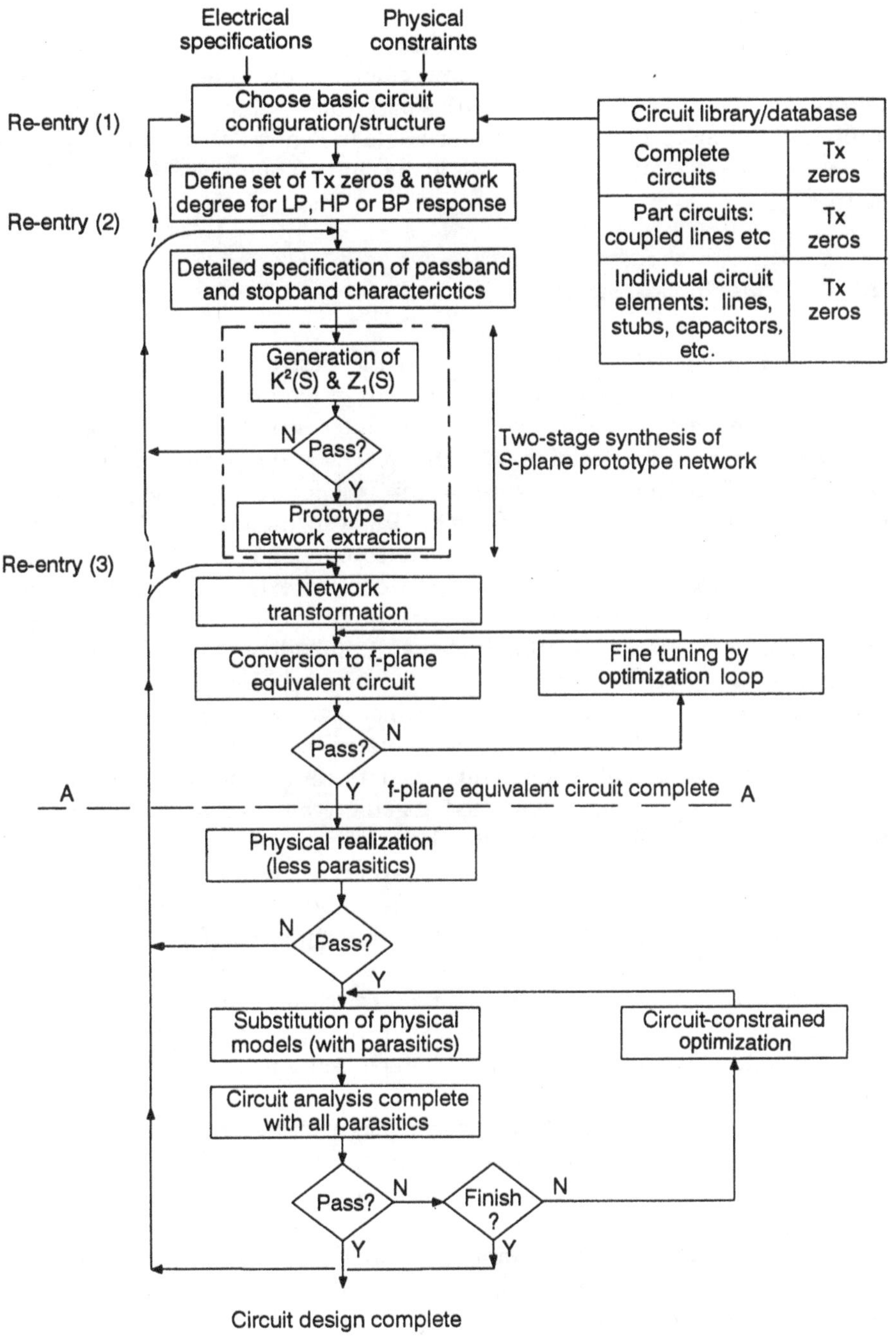

Figure 2.2 The complete procedure for microwave circuit design by exact synthesis.

beginning. Each reentry point gives considerable scope for modification of the circuit topology or element values without changing the basic composition of the network (i.e., the types of elements used). Apart from the inclusion of the prototype synthesis stage enclosed by the dotted rectangle in Figure 2.2, the other significant difference between Figures 2.1 and 2.2 is in the content of the circuit design library/database.

2.2.1 Choosing Physical Structure

As indicated in Figure 2.2, the first task in the design flow is choosing a basic structure for the network/circuit. The information required for this is drawn from a library or database of circuit designs, the contents of which will represent the previous working experience of the designer as well as the published circuit design material to be found in the literature. In this respect the database is the same as that of the nonsynthesis design approach. However, the introduction of network synthesis now allows circuits to be assembled from much more elementary components, and two new categories of these are held in the extended database of Figure 2.2. In addition to complete circuit configurations, the database now holds a collection of the most elementary of circuit elements, including items such as interconnecting lengths of transmission line, open-circuit stubs, short-circuit stubs, lumped capacitors, and lumped inductors, all of which can be used in either series or parallel configurations. In between this elementary level and the top level containing the complete circuits is a collection of subcircuits or compound elements comprising several elementary elements connected together, but not large enough to be regarded as complete circuits. Typically, elements in this second level will comprise capacitively coupled transmission lines of various degrees of complexity. In the event that no complete structure exists in the library that meets the necessary specifications, it is now possible to consider building up a completely new structure with a new set of low-level building elements, with the knowledge that so long as the chosen set of elements conforms to certain basic rules, the subsequent synthesis process is guaranteed to create the complete prototype structure and calculate all its element values. Whatever the combination chosen, no special procedure will be required.

Clearly, past experience still has a role to play in choosing the best combination of circuit elements for a given performance specification. However, as will be explained later in the chapter, there is a set of general guidelines that can be applied to any new circuit design problem and that offer a way forward even when the designer has little or no relevant past experience. In marked contrast to a design approach based on circuit optimization, there is never any uncertainty as to the performance capability of a network, whatever the choice of topology. The synthesis will always compute the exact frequency responses of the network as determined by the particular set of data supplied to the synthesis software.

2.2.2 Identifying the Set of Transmission Zeros

For each element, group of elements, or complete network held in the design library, there must be a specified set of transmission zeros. As will be explained later, transmission zeros are frequencies at which the transmission of energy through a network falls to zero; these not only influence the frequency response, but they also determine the basic set of elements from which a network will be constructed. The transmission zeros form a major part of the input information required to perform the network synthesis. If a suitable complete network exists in the design library, then the choice of the combination of transmission zeros will have been effectively made, leaving only the multiplicity of the combination (i.e., total number of network elements) to the designer's discretion. However, if a suitable network does not exist in the library, then the designer must either modify a set corresponding to another network or choose a completely new set for a completely new network. To do this, the designer must understand the constraints imposed on choosing transmission zero locations and the physical consequences of this choice on the eventual practical circuit. Choosing the set of transmission zeros (see Figure 2.2) is perhaps the most influential step in the whole of the synthesis design process.

In brief, each basic circuit element will have one or more transmission zeros specified at $S = (0, j0)$, $S = (0, j\infty)$, $S = (1, j0)$, or $S = (0, j\omega)$, where S is a complex frequency variable. These correspond respectively to high-pass (HP) elements such as series capacitors or shunt inductors, low-pass (LP) elements such as shunt capacitors or series inductors, unit elements which are the equivalent of transmission lines in a real circuit, and elements such as shunt-tuned or series-tuned circuits. The choice of transmission zeros will determine whether the frequency response of the network is to be band-pass (BP), low-pass, or high-pass. It will also determine such factors as whether the network can produce an impedance level transformation or whether it will be possible to use a mixture of lumped and distributed elements in the subsequent real circuit. It will fix the degree of the network (total number of elementary, nonredundant elements) for which, as will be explained later, at least one important constraint must be observed.

2.2.3 Passband and Stopband Specifications

In the next step of the design process, the designer gives certain numerical information to determine the detailed characteristics of the passband and stopbands. Whether the network is low-pass, high-pass, or band-pass will already have been determined by the selection of transmission zeros, but the width of the passband and stopbands or the levels of attenuation in each will not yet have been specified. Stopband behavior is most strongly affected by the transmission zeros. In selecting the set of transmission zeros, the location of most of them will already be fixed. However, the precise frequency of any that were not located at $S = (0, j0)$,

$S = (0, j\infty)$, or $S = (1, j0)$ must now be given. These zeros at $S = (0, j\omega)$ can force deep nulls at any chosen frequency in the stopbands. They can also be used, for example, to give extremely high selectivity near the passband edges or to hold attenuation below a given threshold throughout one or all of the stopbands.

Specification of the passband response is usually made by choosing passband edge frequencies, a loss value at the passband edges, and a preference for a maximally flat (Butterworth), an equal-ripple (Tchebycheff), or some other type of passband characteristic. Occasionally, it may also be necessary to specify a constant offset loss value and/or a slope parameter if a sloping passband response is required. However, if a highly specialized response is required that does not fall into one of the classical sets of passband responses, then there may be a need instead to specify the location of the entire set of passband reflection zeros. These zeros correspond to frequencies at which transmission loss through the network falls to zero, but unlike the transmission zeros, they are not linked to the types of element from which the network will be composed. They will only affect the element values. The zeros will be specified at complex frequencies $S = \sigma + j\omega$, for which there are few restrictions on the values of the real or imaginary parts. Almost any shape of passband response can be created in this way. Furthermore, it is also possible to exercise some control over the phase characteristics of the network by adjusting the position of the reflection zeros. This is important for designing certain types of group delay equalization networks. The number of reflection zeros that need to be specified depends on the network degree, which is in turn dependent on the number and location of the transmission zeros.

Figure 2.2 shows that there is a reentry point in the flow diagram above the stage at which passband and stopband characteristics are specified in detail. It gives the designer the opportunity to adjust the passband and stopband specifications without changing the basic combination of transmission zeros. The effect on the prototype network is to change frequency responses and the network element values without changing its topology. This might be necessary should the frequency responses of the network be found unsatisfactory after one of the later stages in the design process. It would not be uncommon to iterate the loop between the reentry point (2) and the intermediate stage of the network synthesis until a desired frequency response is achieved. The most useful parameters to adjust during the iteration are usually the passband width and ripple values. Not only do these alter the frequency responses, but they can also change the values of any of the elements in the network that are proving to be difficult to realize. Both passband ripple and passband width are extremely useful degrees of freedom, which, subject to the constraints imposed by the performance specification of the network, should remain available for change as long as possible.

2.2.4 First-Stage Network Synthesis

With all the necessary design parameters now specified, the design process moves into the main synthesis phase in which a prototype network is created. This is done in two stages, both of which are numerically intensive but easily accomplished using a suitable computer program. Reference to both in-house (Philips) software and commercially available synthesis software will be made in Section 2.7.

In the first stage of the synthesis, the program uses the transmission zero locations and passband specifications to compute a pair of polynomials, the ratio of which is called the *characteristic function,* $K(S)$. These are polynomials in the complex frequency variable S whose coefficients are all real. From the characteristic function it is possible to calculate the frequency responses of the input reflection coefficient and the transmission coefficient of the prototype network. These correspond to the responses of the well-known scattering parameters S_{11} and S_{21} for those cases where the network is terminated at both ends by 50Ω resistors. The calculation can be made even though the prototype network has not yet been extracted. Indeed, the resultant frequency responses are *entirely* independent of how the prototype network is eventually extracted. Whatever the order in which the elements appear in the network, the frequency responses will always be the same. This is an extremely useful facility, since the frequency responses can now be adjusted without reference to a network and with an absolute minimum number of control parameters. All the redundant degrees of freedom that might be associated with a network have been eliminated. As shown in Figure 2.2, a pass/fail check can be made on the responses, and, if necessary, the first stage of the synthesis can be repeated with a modified set of frequency parameters until a pass is obtained.

The last step before moving onto the second stage of the synthesis is to convert the characteristic function $K(S)$ into another ratio of polynomials $Z_1(S)$, which represents the input impedance (or admittance) of the network. Optionally, it will be possible to force $Z_1(S)$ to correspond to either a doubly terminated network (i.e., one with resistive terminations at both ends) or, alternatively, a singly terminated network (i.e., one with a resistive termination at only one end). In a singly terminated network, the source is either an ideal voltage source or an ideal current source. Whether a doubly or singly terminated network is chosen, the output load resistance will not necessarily be known until the extraction of the network is complete. Its value may be quite different from that of any source resistance and depends both on the frequency response specifications of the network and on how the network elements are extracted.

The choice between doubly and singly terminated networks will depend strongly on the practical application. There will be marked differences between the two sets of $Z_1(S)$ polynomials, as well as the two sets of network element values.

Another decision that must be made during the generation of $Z_1(S)$ is in connection with the first and second canonical forms of the prototype network. For any prototype network there is always a second network, a so-called network *dual,* with an identical transfer characteristic. The two networks together are sometimes referred to as *first* and *second duals,* respectively. In the case of the first dual, the function $Z_1(S)$ represents the input impedance of the network, whose magnitude tends toward infinity (i.e., an open circuit) in the stopband. In the case of the second dual, the function $Z_1(S)$ represents the input admittance of the network. As an admittance, this also tends to infinity in the stopband, but as an impedance now tends toward a short circuit. The structures of the two dual networks are quite different, but as will be made clear in Section 2.3, it is not difficult to determine the structure of one dual from that of the other. While the specification of a particular dual must be passed to the synthesis software at this stage, the decision over which to use will have already been made at the start of the whole design process, since it is fundamental to the topology of the practical circuit.

2.2.5 Second-Stage Network Synthesis

Once the generation of the $Z_1(S)$ polynomials is complete and the calculated frequency responses meet the required specifications, the synthesis can move into its second stage. In this stage the prototype network is created as a ladder of ideal circuit elements. As in the case of the first stage, the numerical computations of the process are conducted with software. All the circuit designer has to do is indicate the order in which the prototype elements are to be extracted, done by making a successive selection from the list of transmission zeros. The zeros in the list are the same as those used during the first part of the synthesis, but the sequence of the zeros now determines the precise topology of the network. There are few theoretical restrictions on the sequence of the zeros, but there should be little need for experimentation, since the sequence was effectively determined by the selection of the physical circuit made at the beginning of the design procedure. As each element is extracted, the software updates the $Z_1(S)$ polynomials, progressively reducing their degree until the residue is a constant term equal to the value of the load resistor. For most types of networks, the load resistor value depends heavily on the order of the element extraction and, as mentioned earlier, will not be known until after extraction is complete.

In many practical cases, the prototype network that corresponds most closely to the selected physical circuit will contain what are known as redundant elements. Broadly speaking, this means that for each transmission zero, there may be more than one network element involved in its creation. Adding the extra elements has no effect on the electrical behavior of the network, but can have a highly beneficial effect on its physical realizability. As a redundant network, the prototype will contain more elements than is absolutely necessary for the specified frequency responses. Most synthesis software will allow the network to be created in this form; the user simply has to give a little more information to the software regarding the values of the redundant elements. A better alternative, however, is usually to extract the nonredundant form of the network and then introduce the redundant elements after the synthesis by using one of the appropriate network transformations to be described. This saves on computational time and improves numerical accuracy.

2.2.6 Network Transformation

After a prototype network has been created by exact synthesis, it is possible to make changes to its topology and add redundant elements as required without making any changes to its electrical performance. Such network transformations can be applied to all or just part of the network and are, like the synthesis, exact mathematical processes, the details of which will be given in Section 2.4. They are an extremely effective way of changing not only the topology, but also the element values of a network without having to repeat the synthesis. This may be found necessary if, for example, difficulties are subsequently encountered with physical realizability or with the value of the load resistor. Consequently, as shown in Figure 2.2, just above this network transformation stage there is a third reentry point in the design flow which further enhances the flexibility and versatility of the design procedure.

During a typical network transformation, redundant elements may be created by splitting certain reactive elements such as capacitors or inductors into two or more parts, which can then be moved to other locations in the network. For unit elements, redundant versions of these may be created by inserting extra unit elements into the network at the two ends, the characteristic impedances of which must be equal to the adjacent load element values. The redundant unit elements can then be moved further inside the network as required. The actual movement of the elements throughout the network can be calculated by hand, but the process is much simplified with the aid of suitable software. Often the software tools required for the network transformation will be accessories to the main synthesis software package.

2.2.7 Conversion to an *f*-Plane Equivalent Circuit

The final stage in the design process prior to conversion to physical dimensions is concerned with converting the *S*-plane prototype into an *f*-plane equivalent circuit. An *f*-plane equivalent circuit is the form in which the network can be analyzed using a conventional CAD analysis tool such as LIBRA. Except for the absence of parasitics, its response represents the electrical behavior of the real, physical circuit under consideration. For lumped-element *f*-plane circuits, the capacitors and inductors of the prototype remain capacitors and inductors, their element values being determined by a simple linear frequency transformation. For distributed-element *f*-plane circuits, however, the elements of the prototype become transmission lines and stubs whose lengths are all a quarter of a wavelength at a single frequency known as the *commensurate frequency*. Commensurate frequency is an auxiliary synthesis parameter which will already have been specified during the first part of the synthesis procedure. The transformation causes the unit elements of the *S*-plane prototype to become interconnecting transmission lines, the capacitors to become open-circuit stubs, and the inductors to become short-circuit stubs. The element values of the prototype are transformed directly into the characteristic impedances or admittances of the lines and stubs.

For *f*-plane equivalent circuits that contain a mixture of lumped and distributed elements, the conversion from the *S*-plane is a little more complicated. A detailed description will be given later in Section 2.4. In this special case, the network conversion is not quite exact, and while the errors involved are usually small, there is sometimes a need to perform some superficial optimization of the *f*-plane circuit to restore the precise frequency response required. The optimization loop around the conversion stage is clearly shown in Figure 2.2. It needs to be stressed that the optimization is nearly always only a matter of fine tuning. Every design step preceding this conversion into a mixed lumped/distributed circuit has been exact, involving no physical approximations whatsoever. Since the *S*-plane prototype has the exact required frequency response and required topology, the physical approximations that have to be made in this late stage cause only a minor distortion of the frequency responses. They should not be seen as a corruption of the exact procedure, but more a valuable extension that allows an extremely important class of networks to be created. At this time, there is no known truly exact synthesis procedure for generalized ladder networks containing a mixture of lumped and distributed networks.

2.2.8 Physical Realization and Final Optimization

As previously stated, the last phase of the design procedure below the line AA in Figure 2.2 is common to both the synthesis and nonsynthesis design approaches. It involves the conversion of the *f*-plane equivalent circuit into the physical dimensions of the real microwave circuit, followed by the simulation and optimization of the circuit complete with all its parasitics. This can be a rather complicated process (more details will be given in Section 2.5).

2.3 THE PROTOTYPE NETWORK SYNTHESIS PROCEDURES

Having introduced the complete exact synthesis approach to microwave circuit design and identified all the basic stages in the process, we will now explore some of the theory on which the procedures are based. This section deals specifically with the theory behind the synthesis of the *S*-plane prototype, which is the process contained within the dotted outline of Figure 2.2. It is the fundamental core of the whole design process, the other stages serving either to feed the synthesis with the correct combination of data or to convert the output of the synthesis into something that can be physically realized.

The actual mathematical computation for the synthesis will be done by computer software and will therefore be transparent to the circuit designer. However, a general understanding of the mathematical procedures is important because it will help with the manual control of the synthesis software, ensuring that a prototype of the most appropriate type is created. It will explain how the input data for the synthesis relates to the topology of the prototype network and to its electrical performance. It will also explain why certain constraints must be applied to the selection of the input data.

2.3.1 Some General Terms

To begin the theoretical treatment, some of the basic terms and terminology must be defined. The first of these is the term *exact synthesis.* As far as this text is concerned, exact synthesis means the creation by a direct, mathematical procedure of a two-port passive network, which when connected between an alternating current (ac) power source and a resistive load has transfer characteristics that are exactly those defined in the synthesis specification. This is illustrated more clearly in Figure 2.3. Given a particular physical combination of circuit elements from which the two-port is to be composed, an exact synthesis will generate the values of all the circuit elements so as to produce *exactly* the required frequency responses without any recourse to numerical optimization. As will be described later, there is considerable freedom in choosing the internal composition of the two-port, but as far as this definition is concerned, the terminating impedances must be pure resistances. For the load resistor, this must always be finite/nonzero, since the objective of the whole design effort must be to achieve some power transfer between the power source and the load. However, the same limitation does not apply to the power source, which can have any value of internal resistance including zero and infinity. This gives rise to the definition of two additional terms: a network for which the two terminating resistors are both finite is classified as *doubly terminated,* but a network for which the source resistance is either zero or infinite is classified as *singly terminated.*

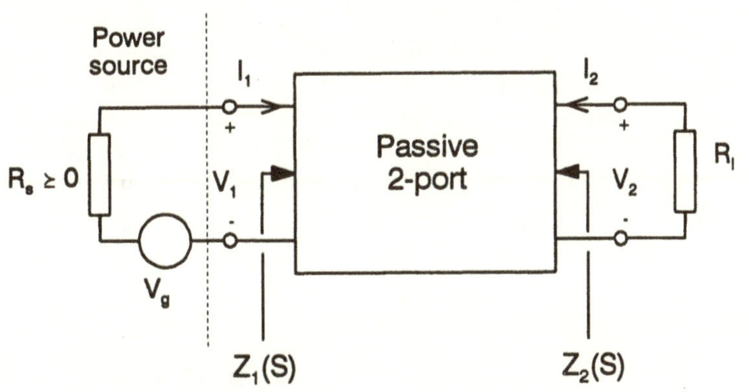

Figure 2.3 The basic concept of passive two-port network synthesis.

It should be noted that the restriction of the exact synthesis procedure to passive networks connected between resistive terminations is only a matter of definition and does not exclude using the techniques to design active circuits or matching networks with complex loads. Active circuits, at least for microwave applications, invariably comprise a small number of active devices such as metal semiconductor FETs (MESFETs) interconnected by a pattern of metal tracks and capacitors. Each of the interconnecting patterns between the active devices can be treated as a passive two-port network, and the reactive elements in the equivalent circuits of the active devices can be treated as part of these passive two-ports. The resistive elements in the equivalent circuits of the active devices become the loads for the two-ports. Hence, complete active microwave circuits can be designed by synthesis so long as the circuit can be partitioned into one or more passive two-ports.

The same is true for the problem of deriving a matching network between a resistive source and a complex load impedance. A good example of this is the case of the transmission line feed, which must be matched to the complex impedance of a radiating antenna. A matching network can be synthesized if the reactive elements of the antenna's equivalent circuit are treated as part of the matching network and its resistive elements as the load. This does depend on the ability to model the antenna as a simple connection of reactive elements plus a single terminating resistor, but this is not usually a problem in practice.

Two of the most fundamental frequency-dependent variables in the synthesis of any two-port network are the input and output impedances. As indicated in Figure 2.3, these impedances, Z_1 and Z_2, are seen looking into the respective input and output ports when the relevant load resistor, R_s or R_l, is in position at the opposite end of the network. That is, with an external voltage source applied to port 1 and all other energy sources set to zero, the impedance Z_1 will be equal to the ratio of the voltage and current at port 1 with R_l in position. The same applies for Z_2 when the ports are reversed. In general, therefore, both Z_1 and Z_2 are complex functions of frequency.

The frequency-dependent parameters of any linear network can be expressed in terms of a complex frequency variable, which reduces the algebraic manipulation of such parameters to one of handling simple polynomials with real coefficients. For pure lumped-element circuits, the variable is the Laplace frequency variable s, where $s = \sigma + j\omega$ and ω is the angular frequency $2\pi f$. Hence, the impedance functions Z_1 and Z_2 are expressed in terms of the ratio of two polynomials in the Laplace variable (s) in the following way:

$$Z_1(s) = \frac{V_1(s)}{I_1(s)} = \frac{P(s)}{Q(s)} \tag{2.1}$$

where $P(s)$ and $Q(s)$ are polynomials in s with real coefficients.

When the network is composed purely of distributed elements such as transmission lines and stubs, as is the case in many microwave circuits, the relevant complex frequency variable is the Richards [1] variable S. More will be said about this extremely useful complex variable in Section 2.3.5, but it is important to note now that in this text the same symbols, σ and $j\omega$, will be used to represent the real and imaginary parts of S as were used for the Laplace variable. However, ω is no longer related to the real frequency variable f by $\omega = 2\pi f$ (i.e., $\omega \neq 2\pi f$). Instead, ω is given by $\omega = \tan\{(\pi f)/(2f_s)\}$, where f_s is the frequency at which all the distributed elements of the network are a quarter of a wavelength long. The frequency f_s is known as the *commensurate frequency*. With the application of this alternative transformation, indicated by the symbolic change of the complex frequency variable to S, the frequency-dependent variables (such as Z_1 and Z_2) for distributed networks are expressed in the same form of polynomials with real coefficients as for lumped circuits. This facilitates the presentation of a single, unified theory for the synthesis of both lumped and distributed networks, the only differences being the substitution of s for lumped circuits and S for distributed circuits and the use of the two alternative mapping functions for transforming into the f-plane. However, since the emphasis in this text is on microwave networks, most of which will involve distributed elements, S should be assumed to be the relevant frequency variable in all further references to the network functions unless stated otherwise.

In the majority of circuit design problems, the main objective will be to transfer as much of the available power from the source as possible into the network load over a specified band of frequencies. There may be a few special circumstances for which there is a need to control the phase of a reflection or transmission coefficient. Although there is no reason in principle why this could not be done, it is an option that will not be covered here. Only the control of the amplitude of the various network frequency characteristics will be addressed by the synthesis.

2.3.2 Definition of Terms for Doubly Terminated Networks

For a doubly terminated network, there is a finite quantity of power available from the signal source, and it tends to be most helpful to work with network functions that are normalized to this available power. Furthermore, it is convenient to think in terms of incident and reflected waves at the two ports of the network rather than the actual terminal voltages and currents. Consequently, two of the most relevant network functions for the synthesis of doubly terminated networks are the *input reflection coefficient* $\rho_1(S)$ and the *transmission coefficient* $\tau(S)$. These are directly related to two other important functions: the *transducer function* $H(S)$ and the *characteristic function* $K(S)$. The set of four are listed together in Figure 2.4 for ease of future reference.

Input reflection coefficient $\rho_1(S)$ is the ratio of the reflected and incident voltage waves at the input port of the network. It relates directly to the input impedance of the network by the expression

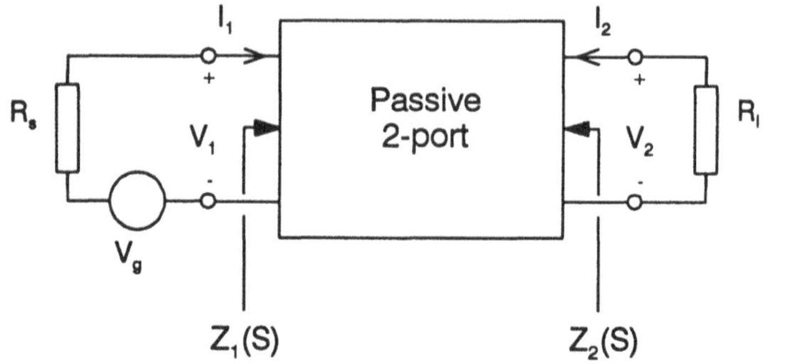

Input Reflection coefficient	$\rho_1(S)$	=	$\frac{Z_1(S) - R_s}{Z_1(S) + R_s}$
Transmission coefficient	$\tau(S)$	=	$2 \times \sqrt{\frac{R_s}{R_l}} \frac{V_2(S)}{V_g}$
Transducer function	$H(S)$	=	$\frac{1}{\tau(S)}$
Characteristic function	$K(S)$	=	$\frac{\rho_1(S)}{\tau(S)}$

Figure 2.4 Definition of terms for a doubly terminated network.

$$\rho_1(S) = \frac{Z_1(S) - R_s}{Z_1(S) + R_s} \tag{2.2}$$

There is also an output reflection coefficient, $\rho_2(S)$, which is related to the output impedance $Z_2(S)$ and the load resistance R_l by a similar relationship to (2.2). In general, $\rho_1(S) \neq \rho_2(S)$. For the purposes of network synthesis, it will be sufficient to deal with only the input reflection coefficient, and hence no further reference to $\rho_2(S)$ will be made in the text. In any subsequent expressions involving the reflection coefficient, the symbol $\rho(S)$ should be assumed to be equal to $\rho_1(S)$ unless stated otherwise. Transmission coefficient $\tau(S)$ is the ratio of the transmitted voltage wave at the load to the incident voltage wave from the source. It is obtained by taking the square root of the power transmitted to the load divided by the available power from the source; that is,

$$\tau(S) = 2\sqrt{\frac{R_s}{R_l}} \times \frac{V_2(S)}{V_g} \tag{2.3}$$

The transducer function $H(S)$ is just the reciprocal of the transmission coefficient:

$$H(S) = \frac{1}{\tau(S)} \tag{2.4}$$

and the characteristic function $K(S)$ is the ratio of $\rho(S)$ and $\tau(S)$; that is,

$$K(S) = \frac{\rho(S)}{\tau(S)} \tag{2.5}$$

The transducer function has been included in the set of definitions only to clarify the connection with $\tau(S)$. Some readers may be more familiar with the transducer function than with the transmission coefficient. The characteristic function is a particularly useful function for the synthesis process, since it embodies the characteristic's of both the reflected and the transmitted waves in its numerator and denominator, respectively.

In the forms in which the above functions have been defined, it should be noted that each will be complex both for S complex and for S equal to $j\omega$ (i.e., purely imaginary). Not surprisingly, this means that they will also be complex at all finite, nonzero frequencies for the equivalent f-plane circuit. Two of the functions are closely related to a pair of the two-port scattering parameters commonly used to represent the electrical performance of microwave circuits. In the special case when the source and load resistors are set to equal to a value of 50Ω, $\rho(S)$ and $\tau(S)$ are equal to the scattering parameters S_{11} and S_{21}. Like the scattering parameters, $\tau(S)$ and $\rho(S)$ can be thought of as *root power functions.* Consequently, while the synthesis only directly controls amplitudes, it is possible when necessary to obtain phase responses by a simple process of evaluation.

The synthesis procedure first creates the basic network functions as squared functions, and these have some interesting mathematical properties that need to be understood. Representing the general case with the function $F(S)$, we can write

$$F(S) \times F(-S) = |F(S)|^2 \qquad \text{for } S = j\omega \tag{2.6}$$

This states that if the real part of S is zero, then multiplying $F(S)$ by $F(-S)$ is equivalent to multiplying $F(S)$ by its conjugate, which then equates to the squared modulus of the function. The resulting polynomial or ratio of polynomials of $|F(S)|^2$ must therefore be purely even, and when evaluated at $S = j\omega$, its value must be real. Of course, this is entirely consistent with the fact that the various network functions are derived from what are fundamentally power ratios, which, by definition, must be scalar quantities. As an even polynomial with real coefficients, $|F(S)|^2$

must have zeros that occur in complex quads in the S-plane or in pairs on the real axis; that is,

$$\begin{aligned}|F(S)|^2 = M &\times (S - S_{z1}) \times (S + S_{z1}) \times (S - S_{z1}^*) \times (S + S_{z1}^*) \\ &\times (S - S_{z2}) \times (S + S_{z2}) \times (S - S_{z2}^*) \times (S + S_{z2}^*) \\ &\times (S - S_{z3}) \times (S + S_{z3}) \times (S - S_{z4}) \times (S + S_{z4}) \\ &\times \quad \text{etc.}\end{aligned} \tag{2.7}$$

where S_{z1} and S_{z2} are zeros at finite values of $j\omega$, S_{z1}^* and S_{z2}^* are their conjugates, and S_{z3} and S_{z4} are zeros for which $j\omega = 0$. M is a constant multiplier.

For a loss-free two-port network, the principle of conservation of energy allows us to write

$$1 = |\tau(S)|^2 + |\rho(S)|^2 \tag{2.8}$$

which is the famous Feldtkeller energy equation for two-port reactance networks. By dividing through by $|\tau(S)|^2$, we find

$$\frac{1}{|\tau(S)|^2} = 1 + \frac{|\rho(S)|^2}{|\tau(S)|^2} = 1 + |K(S)|^2 \tag{2.9}$$

which is a convenient relation for transforming the characteristic function into the transmission coefficient or vice versa. Similarly, a little additional transposition can give a useful expression for $|\rho(S)|^2$ in the form

$$|\rho(S)|^2 = \frac{|K(S)|^2}{|K(S)|^2 + 1} \tag{2.10}$$

To reinforce the interrelationships between these three important network functions, Figure 2.5 contains some graphs based on the frequency responses of a typical equal-ripple low-pass filter. As shown in the figure, the function $|\tau(S)|^2$ ripples just below the value of unity in the passband, falling away in the stopband toward a zero on the $j\omega$ axis, rising again to a local maximum, and finally falling toward zero as the frequency tends to infinity. It never exceeds a value of unity (i.e., zero loss) and therefore can never have any poles anywhere on the $j\omega$ axis. Wherever it falls to zero, $|\tau(S)|^2$, must have at least one transmission zero, but could equally well have several at the same frequency. The function $|\rho(S)|^2$ can be seen to ripple just above zero in the passband, rising in the stopband up to unity, where the transmission of energy falls to zero (i.e., total reflection), before falling again and then rising finally toward unity as frequency tends to infinity. Clearly, $|\rho(S)|^2$ cannot have any poles on the $j\omega$ axis, but it does have zeros corresponding to the frequencies in the passband where all the available power from the source is transmitted to the

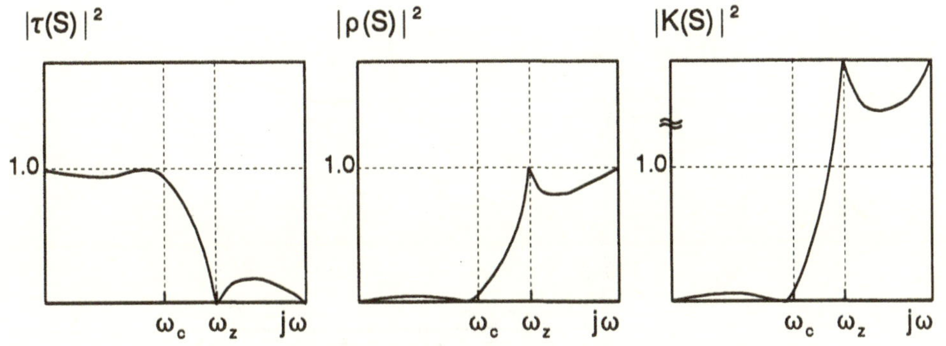

$$1 = |\rho(S)|^2 + |\tau(S)|^2 \quad \text{and}$$

$$|K(S)|^2 = \frac{|\rho(S)|^2}{|\tau(S)|^2} = \frac{1}{|\tau(S)|^2} - 1 \quad \text{for } S = j\omega$$

Figure 2.5 Graphical representations of the basic functions for doubly terminated networks.

load. The function $|K(S)|^2$ also ripples just above zero in the passband, but rises quickly to infinity (i.e., toward a pole) at the position of the first transmission zero in the stopband. It then falls to a local minimum before rising again to infinity as the frequency tends toward infinity. The characteristic function $|K(S)|^2$, therefore, is unique in containing the locations of both the zeros of reflection and the zeros of transmission. The zeros of its numerator polynomial are the reflection zeros, while the zeros of its denominator are the transmission zeros. It is for this reason that the early stages of the synthesis focus on the generation of $|K(S)|^2$.

For further clarification of the allowed distributions of poles and zeros for the three network functions, there are four more sketches in Figure 2.6. Here, a band-pass frequency response has been chosen as an example and in part a) of the figure, the graph shows a likely frequency response for the function of $|K(S)|^2$. This represents the value of $|K(S)|^2$ calculated at points along the $j\omega$ axis of the complex plane. From the stopband responses, it is clear that transmission zeros have been placed at $S = (0, j0)$, at $S = (0, j\infty)$ and at $S = (0, j5)$. It is not possible to determine from the graph how many have been placed at each location nor if there are any other transmission zeros elsewhere in the complex plane. In fact in this example, the network has an additional two transmission zeros at $S = (1, j0)$ which are not visible on the plot. From the passband response, it is apparent that there are reflections zeros positioned at $S = (0, j2.2)$, $S = (0, j3)$ and $S = (0, j3.8)$, giving what is an equal-ripple response extending from $\omega = 2.0$ to $\omega = 4.0$.

Figure 2.6(b) shows the pole/zero distributions in the complex plane that correspond to the same frequency response as that of 2.6 (a) for each of the network

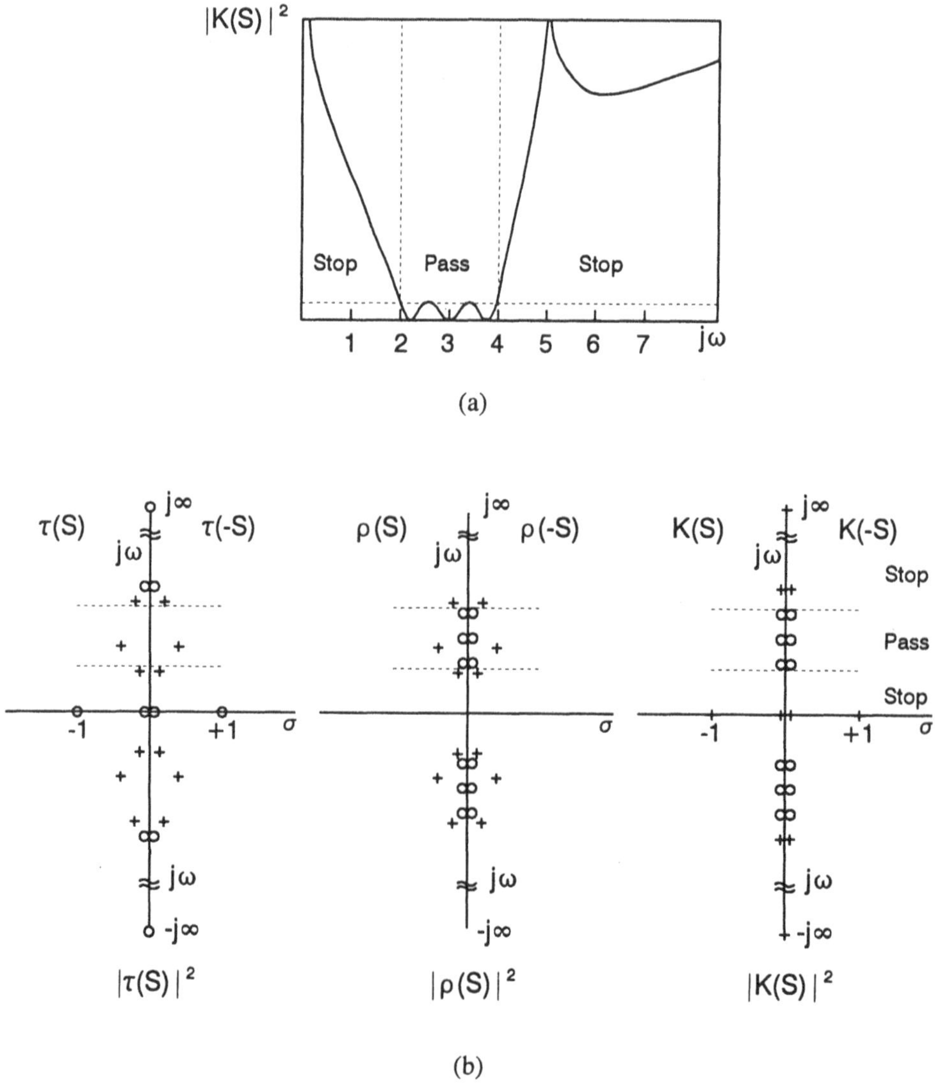

Figure 2.6 Pole-zero distributions for the squared functions of a band-pass network: (a) band-pass frequency response on the $j\omega$ axis; (b) pole/zero distributions (+ = pole, o = zero).

functions $|\tau(S)|^2$, $|\rho(S)|^2$, and $|K(S)|^2$. Poles and zeros are indicated by the symbols + and o, respectively. Their positions in relation to the passband and stopbands are indicated by the dotted lines. It should be noted that the three sets of distributions are only qualitative and are only intended to indicate the allowed distributions of poles and zeros.

Taking $|\tau(S)|^2$ first, most of the zeros have been located on the $j\omega$ axis. It is not essential that they be on the $j\omega$ axis, but this will be the situation in most practical applications. They all reside in the stopband of the frequency response as either complex quads where $j\omega$ is finite/nonzero or in pairs at $S = (0, j0)$ and $S = (0, j\infty)$. They can also be found in pairs on the real axis as shown in the diagram. There are two at $S = (\pm 1, j0)$ corresponding to two unit elements (i.e., the equivalent of two transmission lines) in the prototype network. The poles of $|\tau(S)|^2$ cannot exist anywhere on the $j\omega$ axis. They exist as complex quads or as pairs on the real axis. In this example, they are all in quads and can be seen to cluster around the region of the passband.

For $|\rho(S)|^2$, the rules for the distribution of zeros are the same as for those of the transmission zeros. The only differences in the actual distributions are that the reflections zeros reside within the region of the passband, while the transmission zeros reside within the stopbands. The poles of the reflection coefficient have exactly the same distribution as the poles of $|\tau(S)|^2$ for reasons that will become clearer later. As for $|K(S)|^2$, this appears to be rather different from the other two functions in that all the zeros and all except two pairs of the poles lie on the $j\omega$ axis. The zeros correspond to zeros of reflection, while the poles (note: they are often referred to as the *loss poles*) correspond to the location of the transmission zeros.

With respect to the pole/zero distributions of the root power functions $\tau(S)$, $\rho(S)$, and $K(S)$, Figure 2.6 indicates that the poles and zeros in the left half plane belong to $\tau(S)$, $\rho(S)$, and $K(S)$, while those in the right half plane belong to $\tau(-S)$, $\rho(-S)$, and $K(-S)$. Some variation on this distribution is allowed, but one or two important constraints must be observed. First, the poles and zeros of the squared functions must always be assigned to the root power functions in conjugate pairs, whatever side of the S-plane they originate from. Second, for fundamental reasons of realizability, $\tau(S)$ and $\rho(S)$ must always be assigned the left-hand plane poles of the corresponding squared functions, which are known as the *natural frequencies* of the network. More of this will be discussed later when the impedance function $Z(S)$ is examined in more detail. As far as the zeros of the squared functions are concerned, there are no other hard and fast constraints. However, it is usual to select the zeros for $\rho(S)$ and $K(S)$ from the left-hand planes of the squared functions, since this produces what is called a *minimum phase-shift network.* As the name implies, such a network for a given set of transmission zeros has the smallest possible differential phase shift. The network also has an input admittance that is *minimum susceptance*; that is, there are no poles of $Y_1(S)$ anywhere on the $j\omega$ axis, ensuring that the network never presents a short circuit to the generator. Its dual would have an input impedance that is *minimum reactance.* In addition to these useful properties, it is also the form of the network that tends to have the smallest dynamic range of element values and the least sensitivity to errors in its element values.

From the general property expressed by (2.6), we can write

$$|\rho(S)|^2 = \rho(S) \times \rho(-S) \qquad \text{for } S = j\omega \tag{2.11}$$

which, when numerator and denominator polynomials are identified, becomes

$$|\rho(S)|^2 = \frac{f(S)}{e(s)} \times \frac{f(-S)}{e(-S)} \qquad \text{for } S = j\omega \tag{2.12}$$

where

$$\rho(S) = \frac{f(S)}{e(S)} \tag{2.13}$$

Similarly,

$$|\tau(S)|^2 = \tau(S) \times \tau(-S) \tag{2.14}$$

and also from (2.8),

$$|\tau(S)|^2 = 1 - |\rho(S)|^2 \qquad \text{for } S = j\omega \tag{2.15}$$

Substituting for $|\rho(S)|^2$ we have

$$|\tau(S)|^2 = \frac{|e(S)|^2 - |f(S)|^2}{|e(S)^2|} = \frac{|p(S)|^2}{|e(S)|^2} \quad \text{for } S = j\omega \tag{2.16}$$

and

$$\frac{|p(S)|^2}{|e(S)|^2} = \frac{p(S)}{e(S)} \times \frac{p(-S)}{e(-S)} \quad \text{for } S = j\omega \tag{2.17}$$

and hence

$$\tau(S) = \frac{p(S)}{e(S)} \tag{2.18}$$

With similar algebra, it can be shown that

$$K(S) = \frac{f(S)}{p(S)} \tag{2.19}$$

In this development, we have shown the three network functions to be ratios involving only three polynomials: $e(S)$, $f(S)$, and $p(S)$. The zeros of $e(S)$ are the poles of $\tau(S)$ and $\rho(S)$ and must reside on the left-hand side of the S-plane (i.e., $e(S)$ must be a *Hurwitz polynomial*). The zeros of $f(S)$ are the zeros of $\rho(S)$ and $K(S)$ and may or may not reside on the left-hand side of the S-plane. The zeros of $p(S)$ are the zeros of $\tau(S)$ and the poles of $K(S)$. They will usually be found on the $j\omega$ axis of the S-plane.

The synthesis of a two-port loss-free ladder network begins with the generation of a suitable $|K(S)|^2$ function for a specified frequency response. This is followed by the determination of $Z_1(S)$, the input impedance of the network, and it is from $Z_1(S)$ that the individual elements of the network are subsequently extracted. Deriving $Z_1(S)$ from $|K(S)|^2$ is a relatively simple procedure, but it is constrained by a set of conditions that lie at the heart of network synthesis theory: the set of realizability conditions for any driving point impedance function $Z(S)$. According to Brune [2]:

> *A function Z(S) is realizable as an impedance if, and only if, it is a rational, positive real function of S.*

For $Z(S)$ to be a positive real function, it must possess the following properties:

1. $Z(S)$ is real for all real values of S.
2. $\mathrm{Re}\{Z(S)\} \geq 0$ for $\mathrm{Re}\{S\} \geq 0$.

In other words, $Z(S)$ must be a ratio of polynomials in S with real coefficients such that the real part of $Z(S)$ is positive in the right half of the S-plane. An appreciation of why this must be so can be gathered by considering the behavior of $Z(S)$ along the imaginary axis. The function $Z(j\omega)$ is the actual impedance seen looking into a realizable network at a given frequency ω, where ω is directly related to the frequency of the excitation f by a simple mapping function. In these circumstances, the real part of $Z(j\omega)$ can never be negative if the possibility of negative resistances is to be excluded from the network. It must always be positive, and if it is positive for the real part of S equal to zero, it must also be positive for all positive values of the real part of S. Following from these conditions, it can also be stated that:

> *All the poles of Z(S) are in the left half plane, with any poles on the imaginary axis being simple and having positive residues.*

This is probably the most useful interpretation of the realizability conditions for an impedance and is taken into consideration during the derivation of a realizable $Z_1(S)$ from $|K(S)|^2$. It can be understood with the help of the relation

$$Z(S) = \left.\frac{A_{r1}}{(S - S_{z1})}\right|_{\text{as } S \text{ approaches } S_{z1}} \tag{2.20}$$

where A_{r1} is the residue for the pole at S_{z1}, which is the value of $Z(S)$ at $S = S_{z1}$ after the $(S - S_{z1})$ factor has been canceled from the denominator.

If a pole of $Z(S)$ were to be present in the right-hand plane at values of S approaching the pole, the real part of $Z(S)$ would inevitably tend to a positive or negative value of infinity at some position infinitesimally close to the pole. Hence, since $\mathrm{Re}\{Z(S)\}$ cannot be negative, but must always be greater than zero in the right half plane, there cannot be any poles of $Z(S)$ in the right half plane. As for poles on the imaginary axis, if they are simple (i.e., only one in a particular position), then by considering an approach to the pole along the imaginary axis, it is clear that only an infinite imaginary value of $Z(S)$ can be produced. However, if there is an even number of poles in the same position, then the product of the imaginary components could produce a disallowed negative real result. If there is an odd number of poles, then although $Z(S)$ would tend to an infinite imaginary value, the residue A_{r1} would not necessarily be positive.

Observing the realizability conditions, we can now derive the function of $Z_1(S)$ from the characteristic function $|K(S)|^2$ as follows. Given $|K(S)|^2$, we first calculate $|\rho(S)|^2$ using

$$|\rho(S)|^2 = \frac{|K(S)|^2}{|K(S)|^2 + 1} \tag{2.21}$$

Then the appropriate poles and zeros of $|\rho(S)|^2$ must be assigned to $\rho(S)$ and $\rho(-S)$ with the knowledge that

$$|\rho(S)|^2 = \rho(S) \times \rho(-S) \tag{2.22}$$

and

$$|\rho(S)|^2 = \frac{f(S)}{e(S)} \times \frac{f(-S)}{e(-S)} \qquad \text{for } S = j\omega \tag{2.23}$$

From the relationship between $\rho(S)$ and $Z_1(S)$ expressed in (2.2), it follows that the realizability conditions for $Z_1(S)$ must also apply to $\rho(S)$, and therefore all the poles of $\rho(S)$ must also lie in the left-hand plane. Hence, $e(S)$ is constructed by selecting the poles of $|\rho(S)|^2$ that exist in the left-hand plane. As previously explained, the polynomial $f(S)$ could be constructed by selecting half the zeros of $|\rho(S)|^2$ in any combination, so long as they are taken in conjugate pairs. However, invariably all the zeros will be on the imaginary axis, which eliminates any choice in the matter. When the zeros do not exist on the imaginary axis, they should all be selected from the left half plane zeros of $|\rho(S)|^2$ to obtain a minimum phase network (note: a proof of this is due to Bode [3]).

Transposing (2.2) and assuming a source resistance of unity, we can write $Z_1(S)$ in terms of $\rho(S)$ such that

$$Z_1(S) = \frac{1 + \rho(S)}{1 - \rho(S)} \tag{2.24}$$

and thus

$$Z_1(S) = \frac{e(S) + f(S)}{e(S) - f(S)} \tag{2.25}$$

This is then the input impedance function of a network that is of the first canonical form (i.e., a first dual network). A variety of networks, all with the same transmission characteristics, can usually be extracted from $Z_1(S)$, and all will have an input reflection coefficient that tends toward unity and an input impedance that tends toward an *open circuit* in the stopband. For networks of the second canonical form, the sign of $\rho(S)$ is reversed, giving two alternative expressions for $Z_1(S)$:

$$Z_1(S) = \frac{1 - \rho(S)}{1 + \rho(S)} \tag{2.26}$$

and

$$Z_1(S) = \frac{e(S) - f(S)}{e(S) + f(S)} \tag{2.27}$$

This has absolutely no effect on the transmission characteristics of the network as determined by the conservation of energy relation $1 = |\rho(S)|^2 + |\tau(S)|^2$. However, for networks of the second canonical form, as indicated by (2.26), the input impedance $Z_1(S)$ now tends to a *short circuit* in the stopband as the magnitude of the reflection coefficient tends to unity.

Choosing the appropriate network dual will be an essential part of the circuit selection process at the beginning of the design procedure. For example, the choice of dual determines the types of elements permitted at the two ends of a network. It will be shown that in a network of the first canonical form, any reactance branches at the ends of the network must be series types, and those at the ends of a network of the second canonical form must be shunt types. In some applications, the choice may be determined by the required behavior of the input impedance of the network, whereas in other situations, it may be the realizable range of element values that is the determining factor. As far as transforming from one network dual to the other is concerned, there is no necessity to reextract the elements of the network from $Z_1(S)$. With the exception of unit elements (the S-plane equivalent of a transmission line) and the load resistor, the element values of the first network dual are identically equal to those of the second dual. They simply become the values of the dual types of elements in the same relative positions in the network. For the unit element and load resistor, these values represent admittance values instead of impedance values in the second dual. Hence, if the values of the unit elements and load resistors are to be expressed as impedances in both duals, transforming between duals must involve taking reciprocals.

2.3.3 Definition of Terms for Singly Terminated Networks

Singly terminated networks are another important class of passive networks. They often find applications in active circuits where a matching network is required to operate between a resistive load and the output of an active device (e.g., a FET), where the output of the active device is a close approximation to a pure current source. A power amplifier is a good example, since to achieve maximum output power over a wide frequency range, it is generally more effective to match the drains of the FETs in the output stage with singly rather than doubly terminated networks. More details on this will be covered in Chapter 5, one section of which deals specifically with the practical design of microwave power amplifiers. Another noteworthy example of the use of singly terminated networks occurs in the field of contiguous multiplexers, where a group of networks are usually connected together in parallel at a common junction. Here, singly terminated networks interact more constructively than would be the case for a set of doubly terminated networks, an excellent explanation of which can be found in [4].

An illustration of a singly terminated passive two-port is shown in Figure 2.7, together with some definitions of the most relevant electrical parameters. Since

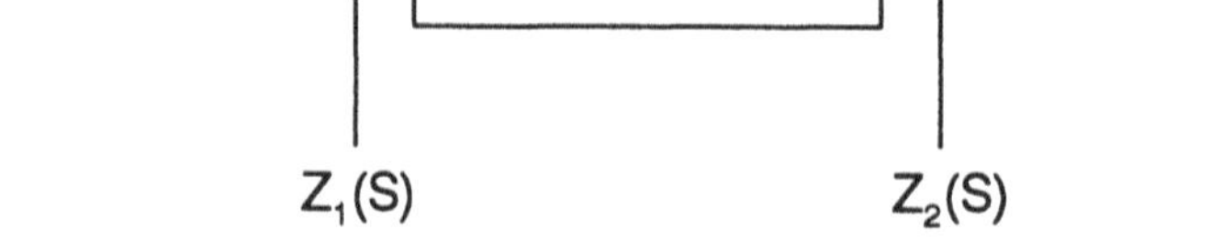

Input reflection coefficient $|\rho_1(S)|^2 = 1$

Input power $P_{in}(S) = I_g^2 \times \mathrm{Re}[Z_1(S)]$

Insertion loss $|L(S)|^2 = \dfrac{R_l}{\mathrm{Re}[Z_1(S)]}$

Figure 2.7 Definitions of terms for a singly terminated network.

the power source now has an infinite internal impedance, the input reflection coefficient of the network is necessarily of unity magnitude at all frequencies. From the conservation of energy relation (2.8), this also means that the transmission coefficient must always be zero. It is indeed true that the transmission coefficient is zero, but because the generator has infinite available power, there is, as might be expected, a finite delivery of power to the network load.

When, as in Figure 2.7, the generator is a pure current source, the input power will be given by

$$P_{in}(S) = I_g^2 \times \mathrm{Re}[Z_1(S)] \tag{2.28}$$

where I_g is the rms value of the input current. Since the network is loss-free, this must also be equal to the power delivered to the load. For singly terminated networks, the most useful parameter for describing the transfer of power to the load is not the transmission coefficient, but the *insertion loss* $|L(S)|^2$. The definition of insertion loss is

$$|L(S)|^2 = \frac{\text{the power delivered to the load with the network absent}}{\text{the power delivered to the load with the network present}} \tag{2.29}$$

from which it follows that

$$|L(S)|^2 = \frac{R_l}{\mathrm{Re}[Z_1(S)]} \quad \text{when } S = j\omega \tag{2.30}$$

where R_l is the network load resistance. Now since the value of R_l is not generally going to be equal to unity and will not necessarily be known until the network elements have been extracted, it is quite possible that the insertion loss of a network could turn out to be much smaller or much greater than unity in the passband. In other words, the network, despite being passive, could actually increase the delivery of power to the network when inserted between the current source and the load. To avoid any misunderstandings brought about by this apparent power gain and to give some consistency with the transmission characteristics of doubly terminated networks, it is therefore desirable to force R_l to be unity by always including an ideal transformer at the output of the network with the appropriate turns ratio. Hence, for the purposes of synthesis, a more relevant expression for $|L(S)|^2$ becomes

$$|L(S)|^2 = \frac{1}{\mathrm{Re}[Z_1(S)]} \quad \text{for } S = j\omega \tag{2.31}$$

in which, for an insertion loss of unity, $\mathrm{Re}\{Z_1(S)\} = 1$. This means that the objective of the synthesis of a singly terminated network must be to create a specified insertion loss frequency response and to do so by controlling the real part of the input impedance function $Z_1(S)$ for $S = j\omega$. This is in marked contrast to the doubly terminated case where power delivery was normalized to the available source power and the control was obtained by controlling the magnitude of the input reflection coefficient.

Notwithstanding the differences in definition between the relevant electrical parameters for singly and doubly terminated networks, the frequency responses and pole/zero distributions of the two sets of parameters are identical. Figure 2.8 shows the same three graphs as Figure 2.5, which referred to doubly terminated networks, but instead of representing the behavior of the parameters $|\tau(S)|^2$, $|\rho(S)|^2$, and $|K(S)|^2$, they now represent the functions $\text{Re}\{Z_1(S)\}$, $(1 - \text{Re}\{Z_1(S)\})$, and $(1/\text{Re}\{Z_1(S)\} - 1)$. We can therefore write the equivalences

$$\text{Re}[Z_1(S)] \equiv |\tau(S)|^2 \tag{2.32}$$

$$(1 - \text{Re}[Z_1(S)]) \equiv |\rho(S)|^2 \tag{2.33}$$

and

$$\left(\frac{1}{\text{Re}[Z_1(S)]} - 1\right) \equiv |K(S)|^2 \tag{2.34}$$

This is a most useful set of equivalences, since with the appropriate substitutions, the procedures used to generate the $|K(S)|^2$ polynomials for a doubly terminated network can also be used to generate the polynomials of $\text{Re}\{Z_1(S)\}$ for a singly terminated network.

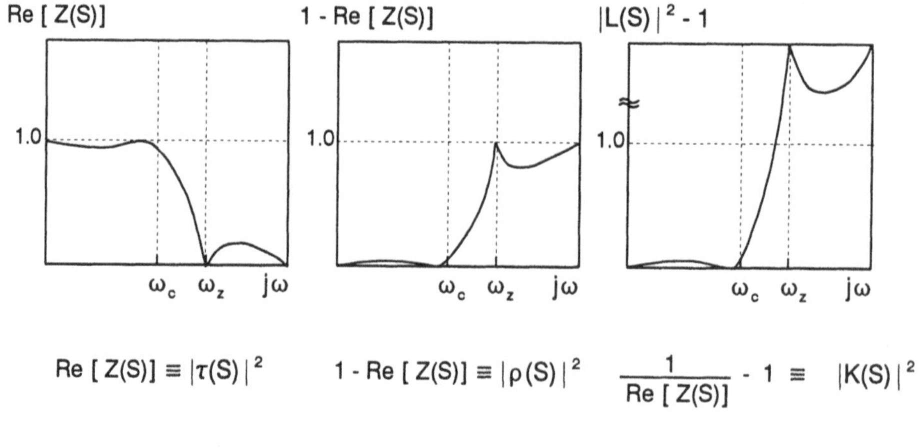

Figure 2.8 Graphical representation of basic functions for a singly terminated network.

Extraction of $Z_1(S)$ from the $\text{Re}\{Z_1(S)\}$ is a relatively simple procedure and relies on the property

$$\text{Re}[Z_1(S)] = 0.5 \times \{Z_1(S) + Z_1(-S)\} \quad \text{for } S = j\omega \tag{2.35}$$

This merely states that the sum of the conjugates of the impedance function $Z_1(S)$ at frequencies on the $j\omega$ axis is equal to twice the real part of $Z_1(S)$. On first consideration, the property seems to suggest that no unique impedance function $Z_1(S)$ exists for a given $\text{Re}\{Z_1(S)\}$, since arbitrary reactances could be added in series with $Z_1(S)$ without altering its real part. However, a unique minimum reactance function $Z_1(S)$ does exist, and the first proof of this is due to Bode [3]. To obtain a minimum reactance $Z_1(S)$, all the poles must lie in the left half S-plane and those of $Z_1(-S)$ in the right half plane. Hence, $Z_1(S)$ can be constructed from $\text{Re}\{Z_1(S)\}$ by a partial fraction expansion, for which $Z_1(S)$ is equal to twice the sum of the terms with left-hand plane poles. That is,

$$Z_1(S) = \frac{2A_{z1}}{(S - S_{z1})} + \frac{2A_{z2}}{(S - S_{z2})} + \ldots \frac{2A_{zn}}{(S - S_{zn})} + R \tag{2.36}$$

Most terms will be in conjugate pairs, but some will correspond to single zeros on the real axis. The constant term R is usually zero, since most networks will have a zero of $\text{Re}\{Z_1(S)\}$ (i.e., a transmission zero) at $S = j0$ and/or $S = j\infty$. It will be positive, however, when there are no zeros of $\text{Re}\{Z_1(S)\}$ at either $S = j0$ or $S = j\infty$, which occurs when a network contains only unit elements. In this case, the numerator and denominator polynomials of $\text{Re}\{Z_1(S)\}$ will be of the same degree. Whatever the precise distribution of poles and zeros, the degree of $Z_1(S)$ will always be half the degree of the function $\text{Re}\{Z_1(S)\}$, which must always be even.

Subsequent extraction of network elements from $Z_1(S)$ will produce a network of the *second* canonical form. If a network of the first canonical form is required, the elements should be extracted from $Y_1(S)$, where $Y_1(S) = 1/Z_1(S)$. Networks of the first kind must be driven by an ideal voltage source and those of the second kind must be driven by an ideal current source.

In reflecting on this process of obtaining $Z_1(S)$ from $\text{Re}\{Z_1(S)\}$, it is rather surprising that a complex quantity can be derived from a function that is entirely real. However, as already indicated, the real function is the summation of a pair of complex conjugates and all that the process is doing is effectively separating out the conjugates from within the resistance function. Hence, far from containing only half the necessary information, $\text{Re}\{Z_1(S)\}$ contains all the information needed for obtaining $Z_1(S)$ in duplicate.

2.3.4 Relevant Transfer Characteristics

There is a virtually unlimited variety of power transfer characteristics that can be created by exact synthesis techniques. However, the characteristics do fall into certain broad classifications. Most practical circuits, whether they be active or passive, have filterlike transmission characteristics, and this means they will possess a set of passbands and stopbands. By definition, *low-pass* networks have a passband that extends from a given cutoff frequency down to zero frequency, *high-pass* networks have a passband extending from a given cutoff frequency up to infinite frequency, and *band-pass* networks have a passband that exists between a pair of specified cutoff frequencies. There is also a *band-stop* classification, but this is not of significant interest for microwave circuits, because sufficiently versatile *f*-plane band-stop responses can already be created using a low-pass *S*-plane prototype. Despite the many variations in passband and stopband shape that can be achieved, any network synthesized *in the complex plane* must fall into one of these four categories. This is true whether the network is subsequently realized in the form of lumped or distributed element circuits.

To illustrate the variety of passband characteristics that can be generated, some likely responses of insertion loss against frequency have been collected together in Figure 2.9. For the sake of brevity, they have all been drawn as band-pass responses with band-edge frequencies of $j\omega_1$ and $j\omega_2$, but the basic passband shapes could apply equally well to low-pass or high-pass responses. The first two responses (Figure 2.9(a,b)) are what could be referred to as *conventional responses*, since they aim, as far as possible, to achieve the flat, zero loss condition of an ideal filter. In the first, the zeros of reflection have been distributed across the passband to produce an equal-ripple, or Tchebycheff, response. Ripple magnitude (L_p) can be specified arbitrarily. For small ripple values, this is probably the best possible approximation to the ideal filter response and tends to result in networks with the smallest number of elements for given slopes of the passband skirts. In the second response, the zeros of reflection have been placed at a single point in the passband to produce a monotonic, or Butterworth, frequency response. Once again a value L_p can be specified, which sets the level of loss at the edges of the passband. For a classical Butterworth response, this is usually 3 dB. Butterworth responses are sometimes to be preferred to Tchebycheff types, since they generally produce a smaller group delay variation.

The third, fourth, and fifth responses in Figure 2.9 tend to be most relevant for the design of matching networks. In the case of the third, an offset L_o has been applied to the passband, fixing the minimum value of the passband loss to a value above zero. This is achieved by a relatively simple modification to the characteristic function for the zero loss case, which involves moving the reflection zeros away from the $j\omega$ axis such that they acquire a positive or negative real part. It may seem a strange objective to deliberately introduce reflection loss into a network. However, as will be explained in Chapter 5, a constant match to a given complex impedance can be achieved over a wider frequency range if the quality of the match is degraded by an appropriate amount. In the case of the fourth and fifth responses, a slope has been introduced across the passband, a positive slope in the case of the fourth and a negative slope in the case of the fifth. Like the third case, it is a simple

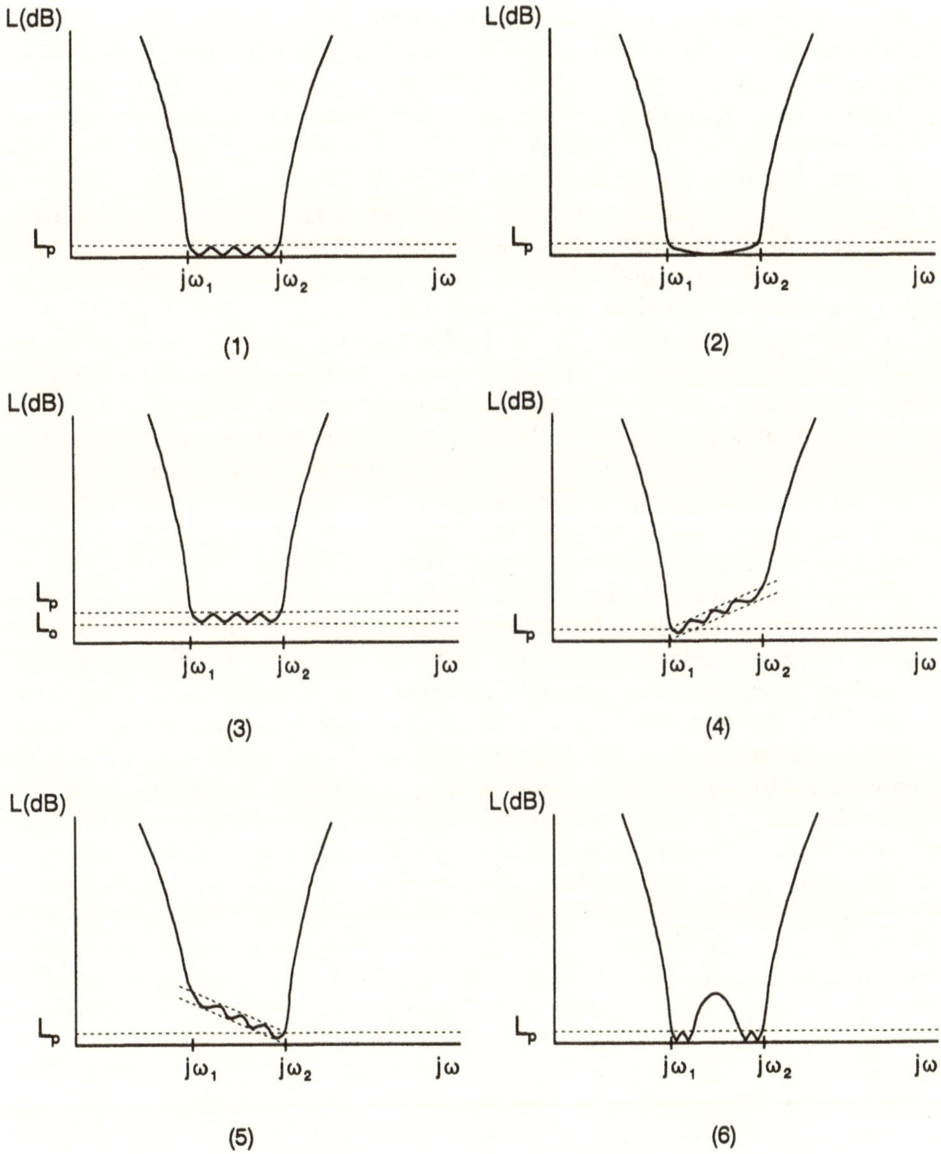

Figure 2.9 Some examples of S-plane passband responses for band-pass filters with similar stopbands.

mathematical procedure to modify the characteristic function $|K(S)|^2$ to produce these slopes, the effect on the reflection zeros being to move them progressively farther away from the $j\omega$ axis with the increasing loss value. The slope of the loss across the passband can be chosen arbitrarily, but a value of ±6 dB/octave is common in connection with active circuit design. These tapered responses, especially the fifth type, find frequent application in reactively matched amplifiers where, for example, an input matching network is used to correct for the natural −6-dB/octave gain slope of the active device to produce an overall flat gain. The network does this by achieving a good match at the top of the frequency band while progressively degrading the match at lower frequencies.

No particular application has been identified for the sixth response (Figure 2.9(f)), but it is included to demonstrate how general the shape of a passband can actually be. Here, a response with what appears to be a pair of passbands has been synthesized by taking a conventional equal-ripple passband response and increasing the size of the center ripple. The center ripple is increased by moving the reflection zeros out toward the passband edges on the $j\omega$ axis. Any peculiar passband shape can be constructed in this way, so long as the basic rules of Section 2.3.2 are observed governing the distribution of the zeros of $|K(S)|^2$.

To examine the variety of stopband responses that can be created, a set of responses have been collected together in Figure 2.10 pertaining to networks with similar passband characteristics. Again for the sake of brevity, the examples chosen are all band-pass, but the options apply equally well to low-pass or high-pass characteristics. The first example (Figure 2.10(a)) is intended to illustrate the situation in which equal numbers of transmission zeros have been placed at zero and infinite frequencies. This will not necessarily produce a precisely symmetrical response, nor will the slope of the two passband skirts necessarily be identical. However, depending on the position and width of the passband, this placement of the transmission zeros will often be the one that produces the closest to a symmetrical response. When an application calls for the skirt on one side of the passband to be steeper than that of the other, one solution is to change the balance of the transmission zeros located at $S = j0$ and $S = j\infty$. The effect is illustrated by the second plot, in Figure 2.10(b), for which only a single zero has been placed at zero frequency, the rest having been placed at infinity. As indicated, the skirt on the upper side of the passband is very much steeper than that on the lower side.

The third and fourth responses (Figure 2.10(c,d)) demonstrate that it is not essential to place transmission zeros at zero or infinite frequency if there are good practical reasons for not doing so. For example, zeros at $S = j0$ are produced by series capacitors or shunt inductors in a prototype network, and if in the chosen medium of construction it proves impossible to realize series capacitors or shunt inductors, there is no point in placing zeros at $S = j0$. In the third response, no transmission zeros have been placed at zero frequency, causing the loss below the passband to tend to a finite limiting value. The actual value of the loss will depend on the number of elements in the network (i.e., its degree), together with the

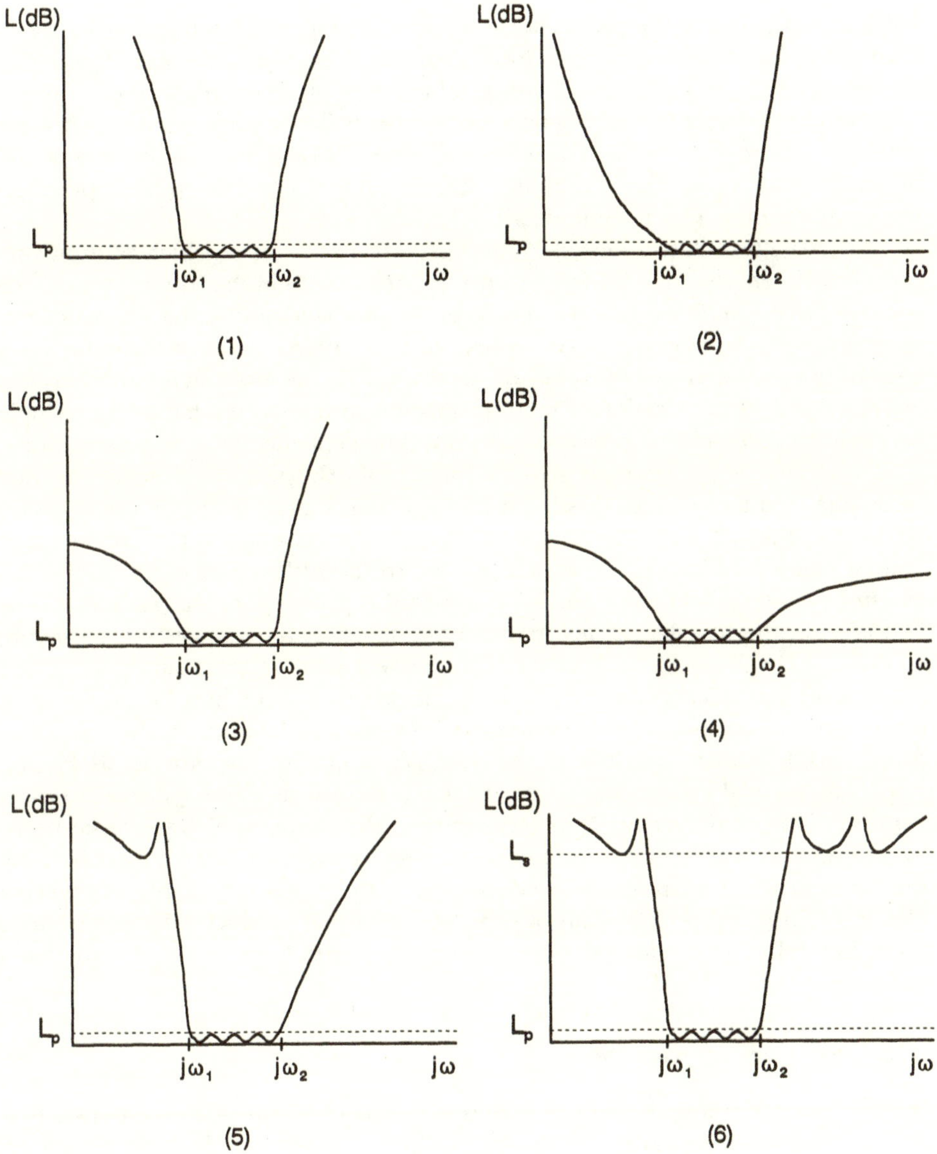

Figure 2.10 Some examples of *S*-plane stopband responses for band-pass filters with similar passbands.

width and position of the passband. In the fourth response, no transmission zeros have been placed at either zero or infinity, and at these two extremes of frequency, the loss tends to two different limiting values. Responses of this type correspond to prototype networks containing only unit elements. The corresponding microwave networks comprise only a cascade of transmission lines and form an important class of impedance-transforming circuits. Chapter 4 will deal with impedance-transforming networks in some detail.

The last two illustrations in Figure 2.10 are responses in which transmission zeros have been placed at finite, nonzero frequencies. As in the fifth response, one or more zeros can be placed on one side of the passband, giving the passband skirt on that side a much greater slope. Instead of a monotonic increase from the lower edge of the passband to a transmission zero at $S = j0$, the loss function now rapidly increases toward the location of the transmission zero and then has a single ripple between the zero and the zero at $S = j0$. As will be explained later, whenever finite, nonzero transmission zeros are specified in a network synthesis, if any are located below the frequency of the passband it is essential that at least one transmission zero exist at $S = j0$, and if any are located above the passband, it is essential that at least one transmission zero exist at $S = j\infty$. In the fifth response, therefore, the transmission zero below the passband forces the presence of the zero at $S = j0$. However, in the absence of any nonzero transmission zeros above the passband, all those at $S = j\infty$ could be removed if this were considered desirable.

In the last response of Figure 2.10, transmission zeros have been placed at finite, nonzero frequencies on both sides of the passband. However, they have also been placed so as to produce an equal-ripple stopband response in which the ripples all have the same minimum value L_s. Known as *elliptic function responses* because of the elliptic integrals involved in the calculation of the transmission zero locations, responses of this type are only really relevant to filter design because they give the highest possible selectivity using the smallest possible number of elements. From a practical point of view, networks that produce elliptic responses are rather difficult to realize physically, and consequently their use is not particularly widespread.

2.3.5 Frequency Transformations

In the previous section, the various shapes of passband and stopband responses that can be synthesized for S-plane prototype networks were explored. These responses, which reside on the $j\omega$ axis of the complex frequency plane, relate to the frequency responses of the f-plane network through certain basic frequency transformations. As well as modifying the frequency responses, the transformations invariably have the effect of changing the values and types of the elements in the prototype network. Because of the topological effects and the limited set of basic physical elements from which a real circuit can be constructed, there are relatively few types of frequency transformations that are relevant to practical circuit design by synthesis.

For lumped-element f-plane networks of the sort that might be used in radio applications at frequencies below 500 MHz, inductors and capacitors are the two most basic types of circuit elements. Several may be combined together to form transformers or more complicated, resonant parallel or series branches within a network, but at the most elementary level, the basic building blocks are inductors and capacitors. Such f-plane networks correspond to networks in the complex plane, which also contain only inductors or capacitors and for which the relevant frequency variable is the Laplace variable s. Depending on the frequency transformation used, there may or may not be a one-to-one correspondence between the inductors and capacitors of the f-plane network and those of the s-plane network. It is, therefore, unwise to attach much physical significance to the elements of the s-plane network; it is better to think of the inductors and capacitors as merely symbolic, mathematical entities.

Examples of the three most common frequency transformations for lumped-element networks are given in Figure 2.11. The first, Figure 2.11(a), is a linear transformation from the s-plane to the f-plane, the ω component of the s-plane variable being replaced by $\omega'(\omega_1/\omega_1')$, where $\omega' = 2\pi f$. Frequencies ω_1 and ω_1' may be arbitrary mapping reference points, but would normally mark the position of the respective passband edges in the two planes. Any s-plane response, whether low-pass, high-pass, or band-pass, is simply scaled in frequency by this transformation without any fundamental change in characteristic. The effect on the elements of the s-plane network is to change only their numerical values. There is no change of element type. The second transformation, Figure 2.11(b), is referred to as a *high-pass transformation* because its effect on a low-pass s-plane response is to produce a high-pass response in the f-plane. Replacing ω with $(\omega_1\omega_1')/\omega'$ changes all the s-plane inductors into capacitors in the f-plane and all s-plane capacitors into inductors. It would be usual to assign ω_1 and ω_1' to the edges of the respective passbands, their values determining the values of the transformed circuit elements. The third transformation, Figure 2.11(c), is referred to as a *band-pass transformation* because its effect on a low-pass s-plane response is to produce a band-pass response in the f-plane. The more complicated substitution for ω, given in Figure 2.11, changes all s-plane inductors into series-tuned circuits and all s-plane capacitors into parallel-tuned circuits. Consequently there will be twice as many elements in the f-plane network as in the s-plane network. Frequency ω_1, which would usually mark the passband edge of the low-pass response, maps into the two frequencies ω_1' and ω_2', the two new edges of the passband in the f-plane.

Note that the numerical stability of the synthesis procedures is normally at its best when the values of ω_1 are in the region of unity rad/s. At values of ω_1 around unity, the dynamic range of the coefficients of the $Z_1(s)$ polynomials is minimized, making the accumulation of residual errors during network extraction easier to control. It is, therefore, strongly advised that whatever the desired f-plane response, a transformation be used that maps the edge or edges of the passband into this preferred region on the $j\omega$ axis of the s-plane.

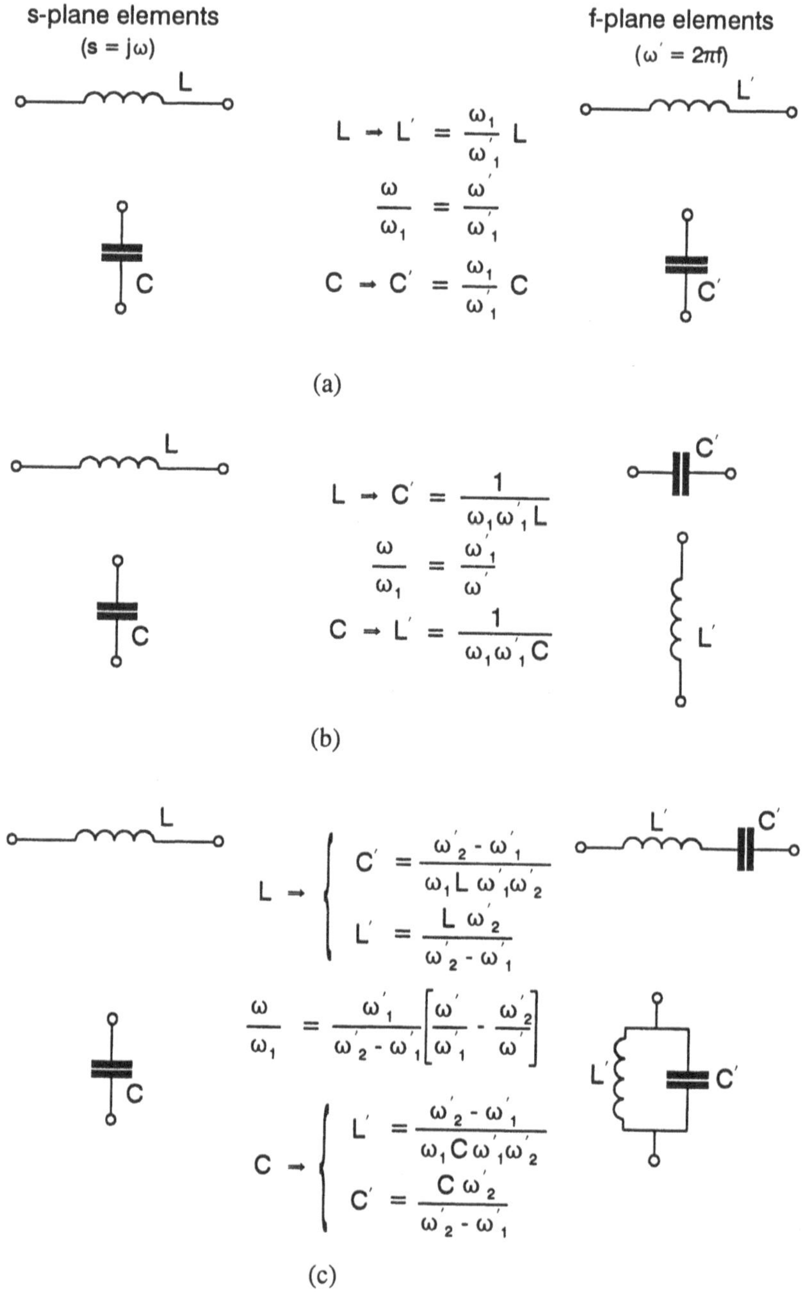

Figure 2.11 Simple frequency transformations between the *s*-plane and *f*-plane for lumped-element networks: (a) low-pass transformation; (b) high-pass transformation; (c) band-pass transformation.

At frequencies above approximately 1 GHz, physical parasitics make pure lumped elements essentially unrealizable. As the physical size of an element becomes a significant fraction of a wavelength, its behavior departs significantly from that of the ideal. Additionally, the electrical length of the physical connections between the physical elements can no longer be ignored and must be taken into consideration. Consequently, most practical microwave components and integrated circuits will either be realized entirely from distributed elements or alternatively from a mixture of distributed and lumped elements. Even in those that use a mixture of distributed and lumped elements, the distributed elements will usually be in the majority. Hence, rather than try to use lumped-element prototypes and correct for the nonlumped nature of their physical equivalents, it is usually much more effective to design microwave circuits starting with a pure distributed prototype. This ensures that the number of lumped-to-distributed or distributed-to-lumped element type conversions that might be required is kept to an absolute minimum. (Type conversions are generally not desirable because of the approximations involved.)

In a pure distributed f-plane prototype network, the basic types of circuit elements comprise open-circuit stubs, short-circuit stubs, and interconnecting transmission lines. As will be seen in Section 2.3.6, the stubs can be arranged into more complicated reactance branches in the network, but the two types of stubs and the interconnecting lines are the most fundamental of distributed-element building blocks. The relevant frequency transformation between the f-plane and the S-plane is the *Richards Transformation* [1], for which S is the complex frequency variable and because of which all the distributed elements in the f-plane network *must* be of equal electrical length.

In Figure 2.12, the Richards Transformation (i.e., $S = j\omega = j\tan((\pi/2) \times (f/f_s)))$, has been applied to low-pass, high-pass, and band-pass S-plane frequency responses. The frequency f_s is known as the *commensurate frequency* and is the frequency at which all the elements of the network have an electrical length of 90 deg. That is, all the elements have a physical length equal to a quarter of a wavelength at the commensurate frequency. In each case, the effect of the transform is to map the aperiodic S-plane response between $\omega = 0$ and $\omega = \infty$ into a repetitive, periodic response in the f-plane. The entire S-plane response is mapped into the f-plane between the pairs of frequencies $f = nf_s$ and $f = (n + 1)f_s$ for all even values of n between 0 and ∞. Its reflection about $\omega = \infty$ is mapped to the region between the pairs of frequencies $f = nf_s$ and $f = (n + 1)f_s$ for all odd values of n between 1 and ∞. The interval of periodicity on the f-axis is $2f_s$. For the low-pass S-plane response, this produces what is classed as a band-stop response in the f-plane, the length of all the elements of the distributed network being a quarter of a wavelength at the center of the stopband. For the high-pass S-plane response, this produces a band-pass response in the f-plane, all the elements being a quarter of a wavelength at the center of the passband. For the band-pass S-plane response, this produces another band-pass response in the f-plane, but unlike the previous f-plane band-pass response, the distributed elements of the network are a quarter of a wavelength

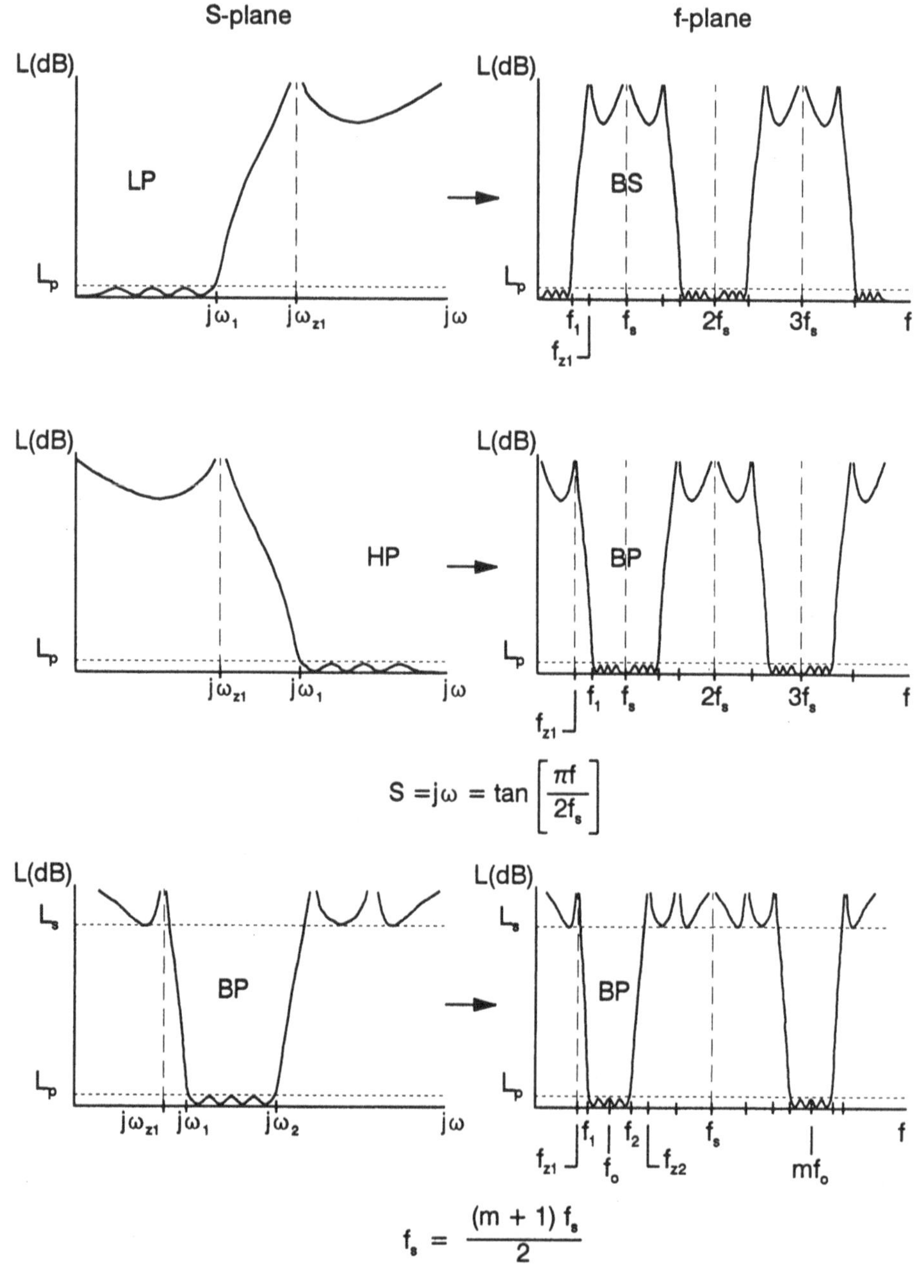

Figure 2.12 The Richards Transformation for low-pass, high-pass, and band-pass *S*-plane frequency responses.

long at the center of the stopband, well above the region of the passband. In all cases, the frequency f_0 is the arithmetic mean of the frequencies f_1 and f_2 (i.e., $f_0 = (f_1 + f_2)/2$), marking as it does the center of the passband for the band-pass f-plane responses and the center of the stopband for the band-stop response. Hence, the frequencies f_0 and f_s are the same for the two cases of the quarter-wave band-pass and band-stop responses.

Band-pass networks derived from high-pass S-plane prototypes tend to be most efficient in terms of the number of elements that are required to produce a passband of a given width and a given selectivity (i.e., the slope of the edges of the passband). This is because the same distributed elements that produce the high-pass half of the passband (i.e., the lower half) also produce the low-pass or upper half of the passband. Below the commensurate frequency they behave like high-pass elements, similar to those of the S-plane prototype, but above the commensurate frequency, the change of sign in the tangent function of the Richards Transformation causes them to behave like low-pass elements. This has some practical advantages, which will become clear in Section 3.1, but a significant disadvantage is that it is impossible to achieve any independent control of the width of the passband and the width of the stopband. Because the second passband on the frequency axis of the f-plane is always centered on a frequency $3f_s$, the width of the stopband immediately above the first passband is fixed by the width of the passband. This means that for wide passbands the stopband can be unacceptably narrow.

In contrast, band-pass networks derived from band-pass S-plane prototypes tend to require twice as many elements as the high-pass case for a given passband selectivity, since there is no "double action" of the distributed elements in producing the complete passband. However, the elements in the band-pass case can be very much shorter than those in the high-pass case, since they are a quarter of a wavelength long in the center of the stopband. Furthermore, there is now some freedom to choose the commensurate frequency f_s independently of the passband specification, which also allows the stopband width to be independently adjusted. Practical constraints on the realizability of the elements will limit the value of f_s, but the ability to adjust f_s is an extremely valuable extra degree of freedom. In the theoretical limit, as f_s tends to infinity, the equivalent inductances and capacitances of the elements of the distributed network become the same as those of a pure lumped network with the same frequency characteristics.

The effects of the Richards Transformation on the elements of an S-plane network are shown in Figure 2.13. Inductors and capacitors in the S-plane network become short-circuit and open-circuit stubs, respectively, in the f-plane. As a one-port element with an impedance $Z = SL$, an inductor is transformed into a short-circuit stub with an impedance $Z = jZ_0 \tan\theta$, where Z_0 is the characteristic impedance of the stub and $\theta = (\pi/2 \times f/f_s)$. The stub length l, which is known as the *commensurate length*, is a quarter of a wavelength at f_s. Similarly, a capacitor with a value C and an admittance $Y = SC$ is transformed into an open-circuit stub of admittance $Y = jY_0 \tan\theta$ and characteristic admittance Y_0. These examples reinforce the point

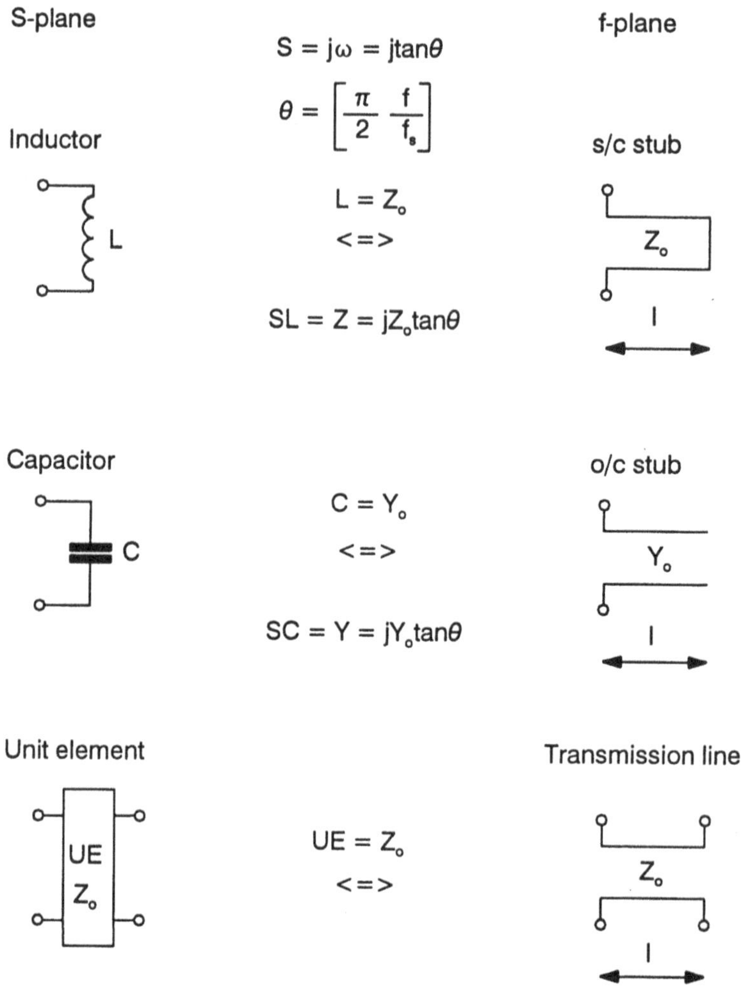

$$\frac{1}{\sqrt{1-S^2}}\begin{bmatrix}1 & Z_oS\\ Y_oS & 1\end{bmatrix} = \begin{bmatrix}A & B\\ C & D\end{bmatrix} = \begin{bmatrix}\cos\theta & jZ_o\sin\theta\\ jY_o\sin\theta & \cos\theta\end{bmatrix}$$

$$\theta = \left[\frac{\pi}{2}\,\frac{f}{f_s}\right] \qquad l = \frac{\lambda}{4}\ @\ f_s$$

Figure 2.13 The Richards Transformation applied to *S*-plane circuit elements.

that elements defined in the complex plane are not necessarily realized in accordance with their symbolic representation. While drawn as inductors and capacitors, these *S*-plane elements actually represent stubs whose values have dimensions expressed in ohms and siemens, respectively.

The effect of the Richards Transformation on a *unit element* (UE) in the *S*-plane is to turn it into an interconnecting transmission line of length l in the *f*-plane. A unit element, therefore, is unique to the synthesis of a distributed network and so was not a member of the set of elements available for synthesizing lumped-element networks. Its behavior in the *S*-plane is unlike any real, physical element, and it is therefore symbolized simply as a two-port box with a value Z_0 or Y_0. Its value corresponds to the characteristic impedance or admittance of a through transmission line of length l. The transformation can be further illustrated by considering the ABCD matrix of a through transmission line of the form

$$\begin{bmatrix} \mathrm{A} & \mathrm{B} \\ \mathrm{C} & \mathrm{D} \end{bmatrix} = \begin{bmatrix} \cos\theta & \mathrm{j}Z_0 \sin\theta \\ \mathrm{j}Y_0 \sin\theta & \cos\theta \end{bmatrix} \tag{2.37}$$

which transforms into a unit element in the *S*-plane with an ABCD matrix:

$$\begin{bmatrix} A' & B' \\ C' & D' \end{bmatrix} = \frac{1}{\sqrt{(1 - S^2)}} \begin{bmatrix} 1 & Z_0 S \\ Y_0 S & 1 \end{bmatrix} \tag{2.38}$$

after dividing throughout by cos θ and substituting S for j tan θ. Since all the elements of any ABCD matrix are ratios of input and output voltages or currents, the $1/\sqrt{(1 - S^2)}$ factor in the matrix for the unit element must create a half-order transmission zero at $S = \pm 1$. The unit element, therefore, does produce a transmission zero, but the zero occurs at $S = \pm 1$ on the real axis of the *S*-plane. In other words, there is no transmission zero on the jω axis of the *S*-plane, which is consistent with the property of a real, through transmission line, which can never in its own right completely suppress transmission.

On a practical note, adding nonredundant unit elements to a network will increase the selectivity of any of the low-pass, high-pass, or band-pass frequency responses, as well as providing the means of separating the reactive elements. However, the increase in selectivity will nearly always be less pronounced than the alternative of adding a reactive element, whatever the type, with its transmission zero or zeros lying on the jω axis.

2.3.6 Generalized Ladder Network Prototypes

The previous sections concentrated on the general properties of two-port prototype networks, defining the external electrical parameters most relevant to their synthesis, and on the range of frequency responses that can be achieved. Attention in this section will now be given to the rules governing the internal structure of the networks.

Starting with only the assumption that the network must be a two-port and that it must be loss-free, it is possible to imagine synthesizing a network of completely unconstrained topology, comprising perhaps a two- or three-dimensional mesh of interconnected *S*-plane elements. However, not only would such general topologies be impossible to synthesize by a single universal procedure, there is no physical justification for allowing such generality. Most practical microwave circuits have one input and one output with a cascade of relatively simple elements in between. Nearly all are ladderlike, and even those that, at first sight, have more the appearance of a two-dimensional mesh can usually be collapsed into some kind of ladder network. Hence, sound practical considerations and the ease with which a universal synthesis procedure can be established combine to force the two-port network to have the topology of a ladder.

The synthesis procedures described in this text are able to synthesize a *generalized ladder network* of the form shown in Figure 2.14. The ladder is a cascade of series and shunt reactance branches (Z_1 and Y_2, etc.), together with unit elements (UE_4 and UE_5, etc.), and with only one or two exceptions, there is complete freedom to choose the sequence in which the elements appear in the cascade. Use of the term *generalized* is intended to signify the extension of the classical lumped-element ladder network to include unit elements. Figure 2.14 illustrates a doubly terminated network, but the ladder is equally relevant to the synthesis of a singly terminated network.

Each reactance branch of the ladder network will be capable of producing at least one *S*-plane transmission zero. These zeros will reside on the $j\omega$ axis somewhere in the stopband response, and as a consequence of the Richards Transformation they will correspond to zeros at a multiplicity of frequencies in the *f*-plane. Each unit element will produce a single zero of transmission at $S = (\pm 1, j0)$ in the *S*-plane, which while not producing an *f*-plane zero, will give some increase in passband selectivity. The reactance branches can take the form of any valid Foster network, the most simple of which would be a single *S*-plane inductor or capacitor. There is no limit to the number of elements that could be used in a complex branch, but it is unlikely that a branch with more than four elements would ever be required. To illustrate some of the options for reactance branches, a set of

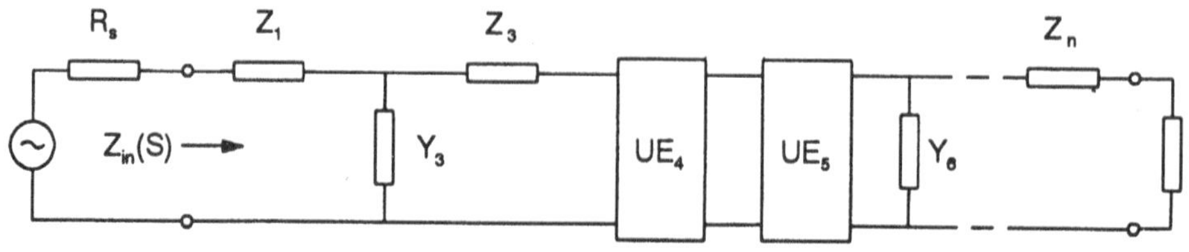

Figure 2.14 The generalized ladder network.

examples is given in Figure 2.15. Six different types, each with a series or shunt implementation, have been tabulated. The table also states the degree of the branch (i.e., the number of simple elements it contains), the corresponding number of transmission zeros it produces, and the location of the transmission zeros.

Branches of type 1 shown in Figure 2.15 are either series capacitors or shunt inductors. As an impedance $1/SC$ or an admittance $1/SL$, they produce a single zero of transmission at $S = j0$ and are generally referred to as *high-pass elements.*

Type No.	Element type		Degree (order) of branch	Number of Tx zeros	Location of Tx zeros
	Series	Shunt			
1	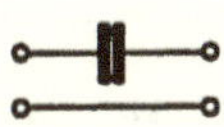	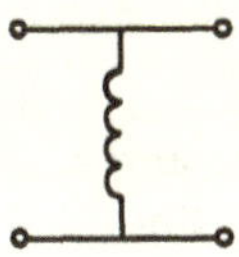	1	1	$S=j0$
2	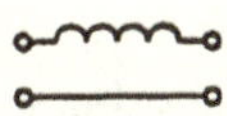	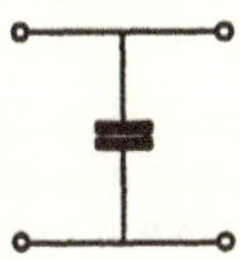	1	1	$S=j\infty$
3	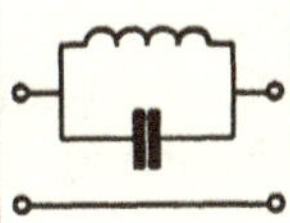	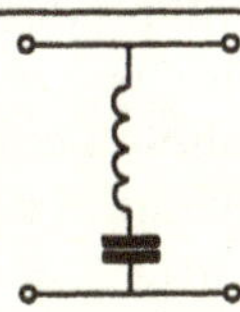	2	1	$S=j\omega_{z1}$
4	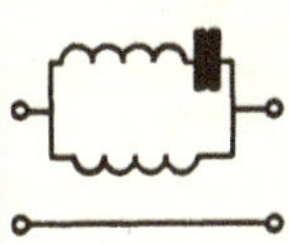	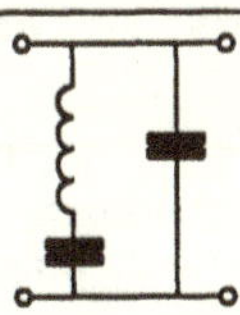	3	2	$S=j\omega_{z1}$ and $S=j\infty$
5	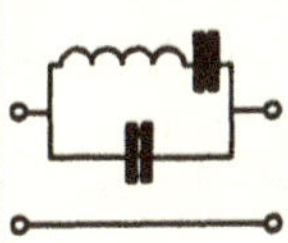	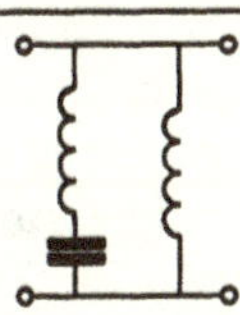	3	2	$S=j\omega_{z1}$ and $S=j0$
6	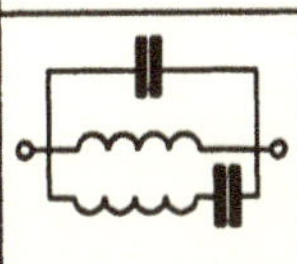	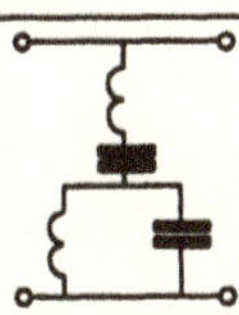	4	2	$S=j\omega_{z1}$ and $S=j\omega_{z2}$

Figure 2.15 A summary of reactance branches for ladder networks.

They make a unity increase to the element count and to the degree of the whole network. (Note: With reference to Section 2.3.2, the *degree* of the network is equal to the degree of either the numerator or the denominator polynomial of the impedance function $Z_1(S)$, whichever is greater.) Branches of type 2 in Figure 2.15 are either series inductors or shunt capacitors. As an impedance of SL or an admittance of SC, they produce a single zero of transmission at $S = j\infty$ and are generally referred to as *low-pass elements.* They also make a unity increase to the network's element count and degree. In the case of type 3 branches, these are series- or parallel-tuned circuits, each containing a capacitor and inductor. Taking the series branch as an example, the impedance of the branch is given by

$$Z(S) = \frac{SL}{1 + S^2LC} \tag{2.39}$$

where L and C are the respective inductor and capacitor values. Clearly, this has a pole at $S = S_{z1} = j\omega_{z1} = j1/\sqrt{(LC)}$, causing a single zero of transmission at $S = (0, j\omega_{z1})$. It is a branch that contributes a *single* transmission zero at a finite, nonzero frequency, but that makes a contribution of *two* to the network degree.

Branches of type 4 in Figure 2.15 comprise three basic components in a series or shunt configuration. They have the appearance of a combination of a low-pass element and a type 3 branch of degree 2. An examination of the impedance function for the series version makes it clear that it will produce two transmission zeros, one at $S = (0, j\omega_{z1})$ and another at $S = (0, j\infty)$. With its three elements, however, it makes a contribution of 3 to the degree of the whole network. The branch of type 5 is very similar to the branch of type 4, but instead of involving a low-pass element, it uses a high-pass element in combination with one of the type 3 branches. This produces zeros at $S = (0, j\omega_{z1})$ and $S = (0, j0)$ and also contributes a value of 3 to the network degree.

The sixth type of branch in the table is a series or shunt Foster network of degree 4. Consideration of the relevant impedance or admittance functions confirms that the element must produce two transmissions zeros at finite, nonzero frequencies of $S = (0, j\omega_{z1})$ and $S = (0, j\omega_{z2})$. With its four basic components, it is of degree 4 and therefore adds 4 to the degree of the whole network.

The effect of adding a unit element in the ladder is to increase the overall degree of the network by 1. This can be verified by considering the input impedance of a unit element terminated in an arbitrary impedance Z_l. In terms of ABCD parameters, Z_{in} is given by

$$Z_{in} = \frac{A'Z_l + B'}{C'Z_l + D'} \tag{2.40}$$

which, after substitution for the matrix elements using (2.38), becomes

$$Z_{in} = \frac{Z_l + Z_0 S}{Y_0 Z_l S + 1} \tag{2.41}$$

In (2.41), both the numerator and denominator polynomials are of unity degree. Therefore, if the unit element were to be added to the front of an existing network with an input impedance Z_l, it is clear that the product of Z_l and S in the denominator of (2.41) would result in a unity increase in the degree of the original network.

Some additional points can be made in connection with the reactance branches. First, the series and shunt versions of each type of reactance branch represent a pair of network duals. As simple networks in their own right, the reactance branches are always to be found in both first and second canonical forms. When embedded in a larger two-port network, converting from one dual to the other is a simple matter of substituting capacitors for inductors and inductors for capacitors, with a one-to-one transfer of element values. The second point concerns the reactance branches of types 3 through 6 in Figure 2.15. While able to give a network a much increased selectivity, they can also be used as a means of making better use of the space available for a given microwave circuit requirement. Networks that contain a large number of elements often run the risk of becoming rather long and thin. By using the more complex reactance branches, it may be possible to reduce the length of the physical circuit in exchange for an increase in width, thereby altering the aspect ratio of the circuit to a value nearer unity.

2.3.6.1 Transmission Line Equivalent Circuits of Reactance Branches

The effects of the various frequency transformations on the elements of a prototype network were discussed in Section 2.3.5. For lumped-element circuits, the relevant transforms change the capacitors and inductors of the *s*-plane prototype into real capacitors and inductors in the *f*-plane. Consequently, all the reactance branches indicated in Figure 2.15 could be realized using real lumped capacitors and inductors. For distributed circuits, the effect of the Richards Transformation is to change the capacitors and inductors of the *S*-plane prototype into open- and short-circuit stubs in the *f*-plane. Therefore, in *theory*, all the branches shown in Figure 2.15 are realizable as combinations of the appropriate types of stub. In practice, however, only the branches of types 1 and 2 would be realizable in the form of simple stubs. The complexity of the interconnections required for the other types would make their realization impossible in any of the planar transmission line media presently in common usage.

One solution to the problem of realizing branches of types 3, 4, 5, and 6 is to employ stubs comprising multiple sections of transmission line. It was proved by Richards [1] that any *S*-plane reactance branch can be realized as a cascade of unit elements terminated in either an open or a short circuit. Branches of type 3, therefore, can be realized as a single stub whose length is twice the commensurate length and whose two sections have, in general, different characteristic impedances. The series version of the branch would be a double-length series stub terminated in a short circuit, whereas that of the shunt type would be a double-length shunt stub terminated in an open circuit. Similarly, branches of types 4 and 5 are realizable as single stubs comprising three sections of transmission line and those of type 6 are realizable as four sections of transmission line. The relevant formulas for converting the degree 2 and degree 4 reactance branches into multisection stubs are given in parts 1 and 2, respectively, of Appendix A. For conversion of the degree 3 branches, the formulas for the degree 4 branches can be used with the appropriate element values set to zero.

Reactance branches of types 4, 5 and 6 can also be realized using a combination of multisection stubs wired in parallel. For example, parallel branches of type 4 can be realized using two stubs wired in parallel, one a single-section open-circuit stub and the other a double-section open-circuit stub. In the case of parallel branches of type 6, these can be realized using two open-circuit, double-section stubs wired in parallel. The relevant formulas for splitting the branches of type 6 (degree 4) into pairs of branches of type 3 (degree 2) are given in part 3 of Appendix A. Which particular stub configuration turns out to be more appropriate will depend on the characteristic impedances of the various transmission line sections and the consequent ease of physical realization. It should be noted that the characteristic impedances of the stub sections are strongly affected by the location of the corresponding transmission zeros.

One final point to be noted with regard to the reactance branches is that despite the preceding discussion on the use of multiple-section stubs, only the shunt branches/stubs tend to be physically realizable in most of the popular transmission line media. The reasons for this will become clear during the coverage of transmission line media in Section 2.5.

2.3.6.2 Creating the Ladder Network from the List of Transmission Zeros

The elements of the ladder network are extracted not by specifying the type of element to be extracted, but by specifying the particular transmission zero that is to be created. In theory, there are virtually no constraints on the order in which *transmission zeros* are created. However, when a particular transmission zero is specified during the sequence of extractions, the precise type of element that is extracted for that zero will not always be the same. Its type will depend on the order in which any preceding zeros were created.

For example, in a low-pass network of the first canonical form, any low-pass element immediately adjacent to the input must be a series inductor to allow the input impedance to tend toward an open circuit in the stopband. It is not possible to create a shunt capacitor until either the inductor has been fully extracted or a single unit element has been extracted. This means that despite the absence of constraints on the order of the zeros, there are constraints on the physical composition of the network, and the designer is not entirely free to specify the order of the *elements.* However, so long as the constraints are understood, this is not a major difficulty, since the designer will always be able adjust the composition of the complete set of transmission zeros to give a permitted network configuration close to if not the same as the type required.

For nonredundant ladder networks containing no unit elements, as a general rule there must be an alternating sequence of any series and shunt low-pass elements, and an alternating sequence of any series and shunt high-pass elements. The presence of high-pass elements does not interfere with the necessary series/parallel sequence of the low-pass elements or vice versa. Any unit elements, however, do alter the sequence for low-pass elements. For each unit element extracted, there will be a change in the type of all succeeding low-pass elements; series inductors become shunt capacitors and shunt capacitors become series inductors. In certain cases where, for example, a sequence of unit elements is interlaced with a sequence of low-pass elements, this can lead to a network in which all the low-pass elements are of the same type (i.e., all series inductors or all shunt capacitors). As far as high-pass elements are concerned, their series/parallel sequence is unaffected by the presence of unit elements. Unit elements themselves remain unit elements at whatever point they are extracted in the network, and there are no restrictions at all on their order of extraction.

2.3.6.3 Constraints on Network Topology and Network Degree

Some important constraints on the order in which elements may be extracted from a network will now be described. To help with the description, six different input configurations have been drawn in Figure 2.16, four of which would not be permitted for the synthesis of a nonredundant band-pass network and only two of which would be permitted. All six configurations involve one high-pass and one low-pass element.

It is perhaps surprising that *unless the proposed network contains an appropriate number of redundant unit elements,* the first four input configurations of Figure 2.16 cannot be synthesized using the techniques described herein, and are therefore not permitted. With reference to Figure 2.16(a), the extraction has begun with an attempt to create a transmission zero at $S = (0, j0)$, resulting in the extraction of a series capacitor. This indicates that the network must be of the first canonical form, whose input impedance will tend toward an open circuit in the stopband. The second element as shown in the figure is a shunt capacitor, which is a low-pass element capable of producing a transmission zero at $S = (0, j\infty)$. Its presence at this position, however, would force the input impedance to tend toward a short circuit in the upper stopband, which is inconsistent with a network of the first canonical form. Hence, given the way in which the $Z_1(S)$ function was derived in

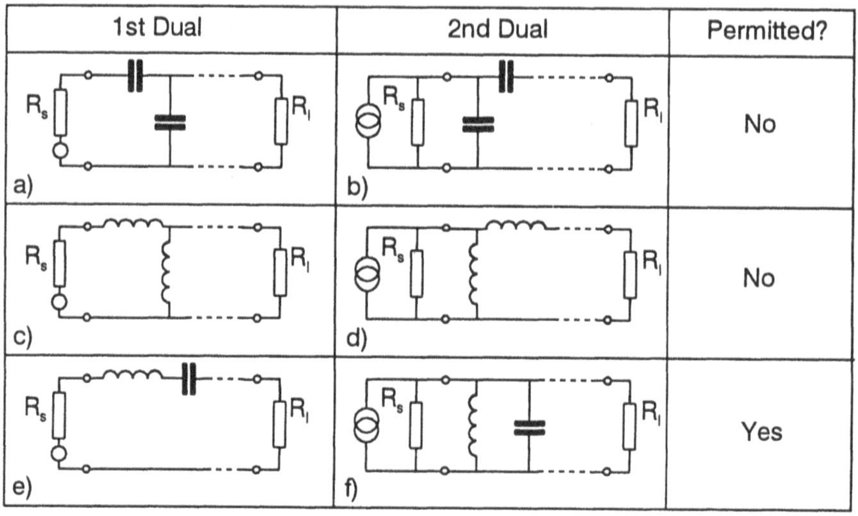

Figure 2.16 Some important prohibited input configurations together with the correct configurations.

Section 2.3.2, it would be impossible to extract the shunt capacitor. Any low-pass element extracted at this position would have to be a series inductor to allow the input impedance to tend toward an open circuit in the stopband. The correct form of the input configuration is, therefore, Figure 2.16(e) and would apply whenever zeros at $S = (0, j0)$ and $S = (0, j\infty)$ are the first two to be created.

Input configurations of Figure 2.16(b–d) are prohibited for the same basic reason as in 2.16(a); that is, within the same canonical form they represent an incompatible combination of series and shunt elements. Figure 2.16(b) is just the dual of 2.16(a); 2.16(c) is similar to 2.16(a), except that the low-pass element has been extracted first to produce a series inductor; and 2.16(d) is the dual of 2.16(c). Being the dual of Figure 2.16(e), 2.16(f) is a perfectly acceptable input configuration for a band-pass network of the second canonical form.

The only way in which the configurations in Figure 2.16(a–d) could exist at the input to a network would be for an odd number of *redundant* unit elements to have been moved through the two input elements from the source using an appropriate network transformation (note: suitable network transformations are soon to be described in Section 2.4.1). This would change the type of low-pass elements, but have no effect on the type of high-pass elements. There will be occasions when this may be advantageous, but it is important to be aware that while redundant unit elements have no effect on the magnitude of the transfer characteristics of a network, they do affect the network's input impedance. More discussion of this will follow in Section 2.3.6.4.

As far as the interior of the network is concerned, the capacitor and inductor combinations shown in Figure 2.16(a–d) are also prohibited unless preceded by an odd number of unit elements. If they are preceded by an odd number of elements, then, conversely, the configurations of types in Figure 2.16(e,f) become prohibited. The difference in the interior of the network is that the unit elements do not necessarily have to be redundant. They can be extracted as part of the extraction of the whole network if transmission zeros have been specified at $S = (1, j0)$.

An important constraint on the degree of a network applies for band-pass frequency responses. Whereas a low-pass or a high-pass network may be of any degree, the *degree of a band-pass network must always be even.* The reason for this can be understood by considering the required distribution of poles and zeros of the squared characteristic function $|K(S)|^2$. As previously explained and illustrated in Figure 2.6, the zeros of reflection must always appear in groups of complex quads. This is because the function must be even, its zeros must be in conjugate pairs, and no zeros, by definition, can be present at $S = (0, j0)$ or $S = (0, j\infty)$ for a band-pass response. Hence, the function $K(S)$ must have an even number of zeros. Without reflection zeros at $S = (0, j\infty)$, the degree of the numerator of $K(S)$ must always be greater than or equal to that of the denominator, which makes the degree of the network equal to the number of reflection zeros. The band-pass network must, therefore, be of even degree. Low-pass and high-pass networks are not similarly restricted, because zeros of reflection may appear in pairs (i.e., not necessarily in quads) at $S = (0, j0)$ and $S = (0, j\infty)$, respectively.

2.3.6.4 Redundant Elements

Nonredundant *S*-plane networks are networks in which the total number of individual inductors, capacitors, and unit elements is equal to the network degree. An alternative way of expressing this is to say that the number of elements in the network is the absolute minimum required to create the specified set of transmission zeros. In these circumstances, each reactance branch will create at least one transmission zero, and each unit element will also create a transmission zero. Redundant elements have the effect of increasing component count as well as circuit complexity and therefore should not be introduced without good reason. However, their introduction is not uncommon in practice, since they are often able to ease topological difficulties and/or reduce the dynamic range of element values.

When redundant unit elements are required, they can be introduced into a network from one or both ends. With reference to Figure 2.14, if an extra unit element with a characteristic impedance of R_l is introduced from the load and transformed through the last network element Z_n, then as far as the network is concerned, the effective load resistance will be unchanged. Furthermore, there will be no change to the network's input impedance, its input reflection coefficient, or the magnitude of its transmission coefficient. The only effect will be to change the phase of the transmission coefficient, and this will usually be of no consequence.

At its position at the end of the network, the extra unit element is clearly redundant (i.e., it does not create any extra transmission zeros). However, even when the unit element is transformed further inside the network, it always remains a redundant element, whatever its subsequent position. This does not mean that the element plays no role in the behavior of the whole network, since to simply cut it out would destroy the network's electrical behavior. But it is, by definition, still a redundant element in the sense that it can be transformed back out of the network toward the load, leaving the network's power transfer characteristics unaffected.

There is no limit to the number of unit elements that can be introduced into a network. They can, however, only be moved through reactance branches and cannot be moved through any existing unit elements in the network. It is unlikely, therefore, that the number of useful redundant elements would ever exceed the number of accessible reactance branches. Their most important role is probably that of element separators within a cascade of reactance branches. They are occasionally preferable in this role instead of nonredundant elements, because, as previously mentioned, they can result in a more realizable range of element values. Another important role that should be mentioned is that of impedance level transformer. The introduction of a unit element and its transformation through a high-pass element at one or both ends of a network has the effect of increasing or decreasing impedance levels. For a symmetrical network, this can be used as a means of altering internal impedance levels to help with physical realization. For an asymmetrical network, it can be used as a means of altering the ratio of the terminating impedances, giving the network the characteristics of a band-limited ideal transformer. The transformation of a unit element through a high-pass reactance branch will be described in Section 2.4.1.

Redundant unit elements can be introduced from the source end of a network, but only when the network is doubly terminated. A redundant unit element of characteristic impedance R_s introduced from the source will not change the magnitude of the reflection or transmission coefficients, but it will change the phase of both. There are several practical circumstances, mostly where matching networks are involved, when the change in phase of the reflection coefficient will not be acceptable. In these cases, unit elements cannot be introduced from the source end of the network. Similarly, redundant unit elements cannot be introduced from the source end into a singly terminated network, as this would require the unit elements to have a zero or infinite characteristic impedance.

Redundant reactive elements are created whenever a single reactive branch is split into one or more parts. For example, a single series capacitor in a nonredundant network will be capable of producing a single transmission zero at $S = \cdot(0, j0)$. Splitting the capacitor into two capacitors in series makes no change in the network behavior, but there is now an extra element in the network that could be moved to some other location by a suitable transformation. The reasons for splitting and moving reactance branches are broadly similar to those for introducing redundant unit elements. Where redundant high-pass elements are involved, it is important to note that they can be moved through any of the network elements except another high-pass element. Similarly, redundant low-pass elements can be moved through any type of element except another low-pass element. As will be explained in Section 2.3.8, splitting reactance branches into redundant components is equivalent to a *partial extraction* of the branches during the second part of the network synthesis process.

2.3.6.5 Determination of Transmission Zeros and Network Degree

There will be occasions when the starting point of a circuit design project is the adoption of an existing circuit that can be modified to achieve the desired performance, rather than starting with an entirely new circuit with a completely new set of transmission zeros. In this situation, the first step is to translate the physical circuit into an *f*-plane equivalent circuit and then make another transformation into a valid *S*-plane prototype. The aim will be to establish a valid *S*-plane ladder network that can be resynthesized, perhaps with some minor topological modification, to meet the required electrical performance. In order to succeed with the derivation of the *S*-plane ladder network, it is essential to be able to:

1. Recognize any redundant elements in the network;
2. Compose the relevant set of transmission zeros;
3. Calculate the network degree;
4. Determine whether it is of the first or second canonical form;
5. Verify that the network falls within the appropriate topological constraints.

With experience, this is a simple task requiring little more than a cursory glance over the *S*-plane network. However, there are occasions when it is more difficult, particularly when the network does contain redundant elements. To help in these more difficult situations, there is a simple, step-by-step procedure that can be followed, which is described below.

The begin with, it is important to try to identify any redundant reactance branches, and this is done by moving all the unit elements to the load end of the network. For this purpose, it is only necessary to take into consideration the changes in the type of the low-pass elements as each one is transformed by a unit element. Any numerical computation of element values is not really relevant, and in any case, the original element values are not likely to be known. This has the effect of forming a group of pure reactance branches toward the input end of the network in which the redundant branches are immediately recognizable. Among this group, it is possible to conclude that any two low-pass elements of the same type separated by anything other than a low-pass element must be part of the same nonredundant low-pass element, and any two high-pass elements separated by anything other than a high-pass element must be part of the same nonredundant high-pass element.

For the unit elements now grouped together at the load end of the network, any that are obviously redundant would need to have a characteristic impedance equal to or very close to the value of the load resistance. Since the information regarding their precise numerical value is unlikely to be available, none would normally be assigned redundant on only this limited evidence. Some clues may be gained from the configuration of the reactance branches at the input to the network or from the apparent degree of the network (i.e., band-pass networks must be of even degree). Similarly, some special prior knowledge about the particular microwave structure in question can be helpful. It may be, for example, that a filter with a given frequency bandwidth is only physically realizable if redundant unit elements are introduced into its prototype. In cases like this, it may be justifiable to assign redundancy to some of the unit elements, but not otherwise. For the sake of completeness, a check on the presence of source-end redundant unit elements should be made by moving all the unit elements back through the network to the source. However, once again it is unlikely that any safe assignment of redundancy could be made without some additional physical justification.

Once the identification of redundant elements has been completed, the network degree can be calculated by adding the number of nonredundant unit elements to the total number of nonredundant inductors and capacitors. Examination of the reactance branches will identify their contribution to the list of transmission zeros, whether these are at $S = (0, j0)$, $S = (0, j\infty)$, or at some finite, nonzero frequency on the $j\omega$ axis. The nonredundant unit elements will each contribute a zero at $S = (1, j0)$. The network will be of the first canonical form if, after the movement of the unit elements, the first reactive element adjacent to the source is a series element. If the first reactive element is a shunt element, then the network will be of the second canonical form. For the network to be valid and one that can be synthesized, there must be a strict adherence to the rule of alternating series and shunt branches for each of the set of nonredundant low-pass and high-pass elements. There should be no occurrence of the sort of prohibited capacitor or inductor combinations that were described in Figure 2.16, unless there is a possibility that an odd number of source-end redundant unit elements might have been introduced into the network. If both low-pass and high-pass elements are present, then it must be assumed that the network has band-pass characteristics and must therefore be of an even degree.

This whole identification and validation process for S-plane prototypes can be further clarified with the help of a simple example. The starting point will be the network illustrated in Figure 2.17, which could represent either a first attempt at the conversion of an existing microwave circuit into an S-plane prototype or just an initial sketch of a prototype for a possible new circuit. To make the problem as general as possible, but at the same time limit its complexity, elements of all the various different types have been included except reactance branches of multiple degree. Hence, elements capable of producing zeros at $S = (0, j0)$, $S = (0, j\infty)$, and

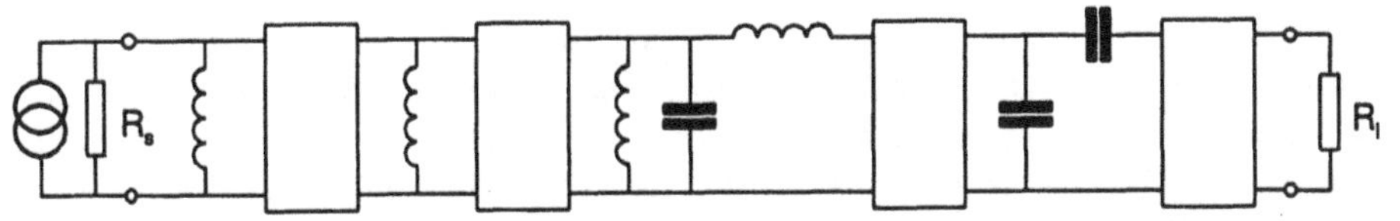

Figure 2.17 An example of an *S*-plane prototype network including redundant elements.

$S = (1, j0)$ are present, but there are none corresponding to any finite nonzero transmission zeros. The network also contains some redundant elements.

With reference to Figure 2.17, it can be seen that the network contains a total of 11 elements, not including the terminations. At first sight, it might be tempting to assume that the eleven elements correspond to a set of 11 transmission zeros and to a network degree of 11. This would not be correct, however, because there are some redundant reactive branches present that complicate the situation and that need to be identified. By a qualitative application of the transforms to be described in Section 2.4, all the unit elements of the network can be moved through the reactance branches to form a unique group at the right-hand side next to the load, as shown in Figure 2.18. This leaves all the reactance branches in another group to the left-hand side, the high-pass elements of which are unchanged in type. Of the low-pass elements, two out of three are unchanged, but the type of the third has changed from a shunt capacitor into a series inductor, because it was transformed through an odd number (i.e., 3) of unit elements. Consequently, at the input to the transformed network, there are now three shunt inductors in parallel, two of which are redundant, the three together forming what is effectively a single inductor. The next shunt capacitor is clearly not redundant, because it cannot be combined with anything else. However, the next two series inductors can be combined into a single series inductor, and, therefore, one of these inductors must be redundant. Finally, despite being transformed by three unit elements, the last series capacitor has not been changed in type and is not redundant.

If it is assumed that none of the unit elements is redundant, then the original *S*-plane prototype must have contained three redundant elements (i.e., two shunt inductors and a shunt capacitor). Hence, after combining the redundant elements, the total number of elements in the nonredundant network is eight and not the original figure of eleven. The degree of the network is therefore 8. Taking note of

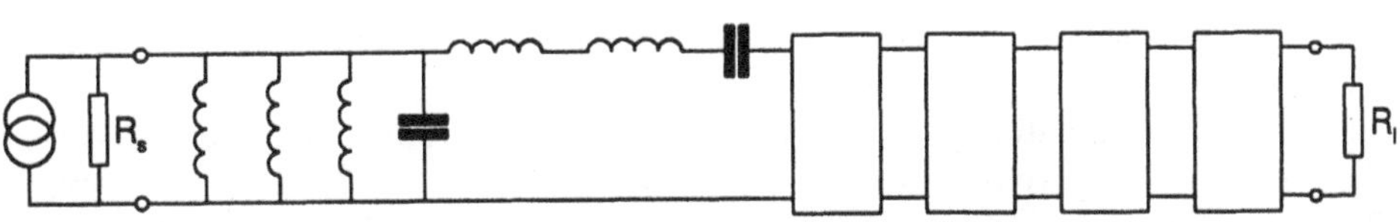

Figure 2.18 Moving unit elements to reveal redundant network elements.

the types of elements, it is clear that there are two transmission zeros at $S = (0, j0)$, two at $S = (0, j\infty)$, and four at $S = (1, j0)$ making another total of eight. Owing to the presence of both low-pass and high-pass elements, it must be a band-pass network, and it must be of the second canonical form, since after the movement of the unit elements, the element next to the source is a shunt reactance branch. Since it is of an even degree and its reactance branches have the required alternating series/parallel sequence, it is reasonable to conclude, therefore, that the network is a valid S-plane prototype.

Some additional insight into the constraints on the composition of a prototype can be gained by considering the effects of certain changes to the circuit of Figure 2.17. If the first shunt inductor had been a series capacitor, this would have forced the network to be of the first canonical form and would have immediately caused a conflict with the first low-pass element, which is a shunt capacitor. It would also have increased the degree of the network to the odd value of 9, and for both these reasons, the network would have been invalid. If the second shunt inductor had been a series capacitor, this would also have increased the degree to 9 and would not have been allowed without some other compensating change to the network being made. Similarly, changing the shunt capacitor nearest the load to a series inductor would have increased the degree to 9, or changing the series capacitor nearest the load to a shunt inductor would have decreased the degree to 7, neither of which would be permitted in isolation.

2.3.7 The Approximation Problem

As previously described in Section 2.2, the synthesis of the S-plane prototype is separated into two-stages, the first dealing with the generation of the input impedance polynomials $Z_1(S)$ and the second with the extraction of the network elements from the $Z_1(S)$ polynomials. As part of the first stage, it is necessary to generate the polynomials of the squared characteristic function $|K(S)|^2$, since these are the polynomials that essentially determine the power transfer characteristics of the network. It is this part of the synthesis process to which the term *approximation problem* applies. By definition, the approximation problem is the task of creating a transfer function of a realizable two-port network, which is as close a fit as possible to some chosen target frequency response. Internal to the synthesis process, the relevant transfer function for the synthesis is $|K(S)|^2$, but the target frequency response is usually expressed externally in terms of the transmission coefficient $|\tau(S)|^2$ or the insertion loss function $L(j\omega)$ dB.

The approximation problem gets its name from the aim to generate a function that is a good approximation to, rather than an exact replica of, some desired frequency response. Indeed, in general, it would be impossible to generate an exact replica of a specified target function and at the same time create a realizable prototype network. Fortunately, this is never necessary in practice, and what happens in most cases is that the transfer function is made to approximate the required shape by forcing it to have certain values at each of a set of discrete frequencies. Generally, the greater the number of target frequency points, the better the approximation, but this also has an impact on the complexity of the prototype network.

The approximation problem can be further illustrated by considering the performance specification of an arbitrary band-pass filter. By definition, an *ideal* filter response is one that is of the general form shown in Figure 2.19, where insertion loss is zero within a specified frequency range f_1 to f_2, but is infinite at all other frequencies. Clearly, achieving this ideal filter response in practice is quite impossible, since this would require a network with an infinite number of elements and zero ohmic losses. However, it is not difficult to create networks with transfer functions that "approximate" the ideal response. The important issues are how good the approximation has to be and what the consequences for the realizability of the network are.

Most target frequency responses of microwave filters are specified using a diagram of the form shown in Figure 2.20. The shaded boundaries in the figure define a set of forbidden zones where no part of the insertion loss function must be allowed to fall. The zone below f_1 specifies the width and depth profile of the lower stopband, the zone in the middle specifies the width and loss profile of the passband, and the upper zone above f_2 specifies the width and depth profile of the upper stopband. Any insertion loss response falling outside these zones, such as the one shown in the figure, would be an acceptable response. To generate a transfer function of the type shown in Figure 2.20, the designer selects a band-pass

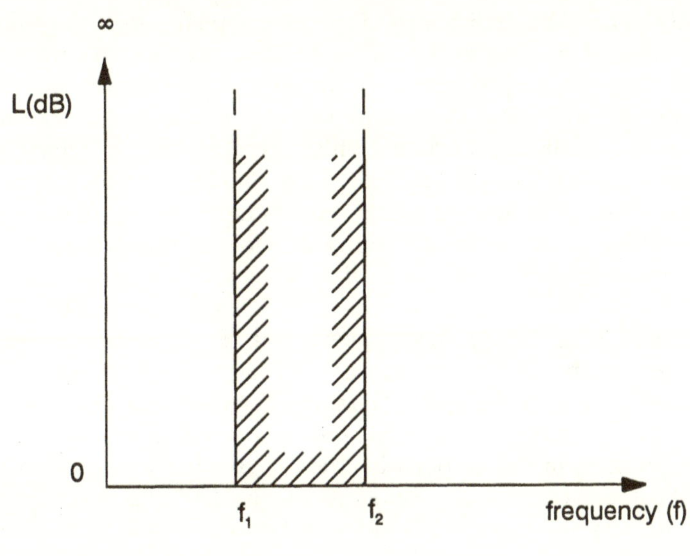

Figure 2.19 Ideal band-pass f-plane frequency response.

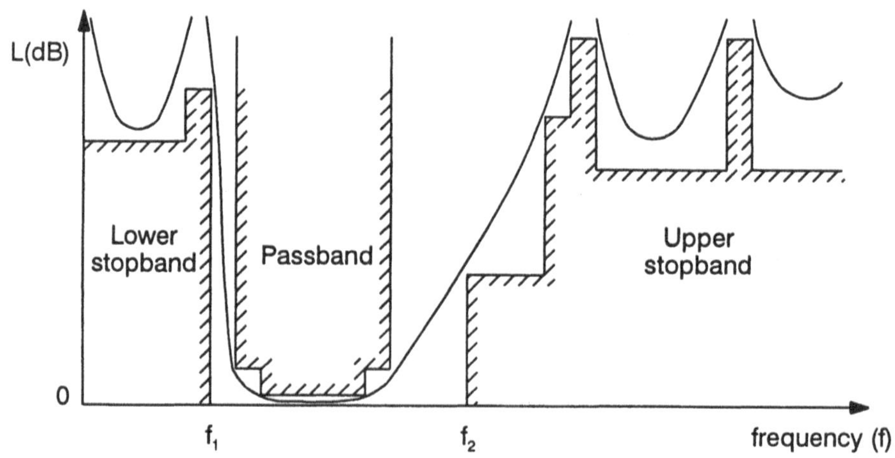

Figure 2.20 A typical band-pass frequency response specification.

response and places the passband edges somewhere in the gaps between the middle and outer two forbidden zones. The number and position of the transmission zeros are then specified to achieve the required stopband profiles. It is sometimes helpful to think of the transmission zeros as pins attaching the transfer function to an imaginary ceiling and the reflection zeros in the passband as pins attaching the function to an imaginary floor. As previously explained, while meeting the target frequency specification, the passband width and ripple values should be used as a means of influencing the realizability of the prototype network. Similarly, the position of the transmission zeros should be chosen not only to give the required performance, but also to give a network with the required mixture of elements.

It should be noted that the complexity of the target specification shown in Figure 2.20 is only really appropriate for filter networks. For matching networks of the sort used in active microwave integrated circuits (MIC), the stopband characteristics are usually of little interest and a much simpler target frequency response would be specified. This has the advantage of allowing the choice of transmission zeros to be based entirely on the desired network topology.

In most of the remainder of this section, the mathematics used by the synthesis software for generating the two most important classes of network frequency characteristics will be described. The mathematics will allow networks with either maximally flat (Butterworth) or equal-ripple (Tchebycheff) passband characteristics to be created, and in both cases, it will be possible to place transmission zeros at finite, nonzero frequencies in the stopbands, thereby creating pseudoelliptic stopband responses. It will also cover the introduction of passband offsets and gradients. The treatment is not absolutely essential reading as far as the application of the synthesis techniques is concerned, but it will be of interest to readers hoping to write their own synthesis software.

To generate a maximally flat passband response, all the transmission zeros, except those at $S = (0, j\infty)$, are assembled as factors of the denominator of $|K(S)|^2$. Any zeros at $S = (0, j0)$ produce simple S^2 factors, while those at positive values of ω on the imaginary axis each produce a group of four linear factors. Transmission zeros at infinity are created by making the degree of the numerator of $|K(S)|^2$ two greater than the degree of the denominator for each of the zeros. If there are no transmission zeros at infinity, then the degree of the numerator is determined by the presence of reflection zeros at $S = (0, j\infty)$. For each one, the degree of the numerator should be made two less than that of the denominator. By definition, all of the reflection zeros for a Butterworth response must be placed at one single frequency on the $j\omega$ axis. For a low-pass Butterworth response, this frequency is $S = (0, j0)$, for a high-pass response it is $S = (0, j\infty)$, and for a band-pass response the frequency is at a value $S = (0, j\omega_0)$, where ω_0 is somewhere close to the center of the passband.

If we define:

a = number of transmission zeros at $S = (0, j0)$
b = number of transmission zeros at $S = (0, j\infty)$
c = number of transmission zeros at $S = (1, j0)$
d = number of transmission zeros at $S = (0, j\omega)$

where ω is finite and nonzero, then for a low-pass Butterworth response, $|K(S)|^2$ is given by

$$|K(S)|^2 = k_p \frac{S^{2n}}{\prod_{i=1}^{d} (S^2 - S_i^2)^2 \times (S^2 - 1)^c} \tag{2.42}$$

where n will be the degree of the resultant network for which $n = b + c + 2d$. There are no restrictions on the value of n. For a high-pass Butterworth response, $|K(S)|^2$ is given by

$$|K(S)|^2 = k_p \frac{1}{\prod_{i=1}^{d} (S^2 - S_i^2)^2 \times (S^2 - 1)^c \times S^{2a}} \tag{2.43}$$

in which all the zeros of reflection are produced by the difference in the degree of the numerator and denominator polynomials. The degree of the network n is in this case $n = a + c + 2d$, and again there are no restrictions on its values.

For a band-pass Butterworth response, $|K(S)|^2$ is given by

$$|K(S)|^2 = k_p \frac{(S^2 - S_0^2)^{2n}}{\prod_{i=1}^{d} (S^2 - S_i^2)^2 \times (S^2 - 1)^c \times S^{2a}} \tag{2.44}$$

where all the reflection zeros are at a frequency $S_0 = (0, j\omega_0)$. The degree of the numerator is always greater than or equal to that of the denominator, as determined by the number of transmission zeros at $S = (0, j\infty)$. This time the network degree will be $2n$, for which $n = ((a + b + c)/2 + d)$, and since n must be an integer, the degree of the network must be even. The sum of transmission zeros at $S = (0, j0)$, $S = (0, j\infty)$, and $S = (1, 0)$ must also be even.

In all the above cases, k_p is a multiplying constant that is used to set the numerical value of $|K(S)|^2$ at certain frequencies, usually the edges of the passband. For Butterworth responses, this invariably corresponds to an insertion loss value of 3 dB.

Generating an equal-ripple (i.e., Tchebycheff) passband characteristic could, in principle, be achieved by the same procedure, but an obvious difficulty arises with the placement of the reflection zeros. Unlike the maximally flat response, the zeros are not located at one frequency, but must be spread throughout the range of the passband to give the required ripple characteristic. To avoid the difficulty, an alternative method for generating $|K(S)|^2$, described in [5], is preferred, during which the correct location of the reflection zeros is computed.

The alternative method makes use of yet another frequency transformation, this time from the S-plane into the z-plane. The low-pass, high-pass, and band-pass forms of the transform are as follows:

$$z = \sqrt{1 - \frac{S_2^2}{S^2}} \quad \text{for low-pass responses} \tag{2.45}$$

$$z = \sqrt{1 - \frac{S^2}{S_1^2}} \quad \text{for high-pass responses} \tag{2.46}$$

$$z = \sqrt{\frac{S^2 - S_2^2}{S^2 - S_1^2}} \quad \text{for band-pass responses} \tag{2.47}$$

where the variable z is complex such that $z = x + jy$ and the constants S_1 and S_2 correspond to the lower and upper edges of the respective passbands in the S-plane. The effect of the transform is to map the passband response in the S-plane onto the entire imaginary axis of the z-plane and to simultaneously map the stopbands onto the real axis. An illustration of a band-pass response is given in Figure 2.21. As a result of the mapping, a surprisingly simple sequence of polynomial manipulation is then able to create the required equal-ripple passband. The procedure will be given below, but for a detailed derivation, the reader should refer to the literature [5].

A polynomial is first constructed in the z-variable of the form

$$E(z^2) + zF(z^2) = \prod_{i=1}^{n} (z + z_i) \tag{2.48}$$

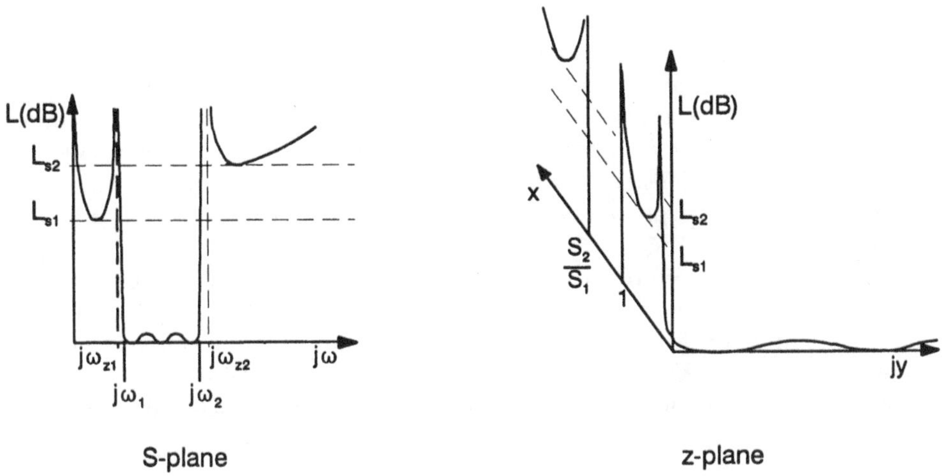

$$z = \sqrt{\frac{S^2 - S_2^2}{S^2 - S_1^2}}$$

where $z = x + jy$ for $x \geq 0$

Figure 2.21 The *z*-transform for band-pass frequency responses.

where z_i are the transmission zeros transformed from the *S*-plane, *E* and *F* are even polynomials, and *zF* is the odd part of the complete polynomial. Then $|K(z)|^2$ is given by

$$|K(z)|^2 = k_p \frac{E^2}{E^2 - z^2F^2} \tag{2.49}$$

where k_p is a constant such that

$$k_p = 10^{\left(\frac{T_p}{10}\right)} - 1 \tag{2.50}$$

where T_p is the magnitude of the passband ripple expressed in decibels. The final step is then to transform $|K(z)|^2$ back into $|K(S)|^2$ using the appropriate *z*-transform (2.45), (2.46), or (2.47).

As presented here, the mathematics for generating either Butterworth or Tchebycheff passband responses are not capable of creating equal-ripple stopband responses. This is quite deliberate, since the sort of mathematics required would be considerably more complicated and would impose certain restrictions on the network topology. Furthermore, equal-ripple stopband responses are only really of any value for high-selectivity filters, which represent a relatively small classification of microwave components. What may be of more concern at first sight, however, is the apparent inability to predict directly the magnitude of any of the ripples in the stopband. Even this is not really of any consequence, because the necessary information becomes available as soon as the frequency responses are computed by the software, which on a modern computer takes substantially less than a second. Hence, an iterative placement of the transmission zeros is a very convenient and highly versatile way to proceed for determining the stopband characteristics of a network.

When a passband is required that is equal-ripple but which is offset from zero loss by a constant value, as in Figure 2.9(c), this is achieved by simply generating the $|K(S)|^2$ function for zero offset and then adding the appropriate constant. Hence, if we write $|K(S)|^2 = P(S)/Q(S)$, where $P(S)$ and $Q(S)$ are the original numerator and denominator polynomials, then the new characteristic function is given by

$$\frac{P'(S)}{Q(S)} = \frac{P(S)}{Q(S)} + k_o = \frac{P(S) + k_0 Q(S)}{Q(S)} \tag{2.51}$$

where k_o is the offset value and $P'(S)$ is the new numerator polynomial of $|K(S)|^2$ with the offset applied. Similarly, when a passband is required that is equal-ripple, but which, on a decibel scale, has a sloping linear offset across its frequency range (e.g., as in Figure 2.9(d,e)), this is easily achieved by again generating the $|K(S)|^2$ for no offset, but adding an offset in the form of a quadratic function in S rather than a constant. For a low-pass frequency response, the modification to $|K(S)|^2$ is indicated by

$$\frac{P''(S)}{Q(S)} = \frac{P(S)}{Q(S)} + k_{g1}\frac{S^2}{S_2^2} + k_o \tag{2.52}$$

where k_{g1} is adjusted to give the required offset gradient, S_2 is the upper edge of the passband, and $P''(S)$ is the new numerator polynomial. High-pass frequency responses require a different modification of the form

$$\frac{P'''(S)}{Q(S)} = \frac{P(S)}{Q(S)} + k_{g2}\frac{S_1^2}{S^2} + k_o \tag{2.53}$$

where k_{g2} sets the offset gradient, S_1 is the lower edge of the passband, and $P'''(S)$ is the new numerator polynomial. For a band-pass frequency response, the offset function is a combination of those of the low-pass and high-pass types, giving

$$\frac{P''''(S)}{Q(S)} = \frac{P(S)}{Q(S)} + k_{g3}\left\{\frac{m_1}{m_2 - m_1}\left(\frac{S}{m_1} - \frac{m_2}{S}\right)\right\}^2 + k_o \tag{2.54}$$

and for which

$$m_0^2 = m_1 m_2 \tag{2.55}$$

Here, m_1 and m_2 are a pair of constants whose values determine whether the slope across the passband will be positive or negative. If the slope is to be positive, then it is necessary to set $m_0 = S_1$ and $m_2 = S_2$, where S_1 and S_2 are the lower and upper edges of the passband, respectively. If the slope is to be negative, then it is necessary to make $m_0 = S_2$ and $m_1 = S_1$. The constant k_{g3} determines the magnitude of the offset slope across the passband. The new numerator polynomial this time is $P''''(S)$. It can be seen that in all three cases of sloping offset, the option for an additional constant offset is also made available by retaining the constant k_o.

An examination of the effect of the offset functions on the distribution of transmission zeros reveals that there is the potential for the creation of extra zeros. The low-pass offset function could create an extra zero at $S = j\infty$, the high-pass function could create an extra zero at $S = j0$, and the band-pass function could create a zero at both $S = j\infty$ and $S = j0$. If disruption of the original set of transmission zeros by the addition of the offset functions is to be avoided, it is therefore essential that the relevant zeros already exist. If they already exist, the offset functions will not create any extra zeros, nor will they change the overall degree of the network function. What the offsets do change is the distribution of reflection zeros, and since the reflection zeros are determined exclusively by the numerator polynomial, only the numerator polynomial of $|K(S)|^2$ is modified. The denominator $Q(S)$ should always remain unchanged.

To conclude this section, a summary will be given of the information that needs to be supplied to the synthesis software in order to solve the Approximation Problem and generate the $Z_1(S)$ polynomials. The information must comprise:

1. The number and location of transmission zeros in the S-plane, paying due regard to the required stopband response and the composition of the prototype network. The zeros must also be consistent with the type of response to be selected (i.e., low-pass, high-pass, or band-pass). They must correspond to an even degree for band-pass responses.
2. The type of passband characteristic required (e.g., Butterworth, Tchebycheff, or something more general) and the overall shape of the frequency response required (i.e., low-pass, high-pass, or band-pass).
3. For Butterworth or Tchebycheff characteristics, the S-plane frequencies of the passband edges and the magnitude of the passband ripple. Also, the magnitude and slope of any passband offsets. Passband ripple and offset values would normally be expressed in terms of transmission loss (decibels).
4. For polynomial generators producing passbands of a general shape, the locations of the reflection zeros as complex values of frequency in the S-plane.

The number of reflection zeros to be specified will be half the degree of the network, rounding down to the nearest whole number for networks of odd degree. The extra zero for networks of odd degree is implicit at either $S = (0, j0)$ or $S = (0, j\infty)$, depending on whether a low-pass or high-pass response is required. In addition to the zeros, the value of a multiplying constant must be supplied. Achieving the desired response is necessarily a process of trial and error. Some software may be able to do the optimization automatically.

5. A specification for a singly or doubly terminated network.
6. A specification for a network of the first or second canonical form.

In principle, the synthesis software should now be capable of computing either the *S*-plane or *f*-plane frequency responses, irrespective of how the network is subsequently extracted, and the coefficients of the $Z_1(S)$ polynomials.

2.3.8 The Network Extraction Problem

The purpose of the second stage of the synthesis process is the extraction of the elements of the *S*-plane prototype from the $Z_1(S)$ polynomials. By virtue of the way in which the $Z_1(S)$ polynomials were generated, the extraction of a prototype in the form of a generalized ladder network with the prescribed transfer characteristics is now *guaranteed.* The specification of the transmission zeros during the solution of the approximation problem will already have determined the number and types of all the elements in the ladder. However, this will not have determined the exact location nor the values of all the elements in the network. It is the placement of the elements and the calculation of their values that form the basis of the *network extraction problem.* As explained in Section 2.3.6, there are some basic rules and constraints that must be observed during the extraction, but overall these have little impact on what is still a wide choice of available prototype topology.

As in the case of the approximation problem, the mathematics of the network extraction problem can be confined to computer software, and element extraction from $Z_1(S)$ is simply a matter of specifying the order in which transmission zeros are to be created. To illustrate the process, an example is given in Figure 2.22 that shows the first few steps of a typical element extraction sequence. Starting at the input to the whole network, a transmission zero is first specified at $S = (0, j\infty)$, and after checking for the presence of a pole of $Z_1(S)$ at $S = (0, j\infty)$, a series inductor is extracted. The network is therefore of the first canonical form. Had a zero of $Z_1(S)$ been found at $S = (0, j\infty)$, then a shunt capacitor would have been extracted and the network would have been of the second canonical form. After extraction of the inductor, a new set of input impedance polynomials $Z_1'(S)$ is calculated, corresponding to a remainder or residual network whose degree is reduced by the degree of the element extracted, which in this case is unity. With one less transmis-

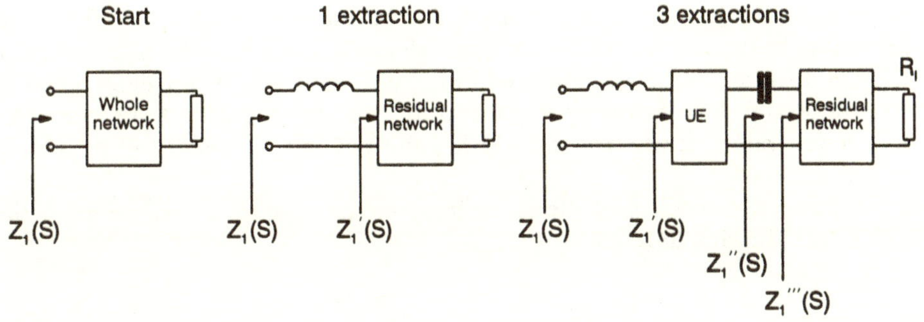

Figure 2.22 Successive element extraction from the $Z_1(S)$ polynomials.

sion zero now available for assignment, a zero at $S = (1, j0)$ is specified, causing the extraction of a unit element and the creation of a second residual network with input impedance $Z_1''(S)$. Then specifying a third zero at $S = (0, j0)$ results in the extraction of a series capacitor and the creation of a third residual network with impedance $Z_1'''(S)$. The process continues in this way until all the elements of the network have been extracted and the degree of the remainder function has fallen to zero. The remainder function at the end of the process is just a constant whose value is equal to the value of the load resistance.

At the start of the element extraction, it will not generally be possible to determine the value of the network load, since this could depend strongly on the order in which elements are extracted. In contrast, however, the input impedance $Z_1(S)$ in the case of a doubly terminated network will always be normalized to a source resistance of 1Ω, whatever the subsequent value of the load. In the case of a singly terminated network, it is the real part of either the input impedance or input admittance that will be normalized to unity, depending on which of the two network duals has been specified.

There are one or two notable circumstances in which it is possible to predict the value of the load resistance. The most obvious is when symmetrical networks are involved for which both the source and load resistances must be of equal value (i.e., 1Ω). Another is when a low-pass network of odd degree is involved, which also requires that both source and load resistances be equal. A third is when a network has all its transmission zeros located at the same frequency, forcing all its elements to be of the same basic type. In this last case, the topology of the network is completely constrained by the transmission zeros, and therefore, in theory, it should be possible to calculate the load resistance with the help of a little additional mathematics.

The mathematical process of element extraction is commonly referred to as *pole extraction,* because, except in the case of unit elements, it involves the removal of a pole from $Z_1(S)$ or $Y_1(S)$ at the frequency of the transmission zero corresponding to the network element being extracted. For transmission zeros at $S = (0, j\infty)$, the element to be extracted must be a low-pass element. There must be a pole of either $Z_1(S)$ or $Y_1(S)$ at $S = (0, j\infty)$, indicated by the difference in the degree between the numerator and denominator polynomials. If $Z_1(S)$ has the pole, the element will be a series inductor, whereas if $Y_1(S)$ has the pole, the element will be a shunt capacitor.

Taking the example of the series inductor (see Figure 2.23), the extraction is based on the simple relation

$$Z_1(S) = Z_1'(S) + SL = \frac{P(S)}{Q(S)} \tag{2.56}$$

where $Z_1'(S)$ is the residue after extraction of the inductance L, and $P(S)$ and $Q(S)$ are the numerator and denominator polynomials of the input impedance of the current network. The value of L is given by

$$L = \left.\frac{Z_1(S)}{S}\right|_{S=j\infty} \tag{2.57}$$

or

$$L = \frac{P_n}{Q_m} \tag{2.58}$$

where P_n and Q_m are the coefficients of the highest order terms of $P(S)$ and $Q(S)$, respectively.

The residue, $Z_1'(S)$, is given by

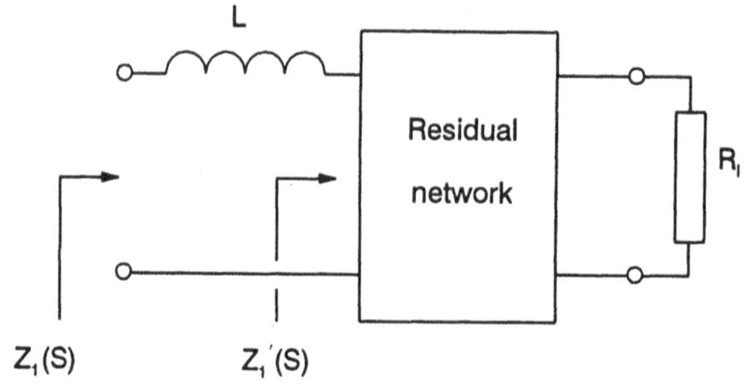

Figure 2.23 Extracting a series L.

$$Z_1'(S) = \frac{P(S)}{Q(S)} - SL = \frac{P(S) - Q(S)\,SL}{Q(S)} \tag{2.59}$$

and the degree of the numerator polynomial will always be reduced by at least 1. If there are more transmission zeros at $S = (0, \mathrm{j}\infty)$ waiting to be created in the residual network, the degree of the numerator will be reduced by 2. When this occurs, the degree of the numerator will be 1 less than the degree of the denominator, and then it is the denominator polynomial that defines the degree of the residue. Hence, the overall degree of the impedance function is always decreased by 1, irrespective of whether there are more transmission zeros at $S = (0, \mathrm{j}\infty)$ to be created or not.

The extraction of a shunt capacitor follows exactly the same procedure as that for the series inductor if, in (2.57) and (2.58), $Y_1(S)$ is substituted for $Z_1(S)$ and C is substituted for L. As far as the resulting residue impedance function $Z_1'(S)$ is concerned, this means that the extraction of a shunt capacitor will reduce the degree of the denominator by either 1 or 2, depending on the presence of more transmission zeros at $S = (0, \mathrm{j}\infty)$, and will always reduce the overall degree by 1.

For transmission zeros at $S = (0, \mathrm{j}0)$, the element to be extracted must be a high-pass element. There must be either a pole of $Z_1(S)$ or $Y_1(S)$ at $S = (0, \mathrm{j}0)$, indicated by a zero value of the first coefficient in the numerator or denominator polynomials. If $Z_1(S)$ has the pole, the element will be a series capacitor, whereas if $Y_1(S)$ has the pole, the element will be a shunt inductor. Extracting a series capacitor (see Figure 2.24) is based on the relation

$$Z_1(S) = Z_1'(S) + \frac{1}{SC} = \frac{P(S)}{Q(S)} \tag{2.60}$$

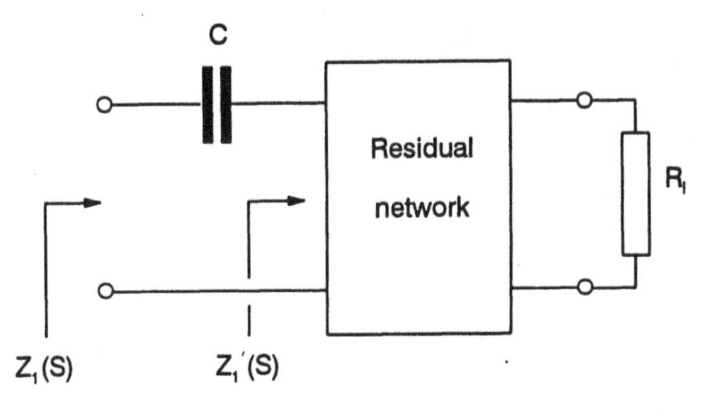

Figure 2.24 Extracting a series *C*.

where $Z_1'(S)$ is the residue after extraction of the capacitance C. The value of C is given by

$$C = \left.\frac{1}{SZ_1(S)}\right|_{S=j0} \tag{2.61}$$

or

$$C = \frac{Q_2}{P_1} \tag{2.62}$$

where P_1 and Q_2 are the first and second coefficients of $P(S)$ and $Q(S)$, respectively.

The residue, $Z_1'(S)$, is given by

$$Z_1'(S) = \frac{P(S)}{Q(S)} - \frac{1}{SC} = \frac{P(S) - \dfrac{Q(S)}{SC}}{Q(S)} \tag{2.63}$$

and there will always be a common factor of S to cancel in the numerator and denominator. The degree of both the numerator and denominator will therefore be reduced by 1. If another transmission zero exists at $S = (0, j0)$, it will be evident in the residue function $Y_1'(S)$ as a pole at $S = (0, j0)$.

The extraction of a shunt inductor follows exactly the same procedure as that for the series capacitor if, in (2.61) and (2.62), $Y_1(S)$ is substituted for $Z_1(S)$ and L is substituted for C.

For transmission zeros at $S = (1, j0)$, the element to be extracted must be a unit element. In contrast with the reactance branches, its extraction does not require any pole of $Z_1(S)$ or $Y_1(S)$ to exist at the frequency of the transmission zero. In fact, any such pole is forbidden. Extracting a unit element (see Figure 2.25) is based on the Richards Theorem [1], and as a first step, the value of the unit element is obtained by evaluating $Z_1(S)$ at $S = (1, j0)$. That is,

$$\mathrm{UE} = Z_1(S)|_{S=1} \tag{2.64}$$

where UE is the value of the unit element.

The residue, $Z_1'(S)$, is given by

$$Z_1'(S) = \mathrm{UE}\left(\frac{S \cdot \mathrm{UE} - Z_1(S)}{SZ_1(S) - \mathrm{UE}}\right) \tag{2.65}$$

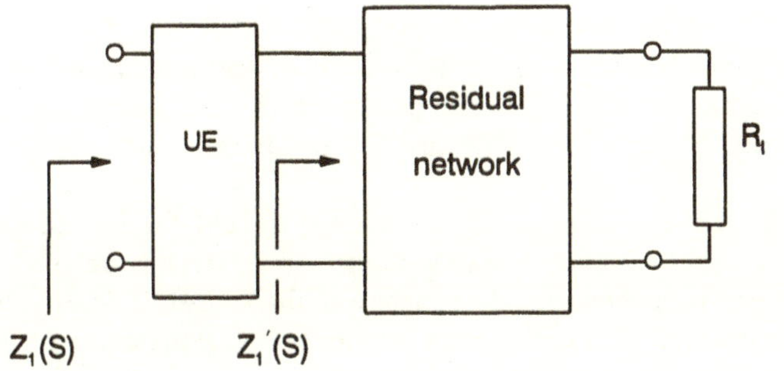

Figure 2.25 Extracting a unit element.

and there will always be a common factor of $(S^2 - 1)$ to cancel in the numerator and denominator. After a successful extraction of a unit element, the degree of $Z_1(S)$ will have been reduced by 1.

In connection with the extraction of redundant *reactance branches*, the concept of *partial pole extraction* needs to be understood. The procedures outlined above for the reactance branches will produce values for elements that cause the complete creation of the relevant pole or zero of $Z_1(S)$. They always cause a reduction in the degree of $Z_1(S)$. However, if instead some other value were to be assigned to the element, the pole of $Z_1(S)$ would only be partially created and would still exist as a pole of $Z_1'(S)$. There would be no reduction in the degree of $Z_1(S)$. Effectively, this means that only part of the element would be extracted, the remaining part being left for extraction later at some other position in the network. Hence, the extracted element would now be a redundant element. For low-pass elements, the element value calculated for complete pole extraction represents the maximum that could be assigned during a partial pole extraction. Assigning a greater value would create a negative element in the residual network and would generally be forbidden. For high-pass elements, the element value for complete pole extraction represents a minimum value for any attempted partial pole extraction. Any value less than this would create a negative element and again would not be permitted.

When a transmission zero is specified at $S = (0, j\omega)$, where ω is finite and nonzero, the relevant reactance branch will be of degree 2. A slightly different approach to pole extraction for these elements is required. If, as in most cases, the approximation procedure were constrained to produce a minimum reactance or susceptance function (i.e., no finite, nonzero poles of $Z_1(S)$ or $Y_1(S)$), then when the appropriate residual $Z_1(S)$ is evaluated at $S = (0, j\omega)$, its value must be finite and nonzero. Because of this, the reactance branch cannot be extracted directly. To extract the reactance branch, the relevant pole of $Z_1(S)$ or $Y_1(S)$ that exists elsewhere in the complex plane must first be moved to the location of the transmission zero at $S = (0, j\omega)$—an operation known as *zero shifting*. In brief, this is done by making a partial extraction of a pole at either $S = (0, j0)$ or $S = (0, j\infty)$. Evaluation of $Z_1(S)$ at $S = (0, j\omega)$ will give the type and value of the element that must be extracted to leave behind a residual $Z_1'(S)$ with the necessary pole or zero. For transmission zeros located below the passband, this will always be a high-pass element, whereas for zeros located above the passband, it will always be a low-pass element.

Once the zero-shifting element has been extracted, the second-order reactance branch can be extracted immediately afterwards. There is an obvious important point to be made here, however. The success of the zero-shifting procedure depends on the required high-pass or low-pass elements being present in what remains of the network to be extracted. If all high-pass or low-pass elements have already been completely extracted from $Z_1(S)$ before the point at which the second-order element is to be extracted, extraction will not be possible without introducing a negative zero-shifting element. Similarly, it is essential that the appropriate transmission zeros be specified at $S = (0, j0)$ and/or $S = (0, j\infty)$ during the solution of the approximation problem.

To extract a series reactance branch of degree 2 (see Figure 2.26) corresponding to a transmission zero below the passband at $S = (0, j\omega_0)$, the procedure would begin by evaluating $Z_1(S)$ at $S = (0, j\omega_0)$ to find

$$\begin{aligned} \mathrm{Re}\{Z_1(j\omega_0)\} &= 0 \quad \text{and} \\ \mathrm{Im}\{Z_1(j\omega_0)\} &= X \end{aligned} \tag{2.66}$$

where X is the reactance of $Z_1(S)$ at $S = (0, j\omega_0)$. For this type of element, X should be negative, confirming that the element required to perform the zero shifting is a capacitor. Its value would be calculated using the simple relation $C_p = 1/(X\omega_0)$. By examination of the $Z_1(S)$ polynomials, it is then possible to establish whether or not a shunt capacitor is available for extraction and to verify that the value of the capacitor is compatible with the value required for the zero shifting.

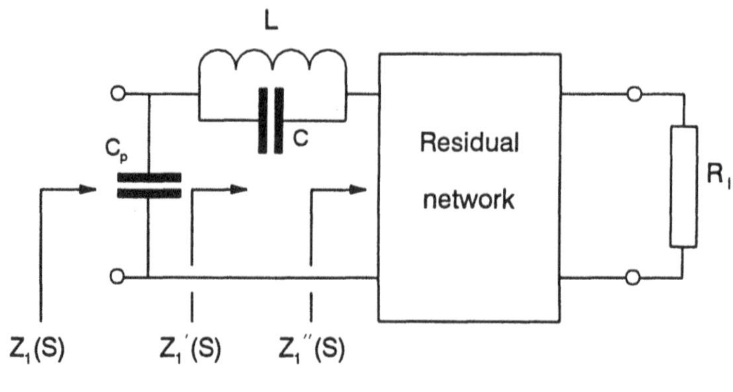

Figure 2.26 Extracting a degree 2 reactance branch.

After the partial extraction of the shunt capacitor C_p, the residue $Z_1'(S)$ should now have a pole at $S = (0, j\omega_0)$. This pole can be extracted to create the series degree 2 branch using

$$Z_1'(S) = Z_1''(S) + \frac{\frac{S}{C}}{S^2 + \frac{1}{LC}} \tag{2.67}$$

where $Z_1'(S)$ is the residue after extraction of the shunt capacitor and $Z_1''(S)$ is the residue after extraction of the degree 2 branch. L and C are the values of the elements of the degree 2 branch. Hence, we can write

$$Z_1'(S) = \left.\frac{\frac{S}{C}}{S^2 + \frac{1}{LC}}\right|_{\text{for } S=j\omega_0} \tag{2.68}$$

and since $\omega_0^2 = 1/LC$, we find

$$C = \left.\frac{S}{P(S)} \times \frac{Q(S)}{(S^2 + \omega_0^2)}\right|_{\text{for } S=j\omega_0} \tag{2.69}$$

and

$$L = \frac{1}{\omega_0^2 C} \tag{2.70}$$

The residue is then

$$Z_1''(S) = Z_1'(S) - \frac{\frac{S}{C}}{(S^2 + \omega_0^2)} \tag{2.71}$$

and it will always be possible to cancel the factor $(S^2 + \omega_0^2)$ in the numerator and denominator polynomials. The degree of the final residue is therefore 2 less than the degree of the initial $Z_1(S)$ before the partial extraction of the shunt capacitor.

When two transmission zeros are specified simultaneously, they can be realized with one of the multielement reactance branches that were shown in Figure 2.15 and described in Section 2.3.6. Apart from the need to work with polynomials of rather higher degree, the extraction process is not substantially different from that for degree 2 reactance branches, and, therefore, no further mathematical details need be given. However, it should be noted that for each finite nonzero transmission zero (i.e., $S = (0, j\omega)$), one zero-shifting element must be extracted so that in the case of degree 4 branches, for example, two must be extracted before the extraction of the reactance branch itself. Furthermore, the zero-shifting elements must both be either series elements or shunt elements. One will be a high-pass element and the other a low-pass element. Partly as a consequence of this, it is essential that the two transmission zeros associated with the degree 4 reactance branch be located on opposite sides of the passband. Extraction of degree 3 or degree 4 reactance branches always decreases the degree of the residual network by 3 or 4, respectively. Some practical examples will be covered later in Section 3.2 of Chapter 3 as part of the study on microwave filters.

To conclude this section, a numerical example will be given of the complete extraction of a prototype network from the $Z_1(S)$ polynomials. A drawing of the relevant network has been given in Figure 2.27, together with the sets of frequency parameters and transmission zeros that were used to generate the $Z_1(S)$ polynomials. Beneath the table of transmission zeros, there is also a listing of the pairs of polynomials representing the input impedance of the residual network after each extraction of a network element. The rows of numbers are the coefficients of the terms of the polynomials in ascending order to the right. For each polynomial, the number of coefficients is always one greater than the degree of the highest order term. It is clearly apparent from the listing how the degree of the polynomials is reduced as the extraction process progresses from start to finish.

Examination of the list of frequency parameters given in Figure 2.27 indicates that the f-plane frequency response of the network is band-pass and that the passband extends from 4 to 8 GHz. A finite nonzero transmission zero has been located at 3 GHz, with the intention of giving good selectivity on the lower side of the passband. Commensurate frequency has been set to 18 GHz, and since 18 GHz is well above the range of the passband, the frequency response of the S-plane prototype must also be band-pass. Using the Richards Transformation ($S = j\omega = j\tan((\pi/2) \times (f/f_s))$), the frequencies of the passband edges and the transmission zero have been transformed to the frequencies ω_1, ω_2, and ω_{z1}, respectively, in the S-plane and the passband ripple has been set to 0.1 dB. As indicated in the table of transmission zeros, there is a total of five zeros, one of which is at a finite, nonzero frequency $(0, j\omega_{z1})$, making the degree of the complete network 6. This is consistent with the rule that for band-pass networks, the degree must always be even.

Referring now to the listing of $Z_1(S)$ polynomials, the first two polynomials represent the numerator and denominator of the input impedance function of the complete network before any elements have been extracted. The numerator degree is 6, the same as the degree of the whole network. The denominator degree is 5, which confirms the presence of a transmission zero at $S = (0, j\infty)$ and indicates

Prototype network

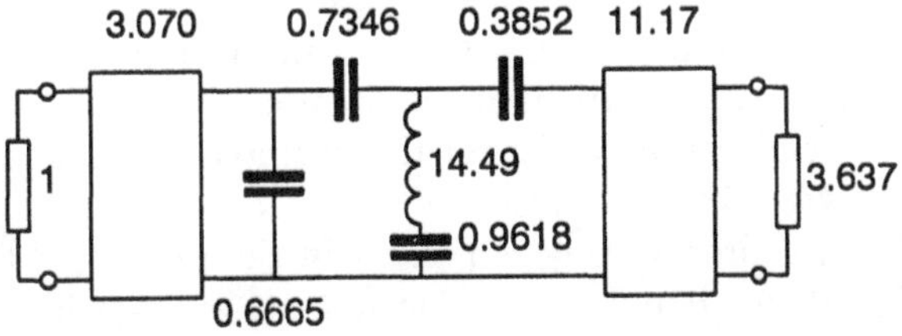

Frequency parameters

(f-plane)	(S-plane)
$f_1 = 4$ GHz	$\omega_1 = 0.3640$
$f_2 = 8$ GHz	$\omega_2 = 0.8391$
$f_{z1} = 3$ GHz	$\omega_{z1} = 0.2679$
$f_s = 18$ GHz	$\tau_p = 0.1$dB

Transmission zeros				
Location	(1,j0)	(0,j∞)	(0,j0)	$(0,j\omega_{z1})$
Numbers	2	1	1	1

$Z_1(S)$ Polynomials:-

```
                        0 EXTRACTIONS
NUMERATOR DEGREE     6
0.4765D-01 0.6993D-01 0.6912D+00 0.6799D+00 0.2571D+01 0.1000D+01 0.2062D+01
DENOMINATOR DEGREE   5
0.0000D+00 0.6993D-01 0.9942D-01 0.6799D+00 0.4705D+00 0.1000D+01
                        1 EXTRACTION
NUMERATOR DEGREE     4
0.7094D-01 0.1041D+00 0.7804D+00 0.6619D+00 0.1500D+01
DENOMINATOR DEGREE   5
0.0000D+00 0.8100D-01 0.1141D+00 0.7580D+00 0.4849D+00 0.1000D+01
                        2 EXTRACTIONS
NUMERATOR DEGREE     4
0.1620D+01 0.2378D+01 0.1783D+02 0.1512D+02 0.3428D+02
DENOMINATOR DEGREE   4
0.0000D+00 0.7704D+00 0.1021D+01 0.5436D+01 0.1000D+01
                        3 EXTRACTIONS
NUMERATOR DEGREE     4
0.5720D+00 0.9877D+00 0.1043D+02 0.1376D+02 0.3428D+02
DENOMINATOR DEGREE
0.0000D+00 0.7704D+00 0.1021D+01 0.5436D+01 0.1000D+01
                        4 EXTRACTIONS
NUMERATOR DEGREE     2
0.7970D+01 0.1376D+02 0.3428D+02
DENOMINATOR DEGREE   2
0.0000D+00 0.3070D+01 0.1000D+01
                        5 EXTRACTIONS
NUMERATOR DEGREE     1
0.1116D+02 0.3428D+02
DENOMINATOR DEGREE   1
0.3070D+01 0.1000D+01
                        6 EXTRACTIONS
NUMERATOR DEGREE     0
0.3637D+01
DENOMINATOR DEGREE   0
0.1000D+01
```

Figure 2.27 A numerical example of a network extraction.

that the network is of the first canonical form. It is also possible to confirm the presence of the transmission zero at $S = (0, j0)$ by observing that the first coefficient of the denominator polynomial is zero. Either of the transmission zeros at $S = (0, j\infty)$ or $S = (0, j0)$ could be created immediately in the form of a series inductor or series capacitor, respectively. However, in the example given, a unit element is the first element to have been extracted.

Extracting the unit element causes a unity reduction in the degree of the residue, but as Figure 2.27 indicates, the denominator polynomial now has a higher degree than that of the numerator. Hence, the transmission zero at $S = (0, j\infty)$ is still present as required, but its creation now requires the extraction of a shunt capacitor, not a series inductor. As far as the transmission zero at $S = (0, j0)$ is concerned, the first coefficient of the denominator polynomial has not been changed, which means the transmission zero is still present and could still be created by the extraction of a series capacitor. This is in agreement with the previously stated effects of transforming unit elements through low-pass and high-pass elements, wherein only the low-pass elements undergo a change of type.

The shunt capacitor of the network is extracted next, again reducing the degree of the residue by unity. However, because there are no more transmission zeros at $S = (0, j\infty)$, both the numerator and denominator polynomials are now of the same degree. Then, to create the transmission zero at $S = (0, j0.2679)$, a partial pole extraction must first be performed before the extraction of the second order element. Because the transmission zero is below the passband, the partial extraction must be of a high-pass element, and given that this network is a first dual, the element must be a series capacitor. After the partial extraction of the capacitor, there is no reduction in the degree of the polynomials and the first coefficient of the denominator is still zero, indicating the continued presence of a transmission zero at $S = (0, j0)$. Then the shunt second-order element is extracted, causing a reduction of 2 in the degree of numerator and denominator polynomials (four extractions completed at this stage). Still, the first coefficient of the denominator is zero, allowing the final creation of the transmission zero at $S = (0, j0)$ by the complete extraction of another series capacitor. The zero disappears from the residue. Finally, a unit element is extracted, reducing the final residue to the ratio of a pair of numbers representing the value of the load resistor. In this case, the value of the load resistor is greater than unity owing to the asymmetry of the network and the slight shift of the high-pass elements (i.e., the series capacitors) toward the right-hand side.

For the sake of completeness, the f-plane frequency responses of the network have been plotted in Figure 2.28. These were derived from the $|K(S)|^2$ polynomials before the network extraction procedure had begun. As expected, the passband width is 4 to 8 GHz and there is a deep null corresponding to the transmission zero at 3 GHz.

As a final comment on this example, the network of Figure 2.27 contains a redundant element; that is, there are is a total of seven elements in the network,

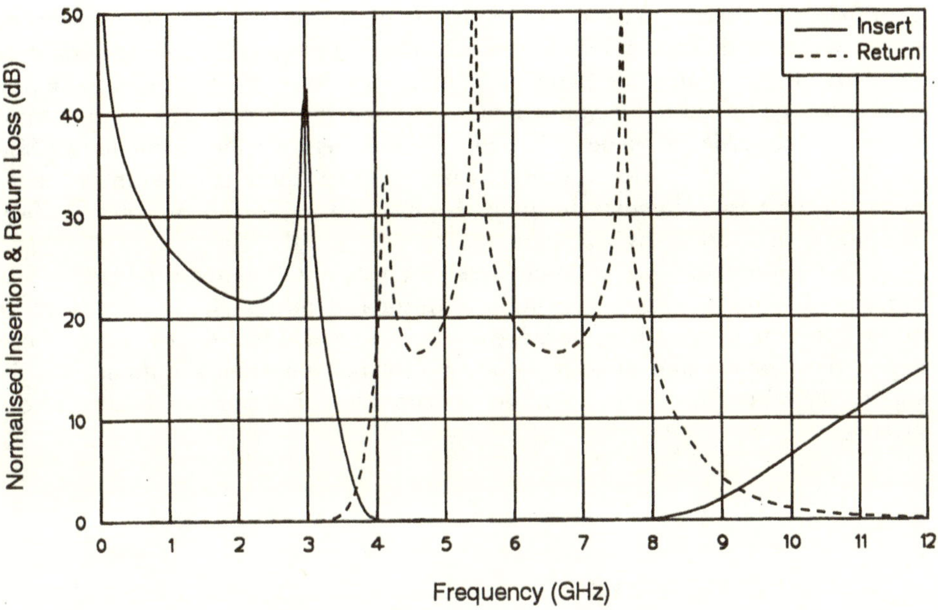

Figure 2.28 Frequency responses derived from $K(S)^2$ polynomials.

while the degree of the network is only 6. One of the series capacitors is by definition redundant because both contribute to the same single transmission zero at $S = (0, j0)$. However, this is a situation in which it is not possible to combine the redundant elements and still obtain a realizable nonredundant network with minimum phase characteristics. Such a network does not exist for the given set of $Z_1(S)$ polynomials. The reader is also reminded that whenever finite, nonzero transmission zeros are to be created (i.e., second-order elements are to be extracted), there must also be at least one transmission zero at $S = (0, j0)$ or $S = (0, j\infty)$.

2.4 NETWORK MANIPULATION AND TRANSFORMATION

With the successful completion of the network synthesis stage of the circuit design process, an *S*-plane or *s*-plane prototype will have been established with the required frequency responses. Now that its transmission zeros and the order in which the elements are to be extracted have been chosen, its topology should be broadly consistent with the form needed for conversion into an *f*-plane equivalent circuit. However, in the majority of cases, it will not be advisable to use the network extraction procedures to create the final form of the network required. A much more effective approach will be to extract the network in its most simple, nonredundant form and then to make the necessary changes using the network transformations about to be described. This is especially true for networks containing significant numbers of redundant unit elements or reactance branches. Similarly, some networks must be forced to be symmetrical or alternatively forced to be asymmetrical to produce a particular value of impedance ratio between the terminating resistances. In these and in the majority of cases in which there is a mixture of high-pass and low-pass elements, the final shaping of the *S*-plane prototype is best achieved by network transformation.

The two transformation procedures most relevant to generalized ladder networks are based on (1) the application of Kuroda Identities [6] or (2) the scaling of two-port admittance and impedance matrices. Kuroda Identities tend to be more appropriate for treating small parts of a prototype, while the admittance matrix scaling technique tends to be more appropriate for treatment of the whole prototype.

2.4.1 Kuroda Identities

As illustrated in Figure 2.29, there are four different Kuroda Identities, which facilitate the movement of one of the four basic types of reactance branches through a single unit element. In all four cases, the values of the unit element and the reactance branch are modified, but the type of the resultant reactance branch depends on whether it was originally a low-pass or a high-pass element. In the case of a low-pass element, there is a change in the type of the element to that of its dual. In the case of a high-pass element, there is no change of type, but the transformation causes the introduction of an ideal transformer instead.

The first and fourth identities shown in Figure 2.29 deal with the low-pass reactance branches. In the case of the first, a shunt capacitor is transformed into a series inductor on the opposite side of the unit element, and in the case of the fourth, a series inductor is transformed into a shunt capacitor. The values of the new elements are given in terms of the old element values and an intermediate parametric variable n. The second and third identities deal with the high-pass reactance branches. In the case of the second, a shunt inductor transforms into another shunt inductor and an ideal transformer with a turns ratio of n. In the case of the third, a series capacitor transforms into another series capacitor and an ideal transformer, but this time the turns ratio of the transformer is in the opposite direction of that for the inductor. The value of the transformer turns ratio is always greater than unity.

Hence, moving low-pass elements through unit elements in any global network will only cause the type of the low-pass elements to change. Moving high-pass elements, however, causes no changes to elements types, but it will change the impedance levels in the network as a result of the introduction of the ideal transformers. Whether the impedance levels increase or decrease will depend on which type

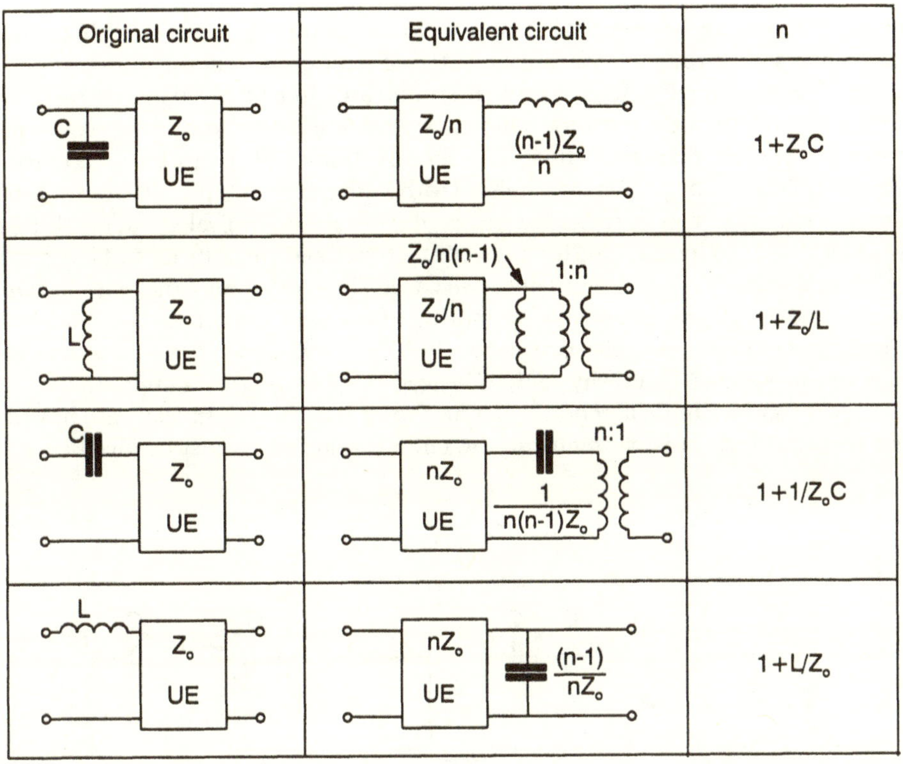

Figure 2.29 Kuroda Identities.

of high-pass element was moved and on the direction in which it was moved. This leads to a pair of important general rules that should be committed to memory:

1. Moving series capacitors to the right-hand side through adjacent elements in a circuit always causes impedance levels to the right to be increased. Moving series capacitors to the left-hand side always causes impedances to the right to be decreased.
2. Moving shunt inductors to the right-hand side through adjacent elements in the circuit always causes impedance levels to the right to be decreased. Moving shunt inductors to the left-hand side always causes impedances to the right to be increased.

A point that must be stressed here is that the Kuroda Identities apply specifically to *S*-plane networks. This means that in terms of the equivalent distributed *f*-plane elements, the lengths of the reactive stubs (equivalent to the *S*-plane inductors and capacitors) and the adjacent transmissions lines (equivalent to the unit elements) *must* all be identical. The transforms cannot be used directly to move an *f*-plane stub through a transmission line of a different length.

As previously explained, Kuroda Identities are often involved in moving redundant elements about a network. One such situation arises when it is required to move part rather than the whole of a reactive branch through an adjacent unit element. This can be particularly helpful with high-pass elements, whereby splitting the element into two parts allows either the resulting equivalent network to be symmetrical or to have a transformer with a predefined turns ratio. In Figure 2.30, a series capacitor C_0 is shown split into two parts. The maximum value of transformer turns ratio would be produced by moving the entire capacitor C_0 through Z_0. However, by transposing the expression $n = 1 + 1/(Z_0C)$ for C, any smaller value of n can be created by moving only C instead of C_0 through the unit element.

After some simple algebra, it can be shown that for the special case in which the resultant section is symmetrical, the chosen value of C must be given by

$$C = C_0 \pm \sqrt{C_0^2 + \frac{C_0}{Z_0}} \tag{2.72}$$

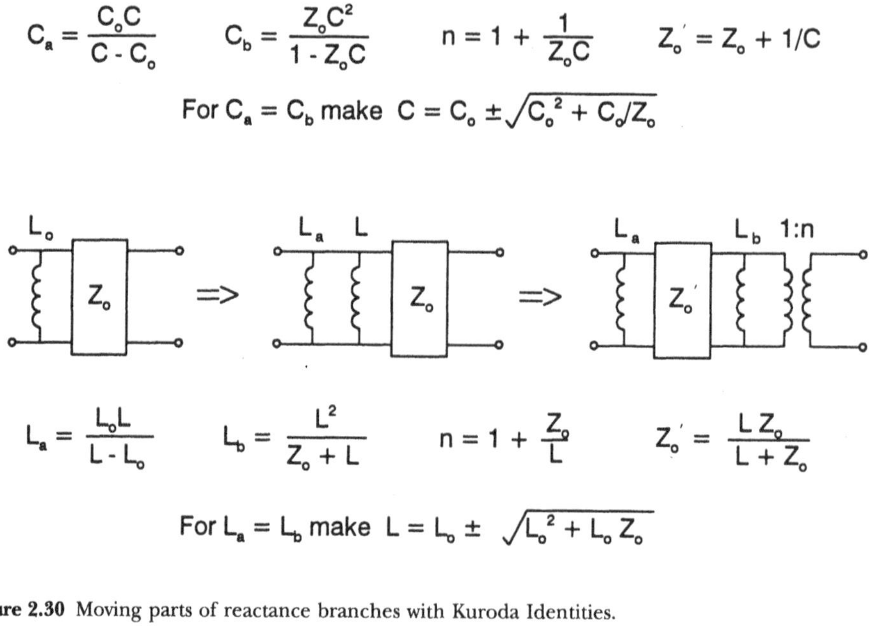

Figure 2.30 Moving parts of reactance branches with Kuroda Identities.

Similar principles apply for moving parts of shunt inductors through unit elements. To obtain a symmetrical section with inductors (Figure 2.30), the relevant equation for the value of inductance that must be moved is

$$L = L_0 \pm \sqrt{L_0^2 + L_0 Z_0} \tag{2.73}$$

There are several noteworthy additions to the basic set of four Kuroda Identities. First, the movement of reactive branches through unit elements is not limited to reactive branches of unity degree. In fact, branches of arbitrary degree can be moved through a unit element, although for branches of degree greater than 2, the transforms become rather complicated. The transforms for degree 2 branches have been published by Levy in [7]. As a general rule, when the transmission zero corresponding to the reactance branch is below the passband, the transform will cause no change in the type of reactance branch. Conversely, when the transmission zero is above the passband, the transform causes the element to change to the form of its dual. This is consistent with the changes that occur for simple high-pass and low-pass elements.

Two important additions to the Kuroda Identities apply for moving high-pass elements through adjacent low-pass elements. This is something that may need to be done almost as frequently as the transform through unit elements. As shown in Figure 2.31, a series capacitor can be moved through a shunt capacitor as if the shunt capacitor were a unit element with a value equal to $1/C_2$. The formulas given in the figure are exactly the same as those given in Figure 2.29, when Z_0 is replaced by $1/C_2$. Similarly, a shunt inductor can be moved through a series inductor as if the series inductor were a unit element of the same value. As in the case of unit elements, it is also possible to move only part of the high-pass elements through the low-pass element. When Z_0 is replaced by $1/C_2$ for the series capacitor, the formulas of Figure 2.30 apply, offering the opportunity to create a transformer with a particular turns ratio or a new section that is symmetrical. Another alternative is to split the low-pass instead of the high-pass element into two parts. The identities 4, 5, and 6 in Appendix A were established in this way, allowing L sections to be converted into T sections or π sections, and T sections to be converted directly into π sections.

It is a simple matter to verify the use of Kuroda Identities for moving high-pass elements through low-pass elements. This can be done by deriving the basic Kuroda Identity and then examining the effect of replacing the unit element with a shunt capacitor. With reference to Figure 2.32, equating the Z-matrices for the two networks before and after the transformation, we can write

$$\frac{1}{S}\begin{bmatrix} \frac{1}{C_1} + Z_0 & Z_0\sqrt{1 - S^2} \\ Z_0\sqrt{1 - S^2} & Z_0 \end{bmatrix}_{\times n}^{\times n} \Rightarrow \frac{1}{S}\begin{bmatrix} Z_0' & Z_0'\sqrt{1 - S^2} \\ Z_0'\sqrt{1 - S^2} & Z_0' + \frac{1}{C_3'} \end{bmatrix} \tag{2.74}$$

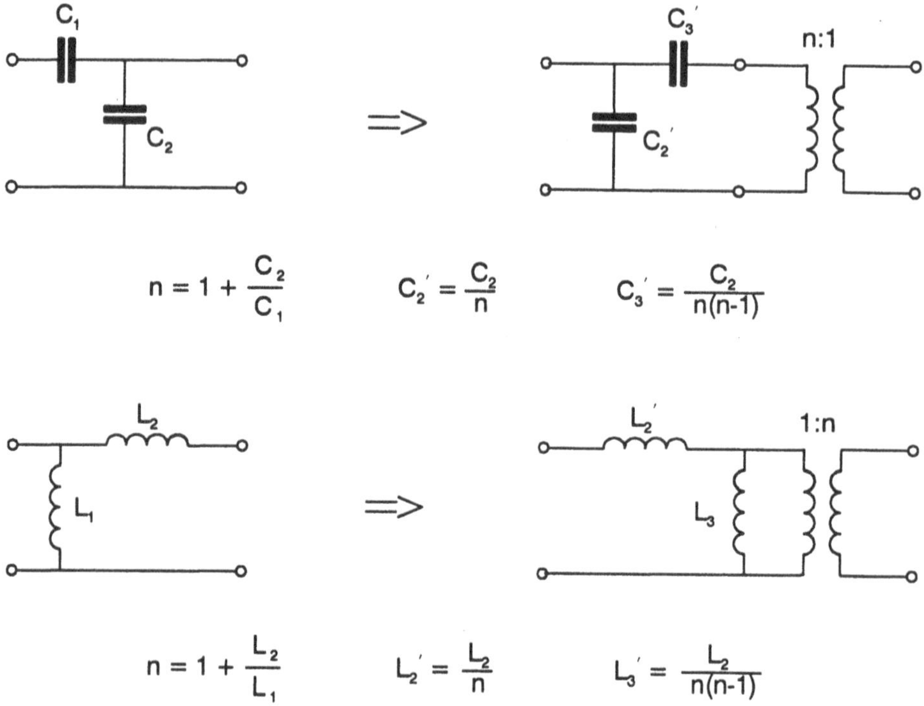

Figure 2.31 Moving high-pass elements through low-pass elements with Kuroda Identities.

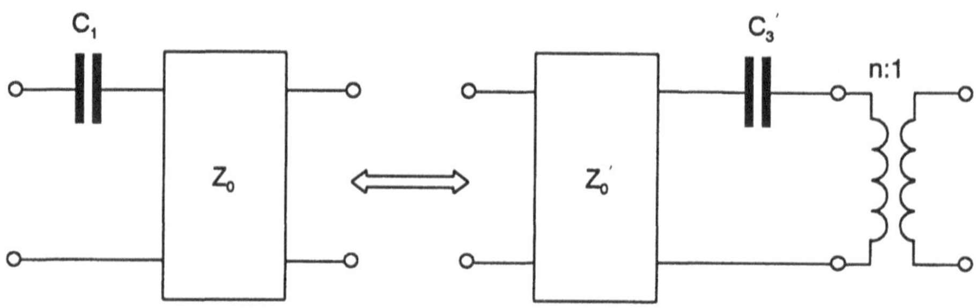

Figure 2.32 Transforming a series capacitor through a unit element.

where the *Z*-matrix on the right of the equation represents the transformed network without the ideal transformer.

Multiplying the second row and column of the original matrix by *n* and equating the terms with those of the second, we can construct the following equations:

$$\frac{1}{C_1} + Z_0 = Z_0'; \quad nZ_0 = Z_0'; \quad n^2 Z_0 = Z_0' + \frac{1}{C_3'} \tag{2.75}$$

Then, in terms of Z_0 and C_1, the expression for n is

$$n = 1 + \frac{1}{Z_0 C_1} \tag{2.76}$$

and then C_3' is given by

$$C_3' = \frac{1}{n(n-1)Z_0} \tag{2.77}$$

These formulas correspond to the formulas given in the table of Kuroda Identities in Figure 2.29. Now, if we write the same pair of Z-matrices for the same pair of transformed networks, but with the unit elements replaced by shunt capacitors (see Figure 2.33), we obtain

$$\frac{1}{S}\begin{bmatrix} \frac{1}{C_1} + \frac{1}{C_2} & \frac{1}{C_2} \\ \frac{1}{C_2} & \frac{1}{C_2} \end{bmatrix} \begin{matrix} \\ \times n \end{matrix} \atop \times n \quad \Rightarrow \frac{1}{S}\begin{bmatrix} \frac{1}{C_2'} & \frac{1}{C_2'} \\ \frac{1}{C_2'} & \frac{1}{C_2'} + \frac{1}{C_3'} \end{bmatrix} \tag{2.78}$$

from which it is clear that the relevant formulas of the Kuroda Identity are unaffected by the absence of the $\sqrt{(1 - S^2)}$ factor. Hence, the identity must apply equally well when the unit element is replaced by a shunt capacitor, whose value is the reciprocal of the characteristic impedance of the unit element.

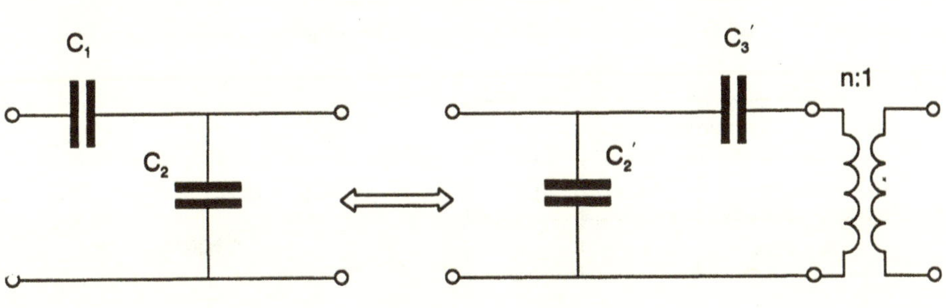

Figure 2.33 Transforming a series capacitor through a shunt capacitor.

A similar proof exists for the second Kuroda Identity and its application to the pair of shunt and series inductors.

2.4.2 Admittance and Impedance Matrix Scaling for Ladder Networks

As has been seen, Kuroda Identities are ideal for transforming individual pairs of network elements. However, when it is necessary to transform a complete ladder network, moving several elements and adjusting several impedance levels simultaneously, scaling of the admittance or impedance matrix is a much more attractive procedure. Admittance or impedance matrix scaling can, in principle, be applied to any type of ladder network, but it is difficult to create a suitable software algorithm to deal with the algebra of the general case. Consequently, algorithms tend to be created for transforming groups of ladder networks of broadly the same type. One of the groups for which matrix scaling is most important comprises networks with transmission zeros at $S = (0, j\infty)$ or $S = (1, 0)$, but only a single transmission zero at $S = (0, j0)$. In such networks, it is usually necessary to split the single nonredundant high-pass element into several parts and move those parts throughout the rest of the network to achieve some desired numerical distribution. The process is not complicated, but is best explained with the help of a working example.

The example chosen is that of a high-pass *S*-plane prototype of the form shown in Figure 2.34. This is an important class of prototype, which when converted into a physical circuit, takes the form of a coupled-line band-pass filter. As indicated by the data given in the figure, the network has been synthesized for an *S*-plane cutoff frequency of 1.7321 and a passband ripple value of 0.1 dB. The corresponding *f*-plane network has a band-pass response, with a passband width of an octave

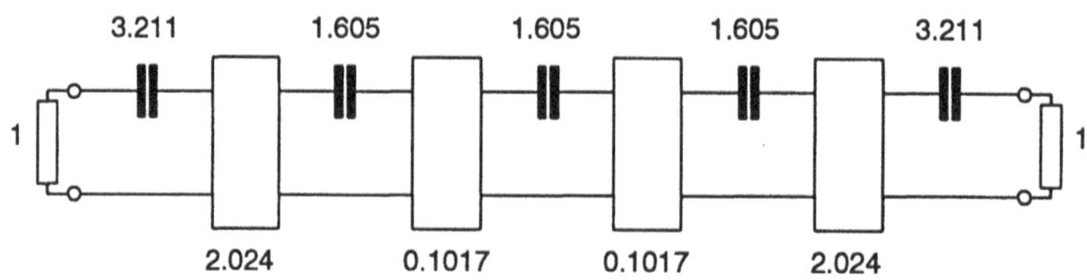

Transmission zeros		
Location	(1,j0)	(0,j0)
Number	4	1

f_1 = 4 GHz ω_1 = 1.7321

f_s = 6 GHz τ_p = 0.1dB

degree = 5

Figure 2.34 Transformed *S*-plane prototype for a coupled-line filter.

centered on the commensurate frequency of 6 GHz. More details of how this prototype is converted into a coupled-line filter will be given in Section 2.7 and in Chapter 3 (Section 3.1).

With only a single transmission zero at $S = (0, j0)$, the prototype of Figure 2.34 is clearly a redundant structure. All the series capacitors could be collapsed into a single capacitor at any chosen position in the network, resulting in a total number of elements of 5. The degree of the network is therefore 5. An important point to be noted is that the capacitors of the redundant prototype are not all of the same value. Those in the inner part of the network all have a value of 1.605, but those at the ends have a value of 3.211, which is twice the value of the internal capacitors. This distribution of the capacitor values is important for a satisfactory physical realization of the coupled-line filter. Achieving the distribution with either Kuroda transforms or by successive partial extraction of the capacitors during the second stage of the synthesis would be extremely difficult. A great deal of iteration would be required. Scaling the impedance matrix is the only realistic approach to the problem.

The equivalent, nonredundant form of the prototype of the coupled-line filter is shown in Figure 2.35. This is the most convenient form in which to synthesize the prototype, since it involves the extraction of only a single capacitor. Extracting the capacitor in the middle automatically makes the whole network symmetrical.

Transforming this network into the network of Figure 2.34 is best achieved by scaling the impedance rather than the admittance matrix. The mathematics of the process begins with the creation of a Z-matrix for the redundant network of Figure 2.34, which must have the same dimensions as the number of meshes in the circuit, which also happens to be equal to the network degree (n). Since there are five meshes, the Z-matrix will be of the form

$$\frac{1}{S}\begin{bmatrix} \frac{1}{C_1}+Z_1 & Z_1\sqrt{1-S^2} & & & \\ Z_1\sqrt{1-S^2} & Z_1+\frac{1}{C_2}+Z_2 & Z_2\sqrt{1-S^2} & & \\ & Z_2\sqrt{1-S^2} & Z_2+\frac{1}{C_3}+Z_3 & Z_3\sqrt{1-S^2} & \\ & & Z_3\sqrt{1-S^2} & Z_3+\frac{1}{C_4}+Z_4 & Z_4\sqrt{1-S^2} \\ & & & Z_4\sqrt{1-S^2} & Z_4+\frac{1}{C_5} \end{bmatrix}\begin{matrix} \times\ m(1) \\ \times\ m(2) \\ \times\ m(3) \\ \times\ m(4) \\ \times\ m(5) \end{matrix}$$

$$\begin{matrix} \times & \times & \times & \times & \times \\ m(1) & m(2) & m(3) & m(4) & m(5) \end{matrix} \qquad (2.79)$$

Multiplying rows and columns by the constants $m(1)$ through $m(5)$ will change all the mesh impedance levels by $m^2(i)$, where i signifies the ith mesh. This will not

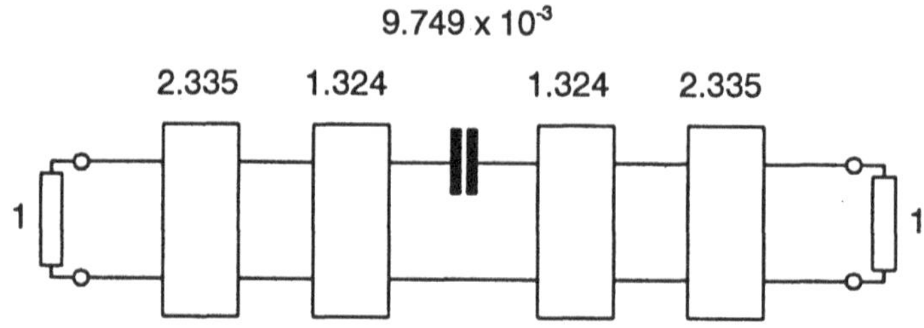

Transmission zeros		
Location	(1,j0)	(0,j0)
Number	4	1

f_1 = 4 GHz ω_1 = 1.7321

f_s = 6 GHz τ_p = 0.1dB

degree = 5

Figure 2.35 *S*-plane prototype for a coupled-line filter with no redundant elements.

change the transmission characteristics of the network. Then if the terms of the original matrix are equated with those of the scaled matrix, a set of $(2n - 1)$ simultaneous equations is created. That is,

$$\begin{aligned} E_1 + Z_1 &= (E_1' + Z_1') \div m(1)^2 \\ Z_1 + E_2 + Z_2 &= (Z_1' + E_2' + Z_2') \div m(2)^2 \\ Z_2 + E_3 + Z_3 &= (Z_2' + E_3' + Z_3') \div m(3)^2 \\ Z_3 + E_4 + Z_4 &= (Z_3' + E_4' + Z_4') \div m(4)^2 \\ Z_4 + E_5 &= (Z_4' + E_5') \div m(5)^2 \end{aligned} \tag{2.80}$$

$$\text{and} \quad Z_i = \frac{Z_i'}{m(i) \times m(i+1)} \quad \text{for } i = 1 \text{ to } n - 1$$

For $i = 1$ to n, the variables E_i are so-called *elastance* values, where $E_i = 1/C_i$ and where the prime (′) denotes the new value of the component in the transformed network. There are $(3n - 1)$ unknowns in these equations. Therefore, n of the unknowns must be selected and given specific values, after which the equations can be solved for the other $(2n - 1)$ unknowns. Clearly, the freedom to set the values of n arbitrary components or mesh scaling factors can give a great deal of design flexibility. However, some care is needed if the resultant network is not to become unrealizable, since as far as the basic equations are concerned, there is nothing to prevent an unwise specification of new element values producing, for example, a negative element value somewhere else in the circuit. When the equations are embodied in a simple computer program, it is not difficult to add the necessary constraints to prevent an unacceptable network from being generated.

For the specific case of transforming Figure 2.34 into Figure 2.35 or vice versa, the boundary conditions that must be applied to the equations are slightly more complicated than just fixing capacitor values. As indicated in Figure 2.34, it is not so much the value of the individual capacitors that matters, but rather that all the internal capacitors become of equal value and equal to half the value of the outer capacitors. In addition, to ensure that the two terminations of the network are equal, there should be no net change in impedance level along the length of the network and the scale factors $m(1)$ and $m(n)$ must both be unity. When these boundary conditions are applied, the simultaneous equations (2.80) effectively become a single, nonlinear error function in a single unknown. The unknown is the value of the internal capacitors in the transformed network and the desired solution is obtained by using a suitable root-finding numerical algorithm.

In this example, impedance matrix scaling has been applied to a network comprising an alternate cascade of unit elements and series capacitors. However, the procedure could also have been applied to a network in which some or all of the unit elements were replaced by shunt capacitors with values equal to the reciprocal of the unit element values. This variation is similar to the variation that was applicable to the Kuroda Identities and described in the latter part of Section 2.4.1. It would also be possible to transform the dual form of the network shown in Figure 2.34 by scaling the admittance instead of the impedance matrix. The dual of Figure 2.34 is a cascade of shunt inductors separated by unit elements, the physical form of which is a line and stub structure representing another important class of microwave band-pass filters.

One other useful capability of the impedance matrix scaling technique is in creating networks similar to that of Figure 2.34, but whose terminating resistances are not equal. By redefining the boundary conditions, the simultaneous equations (2.80) can be solved numerically to give an impedance-transforming network, whose element values taper in magnitude from one end to the other. There are practical limits to be observed, but in theory any specified ratio of terminating impedances could be created in this way. This facility can make the coupled-line filter a highly effective impedance-transforming circuit.

In conclusion, a software tool based on the impedance or admittance matrix scaling of selected prototype structures is a valuable addition to the basic synthesis software and an important part of the microwave circuit designer's toolbox.

2.4.3 Trading Length Against Impedance in Noncommensurate Transmission Line Networks

One of the popular misconceptions about exact network synthesis techniques is that they tend to produce unrealizable networks. In fact, while they can undoubtedly produce unrealizable networks, when properly applied they are no more likely to do so than any other type of circuit design technique. Indeed, most circuits that can be designed by approximate procedures can be designed more accurately and more directly by synthesis. The most important factors determining success are the choice of transmission zeros made at the start of the synthesis design procedure and the application of the most appropriate network transformations toward the end of the process.

The exact transformations described in the previous sections provide enough design flexibility to enable a large proportion of prototype networks to be turned into a realizable circuit. However, there will be occasions, particularly in active circuit design, when some circuit elements remain difficult to realize and some additional design flexibility is required. Very often it will be the characteristic impedance of one or more of the transmission line elements that will be either too high or too low. If this is the difficulty, then it may be possible to obtain the extra degree of freedom required by simply changing the line impedance and making a compensatory change to the line length. This is the so-called *length/impedance trade-off.* Sometimes the trade-off can be taken to the extreme of reducing the length to zero and completely replacing a stub with a lumped element. Not only may this solve the realization problem, but the use of lumped elements such as capacitors instead of open-circuit stubs, can result in a very much more compact circuit. In fact, this is the basis of the approach proposed for designing any circuit containing a mixture of lumped and distributed elements.

It is important to stress that the length/impedance trade-offs are only applicable to a circuit with a low-pass or band-pass S-plane prototype, since they require the length of all the transmission line elements to be substantially less than a quarter of a wavelength throughout the range of the passband. Thus, within the passband, short-circuit stubs always remain inductive and open-circuit stubs always remain capacitive. It is also important to note that the trade-offs generally have little effect on the characteristics of the passband, but they modify those of the stopband. There will be many situations in which this is perfectly acceptable, but when, as in some kinds of special filters, the characteristics of the stopband must not be changed, the length/impedance trade-offs cannot be used.

The length/impedance trade-offs are based on equating the reactances or susceptances of a relevant pair of circuit elements. These may be a pair of distributed elements of different length or, as in the examples given in Figure 2.36, a pair of elements of which one is lumped and the other distributed. In Figure 2.36(a), the reactance of a shunt short-circuit stub is equated with the reactance of a shunt inductor at a common frequency f_1. As the plots of reactance against frequency indicate, the tangent function for the stub and the linear function for the inductor are quite different over the whole frequency range between zero and f_s. However, for networks in which the passband is located substantially below the commensurate frequency f_s, the behavior of the inductor in the passband can be a reasonable approximation to the behavior of the stub.

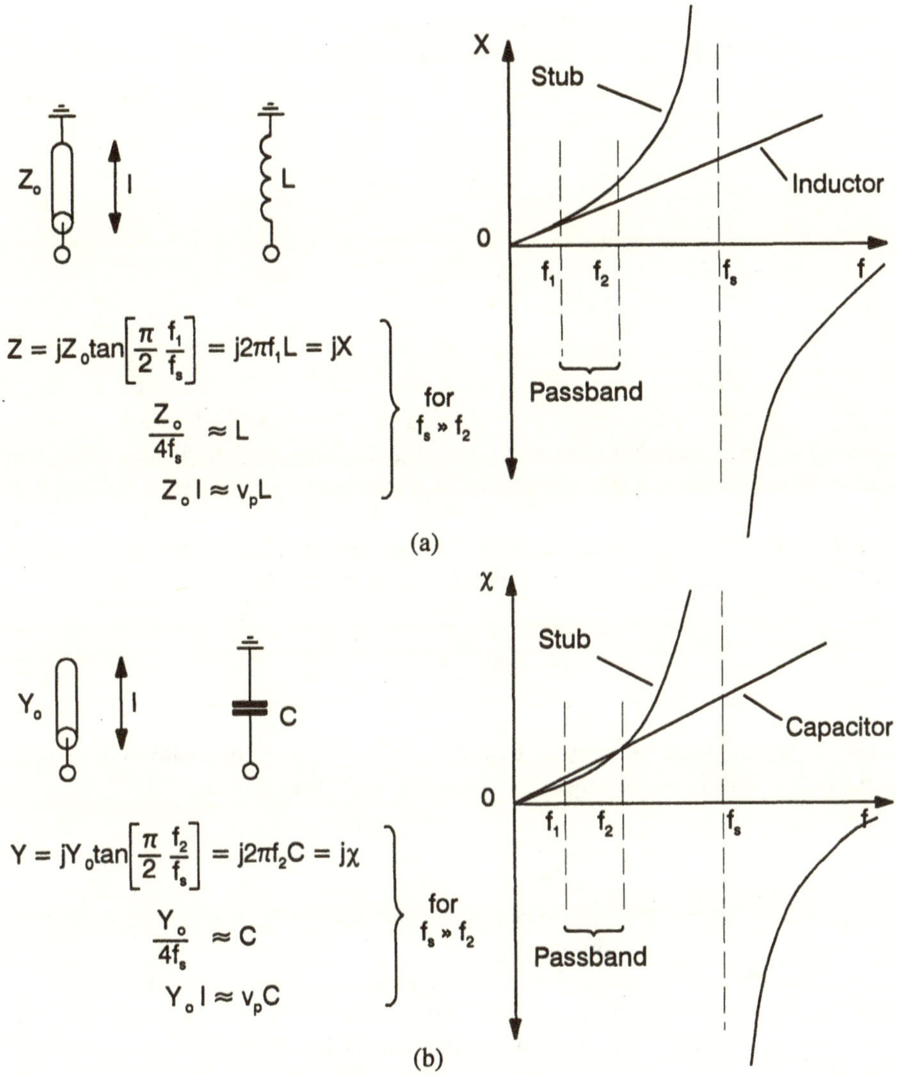

Figure 2.36 Lumped approximations to distributed elements in noncommensurate line networks.

Since both the shunt short-circuit stub and the shunt inductor are high-pass elements and have the most influence on the behavior of the network near the bottom of the passband, it is usual to equate the two reactance functions at the *lower* passband edge. That is,

$$jZ_0 \tan\left(\frac{\pi}{2}\frac{f_1}{f_s}\right) = j2\pi f_1 L \tag{2.81}$$

where Z_0 is the characteristic impedance of the stub and L is the value of the inductor. When f_s is significantly higher than the passband, the equality is reduced to

$$\frac{Z_0}{2f_s} = 2L \tag{2.82}$$

or

$$L = \frac{Z_0}{4f_s} \tag{2.83}$$

Replacing the stub of characteristic impedance Z_0 with an inductor of value L will produce a good approximation to the original passband behavior. The precise quality of the approximation depends on the value of f_s. Stopband behavior, however, as the reactance curves of Figure 2.36(a) indicate, will be substantially modified. By rewriting the reactance expression of the stub in terms of physical length and equating once again with the reactance of the inductor, we find

$$\mathrm{j}Z_0 \tan\left(\frac{2\pi f_1}{v_p} l\right) = \mathrm{j}2\pi f_1 L \tag{2.84}$$

where v_p is the phase velocity of traveling waves in the transmission line medium and l is the length of the stub. This simplifies for large f_s to

$$Z_0\, l \approx v_p\, L \tag{2.85}$$

from which it is clear that the impedance of a stub can be traded against its length for a constant effective inductance.

Using the same principles, a shunt capacitor can be shown to be a good approximation to a shunt open-circuit stub in the region of the passband (Figure 2.36(b)). In this case, the length of the stub is traded against its characteristic admittance so as to maintain the same effective shunt capacitance. An important difference here is that the equality of the susceptances of the open-circuit stub and capacitor is made not at the lower passband edge as previously, but rather at the upper edge frequency of f_2. This is because as low-pass elements, the stub and capacitor are most influential in the upper region of the passband.

Length/impedance trade-offs can also be applied to interconnecting transmission lines (unit elements in the *S*-plane), but this should be generally restricted to lines of high characteristic impedance. In this context, *high* means values that are significantly higher than those of other lines located in the same part of the circuit. Writing the ABCD matrices of a transmission line and a series inductor, we have

$$\begin{bmatrix} \cos\theta & jZ_0 \sin\theta \\ jY_0 \sin\theta & \cos\theta \end{bmatrix} \qquad \begin{bmatrix} 1 & j2\pi f_0 L \\ 0 & 1 \end{bmatrix} \tag{2.86}$$

(transmission line) (series inductor)

from which it can be seen that the two matrices will be approximately equal if Z_0 is large and $\theta = (\pi/2 \times f_0/f_s)$ is small. Since the transmission line is neither a high-pass nor a low-pass element, the frequency at which the comparison is made is usually the center of the passband f_0. Hence, to a first approximation,

$$L \approx \frac{Z_0}{4f_s} \tag{2.87}$$

and

$$Z_0 l \approx v_p L \tag{2.88}$$

which is the same result as for the short-circuit stub and its equivalent inductor.

There is at least one noteworthy category of prototype network (other than those that are not band-pass) for which length/impedance trade-offs are inappropriate.

Length/impedance trade-offs *cannot* be applied to prototypes that are just a cascade of unit elements with no reactance branches. Distributed networks of this kind are often used as wideband impedance transformers. The correct behavior of such networks depends on the presence of the abrupt changes in impedance level at the junctions of the various transmission lines. Without these discontinuities, the network will not function. The point can be reinforced by considering an example in which all the line impedances of an original network are transformed to the same value by a succession of length/impedance trade-offs. After transformation, the network would resemble a single uniform transmission line which could not possibly have the same frequency response as the original network. Furthermore, it would have no stopbands at all if its characteristic impedance happened to be made equal to the value of the terminating resistors.

To restate another general rule, length/impedance trade-offs *cannot* be applied to distributed circuits that have high-pass *S*-plane prototypes. This is because the transmission lines of the *f*-plane network are a quarter of a wavelength in length at the center of the passband, and to reproduce the upper half of the passband, there must be a change in the sign of the reactance of all the elements. However, this does assume that the circuit is required to perform correctly over the entire passband. If the circuit is only required to perform at frequencies below the center of the passband, then it is possible to perform some limited length/impedance trade-offs and lumped-element substitutions as a means of easing a physical realization problem.

An important application of length/impedance trade-offs that has not yet been described is the extension of Kuroda transforms to allow them to deal with noncommensurate line networks. The application will be described with the help of the specific example illustrated in Figure 2.37. Figure 2.37(a) shows a simple network comprising a unit element followed by a shunt inductor. If this were to be realized directly in *f*-plane elements, it would take the form of a through transmission line followed by a shunt short-circuit stub, both of which would be of the same length. For the purposes of the example, the problem that has to be addressed is one of finding a way in which impedances to the right of this network can be increased by a factor of 1.888, which in turn means finding a way of creating an ideal transformer on the right-hand side of the network with a turns ratio of 1.3739. Using the formulas given in Figure 2.30, creating this transformer could be a simple matter of splitting the inductor into two parts and moving the appropriate part to the left through the entire unit element. However, this would introduce a second inductor, which in the *f*-plane network would be a second stub and may not be a particularly attractive prospect.

Instead of moving part of the inductor through the entire unit element, an alternative is to move the entire inductor through only part of the unit element.

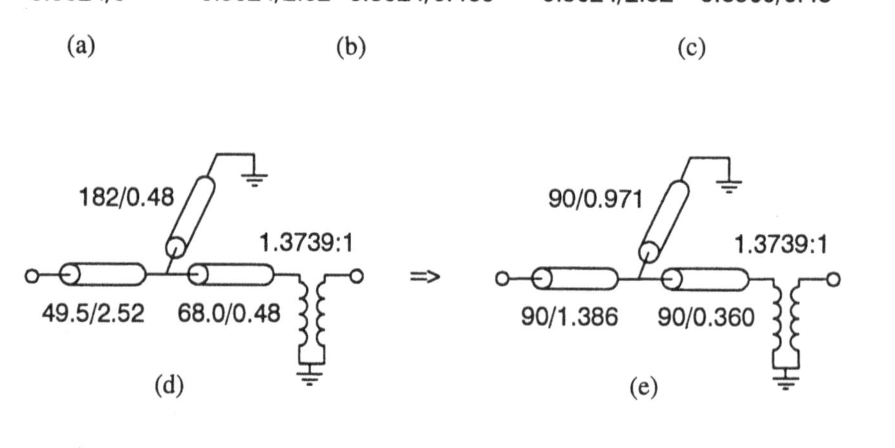

Figure 2.37 Using impedance/length trade-offs and Kuroda Identities in noncommensurate line networks.

This is attractive because no extra stubs will be created. The first step is to recalculate the value of inductance (1.3437) that must be moved into the unit element to create the necessary transformer turns ratio. Then the value of the inductance is reset and the length of the corresponding stub is reduced from a value of 3 to 0.4799 mm. This preserves the product of impedance and length, and therefore the effective inductance remains approximately the same. Now the inductor is moved into the unit element using a Kuroda Identity, the unit element having been split into two separate parts of equal impedance. By making the first part correspond to a length of line equal to 0.4799 mm, the Kuroda Identity can be applied without breaking the rule that the line and stub must be of the same length. As shown in Figure 2.37(c), this creates the required transformer. In Figure 2.37(d), the *S*-plane elements of (c) are converted into transmission line elements after a scale factor of 50Ω has been applied. Here the stub value of 182Ω is too high to be physically realizable in the relevant transmission line medium. Hence, by trading length against impedance for a second time, the impedance of this stub can be reduced to the much more desirable value of 90Ω, and as shown in of Figure 2.37(e), it was considered a good idea to do the same for the other two lines to make all the elements of the network have the same characteristic impedance.

In this example, as well as in the general case, the effect of length/impedance trade-offs within the *f*-plane equivalent circuit will be to cause some disturbance to the frequency responses of the whole network. As previously discussed, the effects will be most marked in the region of the stopband, with usually only a slight effect in the region of the passband. However, for certain networks, particularly those with wide frequency bandwidths of the order of an octave or more, the disturbances to the passband response can be undesirable, and some fine tuning of the elements of the whole network will be necessary by computer optimization. It is important to stress that this optimization of the network is only a case of fine tuning. If the length/impedance trade-offs are carried out correctly, the *f*-plane network will be very close to the required solution. The optimization should be constrained to allow only small changes to the element values, and for each individual distributed element, only the length *or* the characteristic impedance should be adjusted (i.e., not both). On most occasions, it will only be necessary to fit to target frequencies within the range of the passband.

2.5 CONVERSION OF PROTOTYPES TO PHYSICAL STRUCTURES

Following the completion of the prototype synthesis and the creation of an *f*-plane equivalent circuit of the required topology and impedance level, the remainder of the design process comprises the conversion of the *f*-plane prototype into the physical dimensions of a working circuit. This is the part of the procedure that exists below the AA reference line in Figure 2.2 and is also common with the classical design procedure illustrated in Figure 2.1. Despite the apparent simplicity of this part of the procedure, conversion into a satisfactory practical circuit is not always straightforward and is not a task that should be underestimated. Not only must a wide range of physical models be available to enable the essential circuit element dimensions to be calculated, but these models must also be incorporated within the relevant CAD software being used so that an accurate performance prediction can be made that includes the effects of circuit parasitics. In contrast to the preceding network synthesis stages, the conversion into a practical circuit is not an exact procedure. The physical models are by their very nature approximations with limited ranges of validity, and there is considerable scope for destroying an essentially good network design by inept application of the models or careless optimization. However, with the proper attention paid to model validity and optimization constraints, the advantages gained by using the exact synthesis techniques to create the basic prototype network should be not be lost.

It would be beyond the scope of this or any other single publication to present all the information on physical modeling required to design any conceivable kind of microwave component. Furthermore, the main objective of this book is to demonstrate the untapped potential of exact network synthesis procedures for microwave circuit design rather than extend the state of the art on physical modeling. This section will therefore give only a broad overview of the subject of physical realization, making reference to the literature wherever possible for the details of individual physical models. It will include a brief discussion on the design of radio frequency (RF) circuits operating at frequencies below 500 MHz, and then the majority of the section will be devoted to the design of MICs and MMICs operating at frequencies above 500 MHz. Of the many different classifications of microwave components, MICs and MMICs are the groups likely to benefit most from the use of exact network synthesis techniques.

2.5.1 Low-Frequency Circuits (<500 MHz)

For designing circuits at frequencies below 500 MHz, a lumped-element design approach is widely considered to be the most appropriate. This means the design is based on the assumption that the circuit elements and all the interconnections in between are so small that their physical sizes are of no consequence to the electrical behavior of the circuit. In other words, nothing resembling a transmission line appears anywhere in the circuit. In reality, of course, this is an oversimplification, particularly for circuits operating toward the upper end of the 500-MHz frequency range where real lumped elements do introduce significant distributed effects. However, in terms of a general approach, the distributed effects are assumed to be less significant than the lumped behavior of the elements and, at least at the start of the design procedure, are small enough to be ignored. Any corrections for distributed effects are applied later, toward the end of the design process. Although for the purposes of this discussion, 500 MHz has been chosen as the boundary separating lumped- and distributed-element design approaches, this choice was semiarbitrary and the boundary could occur anywhere in the frequency range 250 MHz to 2 GHz.

Because the *f*-plane prototype network of a low-frequency lumped-element circuit will not contain any transmission lines, the corresponding transformed prototype in the complex plane cannot contain any unit elements. The relevant complex frequency variable is the Laplace variable s, where $s = \sigma + j\omega$ and $\omega = 2\pi f$. Hence, the only elements from which the loss-free s-plane prototype can be constructed are capacitors, inductors, and ideal transformers. The capacitors and inductors may be included as either shunt or series elements and may also be incorporated into more complex reactance branches of the sort previously tabulated in Figure 2.15. Their dimensions are expressed in farads and henries, respectively, as a consequence of the linear transformation from the *f*-plane into the s-plane. The ideal transformers can be included to good effect in s-plane prototypes, because when combined with a pair of shunt inductors, one on each side, they can be realized in a physical circuit using a pair of mutually coupled coils of wire. This is only possible at these low frequencies of 1 GHz or less, where the parasitics associated with wire-wound transformers are acceptably small.

The physical forms of passive elements available for RF circuit design depend heavily on the required operating frequency range. Below 100 MHz, capacitors, inductors, resistors, and wire-wound transformers are mostly available as off-the-shelf components from commercial suppliers. The problem of calculating their physical dimensions, therefore, does not really arise. However, it is important to select components with a physical size that is as small as possible for a given electrical value so that the effects of any distributed parasitics can be kept to a minimum. Generally, the larger the components, the larger the parasitics, and the higher the operating frequency, the more pronounced the detrimental effects of the parasitics become.

If compensation for parasitics is necessary, then the relevant element must be represented in the RF circuit using a realistic equivalent circuit model. In the case of capacitors, the dominant parasitics are inductive and resistive, eventually causing both a series resonance at high frequencies and a reduction in Q. A suitable model could then take the form of a series tuned acceptor circuit. In the case of inductors, the dominant parasitics are capacitive and resistive, this time tending to cause a shunt resonance as well as a reduction in Q. There may be other resonances at even higher frequencies, but if these can be ignored, an acceptable model for the inductor could take the form of a shunt tuned rejector circuit. It is important to select capacitors and inductors that have self-resonant frequencies substantially higher than the highest anticipated operating frequency.

At operating frequencies above 100 MHz, some capacitor values may fall as low as 10 pF. For values of 10 pF or less, one method of constructing very high quality capacitors is to use printed circuit technology on ceramic substrates. If the whole RF circuit is to be printed on a ceramic substrate, this has the added advantage that the capacitors can be fabricated as integral parts of the circuit using the same fabrication process. As parallel plate capacitors, they are formed by depositing two layers of metallization separated by a thin-film of dielectric. The relationship between capacitor value and physical dimensions is

$$C = \frac{\epsilon A}{d} \quad \text{farads} \tag{2.89}$$

where $\epsilon = \epsilon_0\epsilon_r$, ϵ_0 is the permittivity of free space, and ϵ_r is the relative permittivity of the dielectric. A is the area of the capacitor plates and d is the dielectric thickness. Fringing capacitances around the edges of the capacitor are usually small enough to be ignored.

Good-quality inductors are more difficult to obtain as standard, prefabricated components and must often be constructed individually from coils of wire. The design of coil inductors is a large and complex subject based on a mixture of theoretical and experimental studies. Determination of coil diameter, wire diameter, and number of turns usually involves consulting design graphs or tables, some useful examples of which can be found in [8].

2.5.2 High-Frequency Circuits (>500 MHz)

In microwave circuits operating at frequencies above 500 MHz, all the component parts will exhibit some distributed characteristics whether or not this is desirable. The physical size of many lumped elements will start to become a significant fraction of a wavelength (e.g., $>\lambda/10$), and when this occurs, their effective electrical value is no longer a constant with frequency. Furthermore, the interconnections between circuit elements that would have been made by wires and ignored at lower frequencies now play a much more significant role and have to be treated as transmission lines. They actually become integral parts of the circuit design and must be properly represented in any prototype network. In microwave circuits, distributed elements tend to be in the majority, and as a consequence, the most appropriate design approach is a *distributed-element* approach. Essentially, this means that the circuit is designed initially as if *all* the elements are transmission lines and any corrections that may be needed to accommodate lumped elements are made at a later stage. All the relevant principles behind this approach have been described in Sections 2.2, 2.3, and 2.4. The relevant f-plane prototype is a ladder network of transmission lines and stubs of various descriptions, their lengths all being equal to a quarter of a wavelength at the commensurate frequency f_s. This transforms into an $\dot{S}$-plane prototype comprising a ladder network of symbolic capacitors, inductors, and unit elements, which is the form in which the basic circuit is synthesized.

Having reached the AA reference line of the design procedure in Figure 2.2, the f-plane prototype for a microwave circuit will be a combination of open- and short-circuit stubs separated by lengths of connecting transmission line. As explained in Section 2.4.3, length/impedance trade-offs may have resulted in some line length adjustments or the replacement of certain stubs with lumped capacitors, but most of the elements are still likely to be transmission lines. In the first stage of the conversion process, which ignores the presence of circuit parasitics, these ideal transmission line elements are translated into the physical dimensions of real transmission lines. Some of them will be translated on a one-to-one basis into the so-called *basic*, or *directly coupled*, physical elements that will be described in Section 2.5.2.5. However, there will be occasions when they are translated in groups into *compound* elements based on one of the many available configurations of capacitively coupled transmission lines. Conversion into capacitively coupled lines will be dealt with in Section 2.5.2.6.

Part of the decision-making process at the beginning of a microwave circuit design will be concerned with the choice of transmission line medium. There are several different transmission line media in widespread use that are relevant to MIC and MMIC design. Four of these, triplate stripline, suspended stripline, microstrip, and coplanar waveguide, will now be described in the following sections. Attention will be given to their basic properties, their advantages, disadvantages, main applications, and the most useful references in the literature. All four of these media are compatible with low-cost printed circuit methods of fabrication and support a dominant transverse electromagnetic (TEM) mode of propagation. Their physical structures are illustrated by the cross-sectional drawings of Figure 2.38. Specific references will be given for each type of transmission line, but for a more general treatment that includes the wider aspects of the modeling of MICs and MMICs, reference should be made to some of the excellent, previously published texts such as [9–12].

2.5.2.1 Triplate Stripline

Triplate stripline, or *triplate* as it is sometimes known, can be thought of as flattened coaxial line. It is a metal track printed on one side of a dielectric sheet that is covered by another dielectric sheet of the same thickness. Both the outer surfaces of the dielectric sheets have either ground plane metallization attached to their outer surfaces or are in contact with metal ground planes provided by an external clamping structure. The dielectric sheets must either be held under pressure or fixed to each other by a suitable low-loss adhesive. Because there is only a single dielectric filling (ignoring air gaps), triplate supports an almost pure TEM mode of wave propagation with no dispersion. Strictly speaking, this only applies when the ground planes and dielectric sheets are of infinite width. However, so long as any side walls are nonconducting and separated by substantially more than the vertical thickness of the structure, dispersion will be insignificant. It is important that the side walls of a triplate structure are nonconducting to avoid the possibility

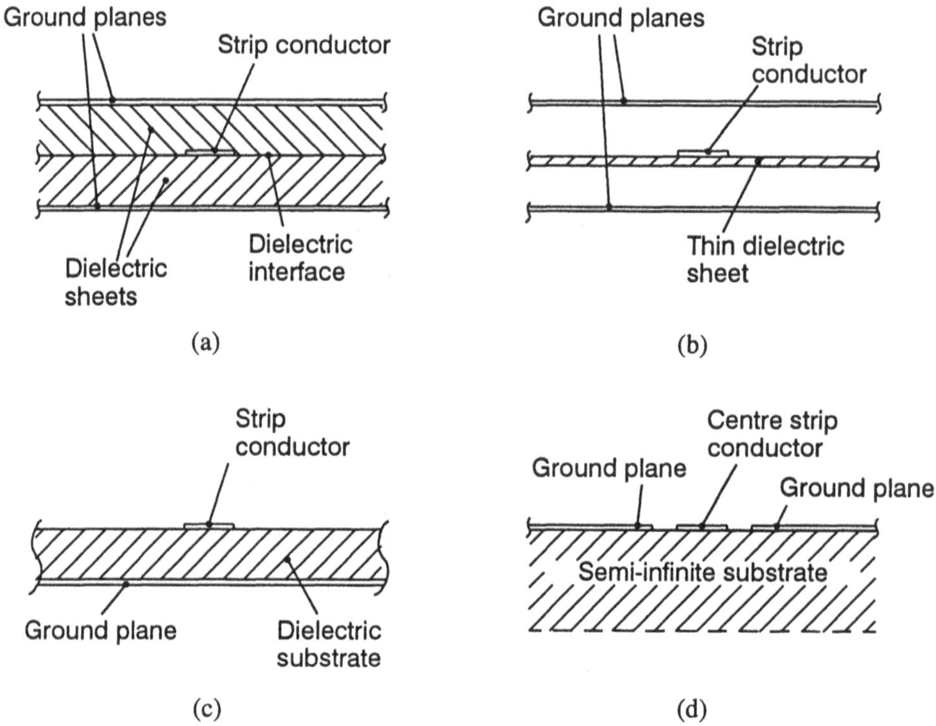

Figure 2.38 Various planar transmission line media for microwave circuits: (a) triplate stripline; (b) suspended stripline; (c) microstrip; (d) coplanar waveguide.

of box modes. If side walls have to be metallized, then additional steps must be taken to ensure the suppression of these unwanted modes.

Being a solid, enclosed structure, it is not particularly easy to incorporate lumped passive or active components into triplate, and hence the medium is mostly used for realizing pure distributed circuits such as microwave filters. It is possible to incorporate lumped elements inside a triplate circuit, but if the elements have significant thickness, clearance holes need to be cut in one or both of the dielectric sheets. This introduces a small risk of exciting parallel plate modes. Another risk of exciting parallel plate modes comes from the inclusion of shunt elements. It is absolutely essential that any shunt elements in triplate are connected to both the upper and lower ground planes if strong parallel plate radiation is to be avoided.

Triplate has the important advantages of being extremely robust and inexpensive to manufacture. Circuits can be fabricated on copper-clad dielectric material using conventional photolithographic etching techniques. Transmission line losses depend on the dielectric material, its thickness and the type of metallization involved, but generally the losses are lower than those of the other transmission line media, with the notable exception of suspended stripline. Resonator unloaded Q values can be on the order of 300. A typical dielectric material is RT Duroid 5880 (Rogers Corporation), which is a glass-loaded PTFE (polytetrafluoroethylene) with a dielectric constant of 2.2. A ceramic-loaded version is available (6010) with a dielectric constant of approximately 10. Practical ground plane spacings tend to be in the range of 1 to 3 mm. Various copper metallization thicknesses can be chosen, mostly in multiples of 17.5 μm. The narrowest lines and gaps that can be etched by conventional techniques are approximately twice the thickness of the copper metallization.

The definitive works on the modeling of triplate stripline can be found in [13–16]. Cohn must be credited with making the most significant contribution to the problem of calculating characteristic impedance from physical dimensions or vice versa. He used conformal transformations to give exact formulas for strips of zero thickness and then showed how further transformations could be used to apply corrections to the line widths to account for nonzero strip thickness. Bahl in [16] brings together Cohn's formulas into a complete set that can be used for the analysis or synthesis of strip dimensions. The range of impedances that can be realized in triplate depends on the ground plane spacing, the dielectric constant of the filling material, the capabilities of the etching process, and the frequency of operation. Typically, the range is between 25 and 125Ω for lines whose length is on the order of a quarter of a wavelength at frequencies below 20 GHz. (Note: Length is relevant here because the width of a transmission line should not be greater than its length.)

2.5.2.2 *Suspended Stripline*

Suspended stripline is like triplate stripline without most of the solid dielectric. In an effort to reduce dielectric losses, the center conductor is printed on a thin dielectric film that is stretched and held under tension in a central position between the ground planes by a clamping action of the box walls. Sometimes, large circuits need additional clamping in the form of matching sets of posts extending up and down from the two ground planes to ensure that the dielectric film is held flat. Wave propagation is predominantly a TEM mode when the dielectric film is thin compared with the ground plane spacing. A ratio of ground plane spacing to film thickness of more than 10:1 is common, wherein the effective dielectric constant is close to unity. A frequency-dependent variation in effective dielectric constant (i.e., dispersion) can, however, be observed at high microwave frequencies when reductions in ground plane spacing to prevent parallel plate modes make the thickness of the dielectric film more significant. The necessary presence of the box walls for clamping purposes can also give rise to undesirable waveguide modes. Fortunately, these are easily suppressed by the metal posts used to clamp the circuit, especially if the posts make direct electrical contact through the dielectric film.

With its lower losses, suspended stripline is especially attractive for making ultralow-loss filters. Unloaded Q values of circuit resonators can be expected to be in excess of 600. The absence of dielectric filling above the circuit makes the incorporation of surface-mounted lumped components more convenient than with triplate, but the need to make contact with both ground planes for shunt elements still presents a difficulty. It is not easy to gain access to the circuit for repairs or for circuit adjustment while the upper ground plane is in place and there is a risk that relatively thick lumped elements will disturb the symmetry of the structure, causing parallel plate modes to be excited. A further advantage of suspended stripline is that it is relatively easy to interface with a waveguide. One disadvantage derives from the absence of a solid dielectric. While an air dielectric reduces line losses, the unity dielectric constant makes suspended stripline circuits comparatively large for a given operating frequency range.

RT Duroid 5880 with a thickness of 125 μm is a common choice of material for the dielectric film. The range of realizable track widths and impedances is similar to that of triplate except that slightly higher impedances can be realized, owing to the lower dielectric constant. For the most part, Cohn's formulas for triplate can be used for calculating the dimensions of suspended striplines, but Weil [17] and Smith [18] have published two papers dealing specifically with suspended line. Hoffmann [12] also gives an excellent treatment of suspended line.

2.5.2.3 Microstrip

As illustrated in Figure 2.38(c), microstrip line resembles triplate stripline without the top layer of dielectric and ground plane. Because microstrip is an asymmetrical structure, fields associated with the propagating waves are no longer distributed equally above and below the center strip, but are instead contained mostly by the solid dielectric. Some field energy inevitably exists in the air above the strip, and this modifies phase velocity and distorts what would otherwise be a pure TEM mode of propagation in a single dielectric medium. Microstrip line therefore has an effective dielectric constant whose value lies somewhere between the unity value for air and the larger value for the solid dielectric. Its value is a function of strip width.

In addition to a TEM mode, microstrip supports dispersive modes of propagation trapped between the top surface of the dielectric and the bottom ground plane. The lowest order of these has a cutoff frequency that also extends down to zero frequency, and consequently some dispersion will always have to be tolerated. However, energy contained in the surface modes can be minimized by keeping the thickness of the dielectric substantially less than a tenth of a wavelength. In these circumstances, the dominant propagating mode for microstrip is *quasi-TEM,* which means that for most circuit design purposes, the effects of dispersion can be ignored.

Typically, dispersion is not significant in MICs and MMICs operating at frequencies below 20 GHz. It can be significant above 20 GHz, and will be apparent by the increase in the effective dielectric constant that occurs as more energy shifts into the surface modes.

Microstrip has grown in popularity because of the ease with which lumped, particularly active, components can be attached to the surface transmission line. Being an open structure, there is ready access to the top surface for the attachment of components, the trimming of circuit elements, and postfabrication maintenance. Complete MICs can be fabricated comprising, if necessary, upward of several hundred lumped capacitors, resistors, diodes, and FETs, all interconnected by a network of microstrip lines. The earliest of these were fabricated in what is now known as *hybrid technology*, where the attachment of the lumped elements takes place as a second part of the process after the patterning of the microstrip circuit on its dielectric substrate. Later, the concept was extended to MMICs in which the microstrip lines and all the lumped components were fabricated at the same time on a single semiconductor substrate of GaAs. MMICs are now very much the center of attention in advancing the state of the art for microwave circuit design. They are revolutionizing the size, cost, and performance of microwave components and systems. Their success has depended heavily on the availability of microstrip transmission line.

There are at least three relevant fabrication technologies associated with microstrip line. The first is what is referred to as *plastic hybrid technology*, wherein microstrip lines are etched onto one side of a copper-clad plastic substrate material. A typical material would be RT Duroid 5880, the same glass-loaded PTFE material used for triplate. Dielectric thicknesses range from 0.125 mm to about 3 mm and copper thicknesses are supplied in multiples of 17.5 μm. The principle advantage of using plastic substrates is the relatively low cost of processing. While they are acceptable for many commercial applications, their mechanical and thermal stabilities are often inadequate for military applications. Lumped components can be attached to the circuits using conducting epoxy adhesives. Wire connections can be made by ultrasonic bonding, but it can be difficult to attach either wires or components to the copper tracks using thermocompression techniques. Thermocompression bonding often causes serious deformation of the plastic substrate.

Thin-film hybrid technology is probably the most widely used method of fabricating MICs. Fired aluminum oxide (Al_2O_3), known as *alumina*, is the most common of the ceramic substrates for thin-film circuits, available in standard thicknesses ranging from approximately 0.125 to 0.625 mm. It is supplied polished to provide the necessary smooth surface finish or is alternatively coated with the chosen metallization (usually gold). If ready-coated, the microstrip lines are etched from the metallization by subtractive etching and conventional photolithography. If, however, the alumina is nonmetallized, the microstrip circuit is usually created by a "plating-up" process that tends to give better dimensional accuracy. In this a thin layer of a few hundred angstroms of nickel and chromium (nichrome) is evaporated onto the alumina to give good adhesion. Then 1,000Å of gold is evaporated onto the surface, the small thickness of which gives the process its name. Photolithography is then used to open windows in a resist material, inside of which the rest of the thickness (usually 3 μm) of the gold is deposited by electroplating. A light etch removes the thin-film gold and seed layer as the last step in the pattern definition process.

Ceramic substrates are hard and have more stable properties than plastic substrates. Components can be attached by thermocompression bonding or soldering without any risk of damage to the substrate. They also tend to have high dielectric constants, giving a reduction in the size of the microwave circuits. Alumina has a dielectric constant of 9.9, and the effective dielectric constant for a 50Ω line is approximately 6.5. A further advantage is the ability to fabricate thin-film resistors as an integral part of the circuit. The resistive elements are created by etching away surface gold to expose the nichrome seed layer. Particular values are obtained by calculating the correct aspect ratio for the exposed nichrome, given the knowledge of its resistivity.

The third technology relevant to microstrip line is MMIC technology on GaAs. As a complete process, this is too complicated to describe here, but the techniques used as part of the process for producing the microstrip lines are largely the same as those used for thin-film hybrid circuits. The difference is that the process has many more steps to create all the other lumped components, including the active devices on the semiconductor substrate.

MMIC fabrication has become sufficiently complicated to require the provision of large dedicated foundry facilities, the cost of which prevents most companies from having their own capability. Consequently, specialist companies throughout Europe and the United States have installed MMIC foundries that now offer a fabrication service to third parties. Given the recipe or design rule book for the particular foundry process, circuit designers can design their circuits from a wide range of active and passive elements in the knowledge that their circuit will be fabricated by the foundry using a process with a guaranteed consistency. In its semi-insulating form, the dielectric constant of GaAs is 12.9 and substrate thicknesses are typically 100, 200, or 400 μm. The metallization thickness is usually 3 μm. Practical microstrip line widths are in the approximate range of 5 to 100 μm.

Whatever the particular technology used for fabrication, microstrip lines tend to have more loss than triplate lines because of the higher current densities on one side of the conducting strip. Unloaded Q values of microstrip resonators are nearer 100 than the value of 300 for triplate. Such losses make microstrip rather unsuitable for high-performance filter components. Furthermore, microstrip is a rather leaky structure, since any bend or step discontinuity produces both free-space radiation and radiation into the substrate. This means that it is difficult to achieve an isolation between any of the elements of a microstrip circuit of more than approximately 40 dB. As a transmission line, microstrip does not have a particularly good performance as far as its transmission properties are concerned, but its advantages for MICs and MMICs far outweigh the disadvantages.

Because of its asymmetry, microstrip is more difficult to model than some of the other types of transmission lines, but the problem has probably received more study and resulted in more publications than almost any other aspect of microwave circuit design. Large numbers of papers are available in the literature for calculating impedance, attenuation constant, and effective dielectric constant of microstrip lines on almost any substrate. Many also provide the theory for calculating the effects of metal housings on the electrical parameters. Some of the most notable references are [19–22]. Bryant and Weiss [19] produced the first accurate impedance and effective dielectric constant data for microstrip lines of zero thickness. The accuracy of the data is widely acknowledged and has been frequently used as a basis for testing many of the more recent analytical formulas. However, the tables of data are rather inconvenient to use, are compiled for only certain substrate materials, and the numerical procedure used for their calculation is not particularly portable. In response to the strong demand for analytical formulas that could be used to calculate electrical parameters from physical dimensions or vice versa, Wheeler [20], Hammerstad [21], and Jensen [22] have provided some of the best solutions. They have also been able to accommodate strip thickness in their formulas. The range of realizable impedances in microstrip line is approximately 20Ω to 125Ω for plastic hybrid technology, decreasing to between 25Ω and 90Ω for MMICs on 100-μm-thick GaAs.

2.5.2.4 Coplanar Waveguide

Except for the thickness of the strip conductors, coplanar waveguide is a truly planar printed circuit transmission line medium. As indicated in Figure 2.38(d), both the signal strip in the center and the two ground plane conductors on either side are now present on the top surface of the solid dielectric. Pure coplanar waveguide has, by definition, no ground plane conductor on the back surface, is of infinite width, and has a substrate of infinite thickness. Like microstrip line, the electromagnetic fields associated with the propagating waves occupy both the air above the strip and the solid dielectric beneath. Most of the fields are contained by the dielectric, but the effective dielectric constant relevant to wave propagation has a value somewhere between that of air and the solid dielectric.

There are two dominant modes of propagation, both of which are quasi-TEM. One is often referred to as the *odd mode,* since this corresponds to the situation in which the electric fields on the two sides of the central strip point in opposite directions. In terms of electric potential, the two ground plane conductors are at the same potential, while the center strip is at a potential above or below as determined by the instantaneous value of the excitation. This is almost always the mode used for practical microwave circuits. The second mode is the *even mode,* for which electric fields on opposite sides of the center strip point in the same direction. This time the two outer ground plane conductors (not really ground planes any more for this mode) are at equal and opposite potentials with respect to the center strip for perfect geometrical symmetry. Essentially, the even mode is like a slot-line mode that could exist even in the absence of the center strip. The odd and even modes of coplanar waveguide have quite different characteristic impedances for a given physical geometry.

As a transmission line medium for use in MICs, the planar structure of coplanar waveguide would seem to offer considerable advantages over microstrip. As well as being able to realize series transmission line elements, coplanar line now allows lumped active or passive components to be connected in a shunt by simply placing them across the gaps between the center conductor and the ground planes. Compared to microstrip line, where shunt components have to extend some considerable distance vertically down through a hole in the dielectric, the connection is very much more convenient, it is more compatible with the physical size of discrete microwave devices, and should give rise to fewer undesirable parasitics. With care, these advantages can be realized, but as will be explained in the next paragraph, there are serious difficulties with coplanar waveguide that have so far impeded its use in MICs.

The problems with using coplanar waveguide for MICs stem from its willingness to support undesirable modes. In an MIC, the interconnecting transmission lines must incorporate bends and junctions of various descriptions. In coplanar line, bends and junctions will both radiate into free space and cause strong mode conversion. Similarly, any shunt connections to the center strip that are not symmetrical or any slight asymmetry anywhere in the coplanar line will cause excitation of the unwanted even mode, with disastrous consequences for the operation of the MIC. Some suppression of the even mode is possible by incorporating short-circuit links between the ground planes at regular intervals along the lines. These might take the form of bond wires in an MIC or even metallic air bridges in an MMIC. However, this can be highly inconvenient and will not always provide adequate suppression. It is also necessary to use the bond wires to interconnect isolated islands of ground plane that can occur in even the simplest of MICs realized in coplanar waveguide. Yet another source of undesirable modes originates from the practical need to house coplanar MICs and MMICs in metal boxes and to use a finite thickness of dielectric material. For mechanical integrity, most need to be fixed to the floor of a metal housing. Unfortunately, this allows microstrip modes to be created almost as readily as the main coplanar waveguide mode itself. Making the dielectric as thick as possible discourages their excitation and painting the back surface of the dielectric with resistive paint will help with their absorption, but complete elimination of microstrip modes is still extremely difficult to achieve.

Coplanar waveguide is only rarely used on its own in an MIC, and even then it is confined to relatively simple circuits. However, it has been used in some specialized circuits where its multimode nature can be used to some advantage. A good example is the millimeter-wave balanced mixer circuit shown in Figure 2.39, which employs several different transmission line media, including coplanar waveguide. A comprehensive account of the mixer design can be found in [23]. A novel aspect of the design is the deliberate exploitation of both the even and odd modes of propagation in a short section of coplanar waveguide.

With reference to the circuit in Figure 2.39, a millimeter-wave input signal at 94 GHz approaches the circuit from the left in waveguide and passes through a tapered transition into a slot-line. The signal propagates along the slot-line until it encounters a short length of coplanar waveguide, where it produces an even mode excitation. Because the coplanar waveguide section is approximately a quarter of a wavelength long and the termination at the far end is an effective short circuit to the even mode, the impedance at the input plane is a virtual open circuit.

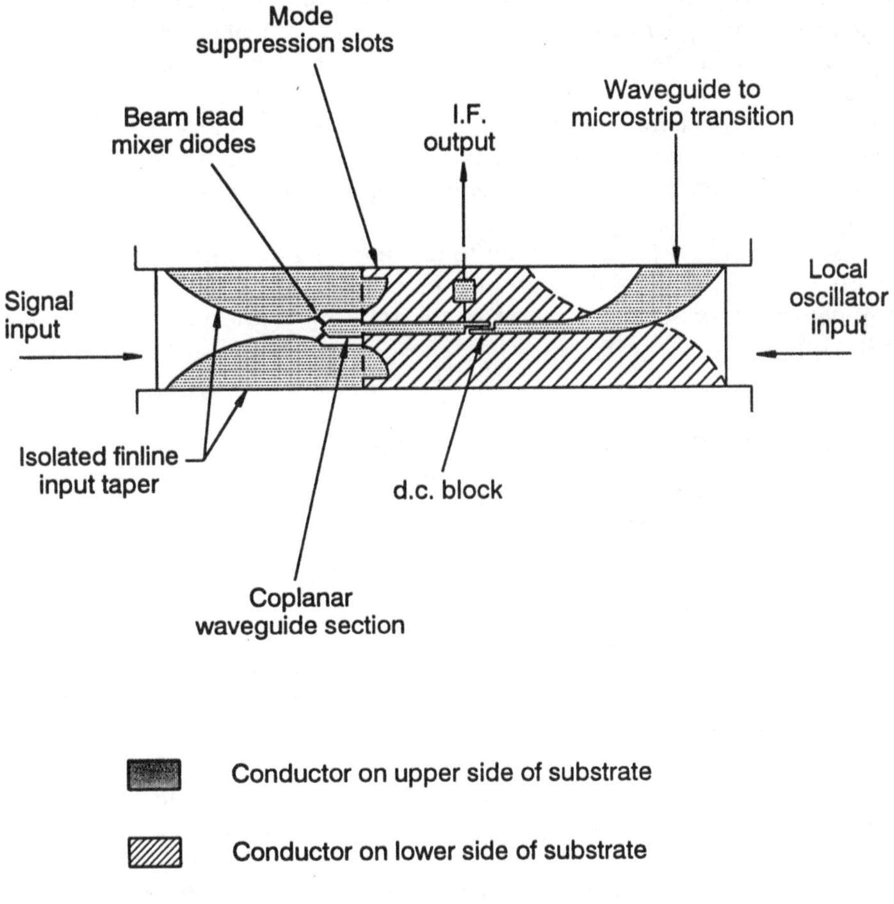

Figure 2.39 A millimeter-wave balanced mixer illustrating the versatility of coplanar waveguide.

Located in this same input plane is a pair of back-to-back mixer diodes connected to the center conductor of the coplanar waveguide section. These cause a balanced mixing action when pumped by a large local oscillator (LO) signal arriving from the opposite direction in the odd mode. The LO signal enters the circuit from the right, where it first makes a transition into microstrip before making a second transition into coplanar waveguide. With proper attention to symmetry, there can be more than 25 dB of isolation between the odd and even modes of the coplanar waveguide section (i.e., 25-dB isolation between the LO and RF ports of the mixer). The MIC was fabricated on a 125-μm-thick RT Duroid 5880.

Transmission losses in coplanar waveguide are similar to those of microstrip. Unloaded resonator *Q* values would typically be on the order of 100. The materials from which the medium can be fabricated are also similar to microstrip, which includes GaAs for MMICs. As more effective techniques for mode suppression become available through MMIC technology, coplanar waveguide can be expected to find more widespread use in the near future. Its potential is greatest in the area of low-power millimeter-wave receiver circuits. Pattern definition by photolithography can achieve minimum strip and slot dimensions of typically 25 μm for thin-film technology on alumina or below 10 μm for MMIC technology on GaAs. Characteristic impedances are realizable in the approximate range of 35Ω to 100Ω for 0.5-mm-thick alumina. An effective dielectric constant is about 5.4 (alumina $\epsilon_r = 9.9$).

References [24–27] are relevant to the modeling of coplanar waveguide and the calculation of physical dimensions from characteristic impedance. Particularly thorough treatments can be found in the texts by Hoffmann [12] and Gupta et al. [10].

2.5.2.5 *Basic (Directly Coupled) Physical Elements*

The simplest set of physical elements from which a new circuit can be constructed will be referred to as *directly coupled* elements. These are individual components that relate to single elements in the *f*-plane prototype and that couple to their nearest neighbors in the network via direct electrical connections. They contrast with the more complicated set of physical elements to be described in Section 2.5.2.6, which are based on capacitively coupled transmission lines.

Lumped elements are the first of the two main subgroups of directly coupled circuit elements. For MICs and MMICs, the passive lumped elements are restricted to capacitors, resistors, and spiral inductors, each of which can be found in several different physical forms. Lumped element transformers would not usually be considered suitable, because of their bulk and their excessive parasitics. The most common forms of lumped capacitors, resistors, and inductors are illustrated in Figure 2.40. The first is an MIM (metal-insulator-metal) capacitor, which, as the name suggests, comprises two rectangular metal plates separated by a thin layer of dielectric. In

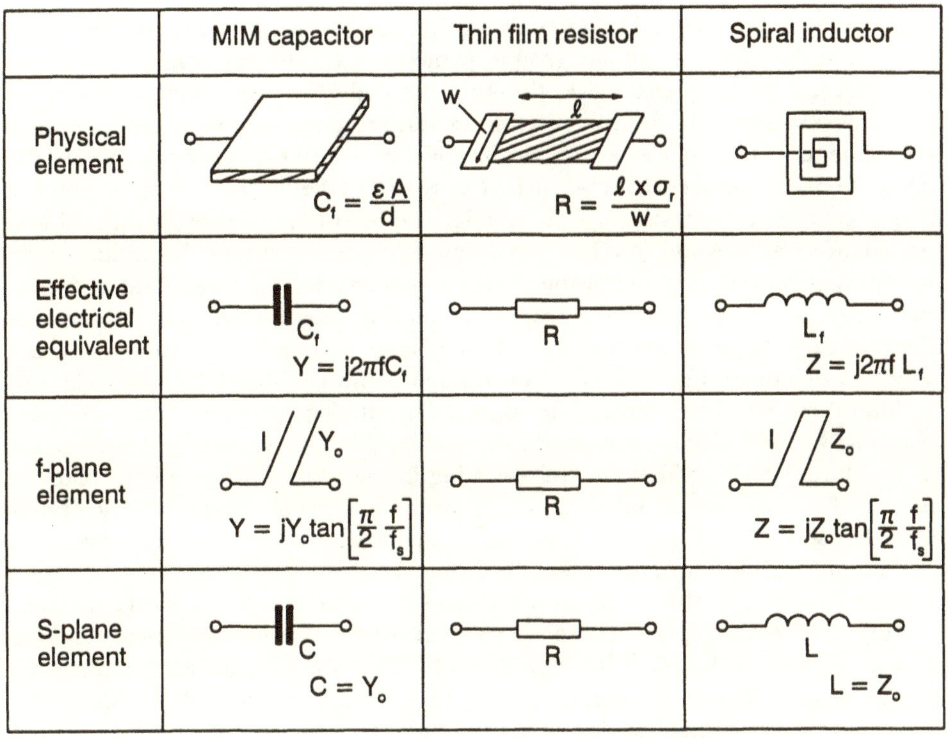

Figure 2.40 A set of directly coupled lumped elements commonly used in MICs and MMICs.

hybrid MICs or MMICs, the typical metallization is evaporated gold for the bottom plate and a rather thicker layer of electroplated gold for the top plate. The dielectric is usually silicon nitride deposited to a thickness of 0.15 μm. With such a thin dielectric layer, fringing capacitance around the edges of the plates can be ignored, except the very smallest of capacitors (e.g., <0.2 pF). Hence, a sufficiently accurate capacitance value can be calculated using (2.89). Typically, values are realizable between 0.1 and 10 pF. The second element is a resistor that comprises a pair of ohmic metal contacts separated by a thin film of resistive material. In MICs the material would normally be an evaporated layer of nichrome, but in MMICs is more likely to be a layer of ion-implanted GaAs. Resistance values are obtainable in the approximate range 5Ω to 5,000Ω, as determined by the aspect ratio and the resistivity of the active material. The third element is a spiral inductor comprising a narrow strip of gold metallization printed in the shape of a spiral. A connection to an outside edge provides one terminal of the inductor, while the other is provided by a radial connection to the center point involving a series of air bridges. Spiral inductors are really only applicable to MMIC technology and are even then unattractively large. Inductance value depends on the number of turns, the spacing of the turns, and the proximity of any ground plane beneath the substrate.

Because of their small size, thin-film lumped MIM capacitors and resistors are relatively free of undesirable parasitics and are good approximations to ideal components at frequencies as high as 20 GHz. Above this frequency or where the components are especially large, it is necessary to take account of their physical size by creating an equivalent circuit model that incorporates short lengths of low-impedance transmission line on each side of the ideal element. Spiral inductors, however, are heavily affected by the presence of parasitics; the long spiral strip gives rise to high series resistance, the proximity coupling between the turns produces an interelectrode capacitance, and the relatively large surface area creates significant shunt capacitance to ground. If spiral inductors must be used, it is important to include these parasitic elements in any equivalent circuit model. Even when the parasitics are included, it is difficult to create a model that is valid for a wide range of structures or for a wide range of operating frequencies. This and their relatively poor performance are strong grounds for avoiding spiral inductors wherever possible. They should only be considered when large values of inductance are required and the operating frequency is less than 10 GHz.

In Figure 2.40, the second row indicates the effective electrical equivalent of the three different kinds of components. This does not necessarily mean that the components could be adequately modeled by these simple equivalent circuits. As previously described, more complicated models would be required, particularly in the case of a spiral inductor, for an accurate representation of the behavior of the real component. However, it is into these simplified forms that their equivalent circuits must be reduced to allow proper association with the elements of an f-plane prototype network. When converting an f-plane prototype network into a physical circuit, the values of any ideal lumped elements already in the prototype will correspond to these effective values of the physical elements.

With the exception of the resistor, the third row of Figure 2.40 contains the distributed elements to which the lumped elements of the first and second rows may correspond. If the f-plane network was a purely distributed commensurate line network, then as previously discussed in Section 2.4.3, the series open-circuit stubs and the series short-circuit stubs can be converted into the lumped elements shown if the stubs are electrically short in the region of the passband. Equating admittances at the appropriate frequencies converts the characteristic admittance of an open-circuit stub into the value of a lumped capacitor, or alternatively, equating impedances converts the characteristic impedance of a short-circuit stub into the value of a lumped inductor. The fourth row of the figure indicates the S-plane equivalent circuit elements if the f-plane prototype network is distributed.

Turning now to the second subgroup of directly coupled elements based on pure distributed elements, these are basically sections of transmission line that are employed either as interconnecting lines (i.e., unit elements in the S-plane) or alternatively as stubs. Without taking practical constraints into account, there are at least five different types of distributed elements that could, in principle, be incorporated into an *f*-plane prototype, as indicated in the second row of Figure 2.41. These comprise interconnecting lines, shunt short-circuit stubs, shunt open-circuit stubs, series short-circuit stubs, and series open-circuit stubs. In addition to these, there is also the possibility of incorporating multiple-section stubs of the types that were described in Section 2.3.6.1. Within the exact *f*-plane prototype, all the elements will be of the same length, but owing to the possibility of length/impedance trade-offs, the lengths of the distributed elements in the physical circuit need not be the same.

Unfortunately, as indicated in the first row of Figure 2.41, only three out of the possible five *f*-plane elements are physically realizable in any of the four transmission line media that were discussed in previous sections. Only interconnecting lines and the shunt stubs can be realized as separate, directly coupled elements, and it is generally impossible to realize either of the two types of series stub. As suggested in Section 2.5.2.4, there may be some scope for realizing series distributed elements in coplanar waveguide, but there would be inevitable drawbacks for the rest of the circuit. The reason for the exclusion of series elements has to do with the necessary

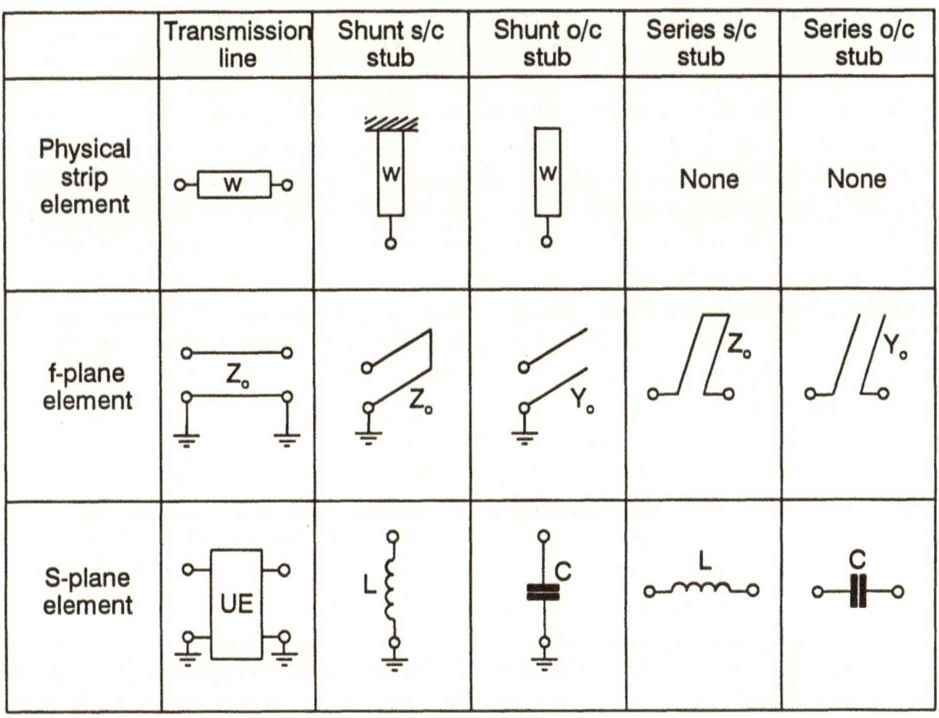

Figure 2.41 A set of directly coupled distributed elements available for MICs and MMICs.

presence of a ground plane in all the transmission line media, whether on the back side of the substrate or on the front side, as in the case of coplanar waveguide. With the ground plane present, any series element cannot avoid having shunt capacitance to ground. This means that the only way to realize series distributed elements is to combine them with other types of elements, such as interconnecting lines or shunt stubs. These other elements would have to be adjacent to the series element and deliberately created in the *f*-plane prototype network. As will be seen in the next section, groups of elements containing series stubs are realizable in the form of capacitively coupled transmission lines.

The widths of directly coupled line and stub elements are calculated from their characteristic impedances using the relevant physical models referenced in Sections 2.5.2.1 through 2.5.2.4. Before the effects of circuit parasitics are taken into consideration, their lengths are calculated by dividing the lengths of the elements in the *f*-plane prototype by the appropriate velocity factor. The velocity factor is equal to the square root of the effective dielectric constant (ϵ_{reff}) for an element of a given width in the chosen transmission line medium. The effective dielectric constant is calculated as part of the calculation of strip width. Later, there may be cause to adjust the length of the lines to take account of lumped parasitics at one or both, ends but at this particular stage in the procedure, it will be more important to establish approximate line dimensions so that an early impression can be gained as to the physical realizability of the circuit. The limits on the line dimensions will depend on the technology involved. However, it is always important to ensure that the length of any line is greater than its width and preferably more that twice its width. Line width should also always be less than a quarter of a wavelength, and in microstrip, less than the thickness of the substrate.

The effect on the *S*-plane prototypes of the restricted availability of physical elements is generally to exclude series *S*-plane capacitors and inductors unless capacitively coupled line sections or lumped-element components can be used in the real circuit. If capacitively coupled lines are acceptable in a circuit, then so long as the appropriate coupled-line section is selected, the restrictions will not apply. However, without capacitively coupled lines, the restriction on series capacitors can only be lifted if lumped capacitors of sufficiently good performance can be made available. This also depends on the *S*-plane prototype being low-pass or band-pass. It cannot be lifted for high-pass prototypes, since these always transform into pure distributed, commensurate line networks in the *f*-plane.

The restrictions on series *S*-plane inductors can rarely be lifted, because of the absence of good-quality-lumped element inductors at microwave frequencies. However, the restriction is not really relevant for series *S*-plane inductors, because in the presence of unit elements, Kuroda transforms can always be used to move them into other positions in the network, where they become shunt capacitors. There are few occasions when the need for series *S*-plane inductors cannot be avoided.

2.5.2.6 Physical Elements based on Coupled Lines

Physical elements based on coupled lines are necessarily distributed structures. They may be realized in any of the four transmission line media previously described and as a basic concept comprise a pair, or possibly several parallel strip conductors, placed in close proximity to each other. In the most simple of configurations, the conductors are printed on the same surface of a dielectric substrate with a small gap in between. Couplers of this type are easy to fabricate, but do not produce particularly strong coupling. Stronger coupling can be achieved if the conductors are overlapped and separated by a thin layer of dielectric material. Coupling is then across the broad dimensions of the strips rather than the edges, giving a much greater interstrip capacitance. Alternatively, strong coupling can also be produced by using an interdigitated arrangement of the strips like that used in Lange [28] couplers. Sketches of microstrip versions of these three coupling structures, the edge-coupled transmission lines, the overlapped coupled lines, and the interdigitated coupled lines are shown in Figure 2.42.

As explained in the previous section, capacitively coupled lines are used to realize groups of *f*-plane prototype elements rather than individual elements. They are able to do this, because as physical structures, they themselves comprise more than one transmission line element. They are able to give an important increase in topological flexibility to an *f*-plane network, enabling series stubs and a much wider range of transmission line impedances to be realized than would otherwise be possible. They can also result in a more compact physical circuit.

There are some disadvantages of coupled lines to be noted, however. As physical structures, they are considerably more complicated to model than individual transmission line elements, which increases the risk of a finished circuit not working to specification. Compared to a circuit not containing coupled lines, one with coupled lines is likely to need more design iteration. It is likely to have more

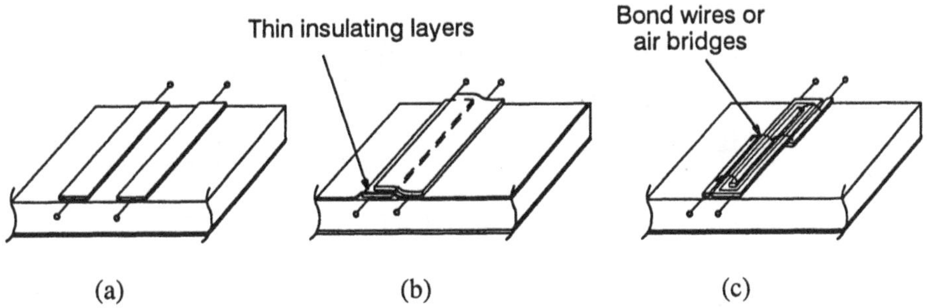

Figure 2.42 Sketches of (a) edge-coupled, (b) overlapping, and (c) interdigitated coupled-line structures in microstrip.

severe dimensional tolerances associated with the gaps between the strips and have higher ohmic losses. The higher losses are due to current crowding along the inside edges of the strips. Circuits containing capacitively coupled lines are most appropriate for relatively narrow instantaneous frequency bandwidths (e.g., <10%). There are some exceptions to this trend, as will be seen in Chapter 3 on wideband filter design. However, because the coupling between adjacent distributed elements relies on proximity capacitance, coupling is much weaker than for directly coupled elements, and if wideband operation is a requirement, directly coupled elements are usually to be preferred.

The simplest four-part equivalent circuit of a pair of capacitively coupled lines is the tunnel shaped arrangement of three ideal transmission lines shown in Figure 2.43. The lines on the sides of the tunnel represent the physical transmission lines formed by each individual strip with respect to ground, and the line across the top of the tunnel corresponds to the balanced line formed by the distributed mutual coupling between the strips.

If we ignore slight differences in propagation constants that would exist in a real coupler, the three lines of the equivalent circuit are of equal electrical length. Their characteristic impedances are Z_a, Z_b, and Z_{ab}, respectively, but for the purposes of calculating distributed capacitances, it can be more convenient to work with their characteristic admittances, Y_a, Y_b, and Y_{ab}. The expressions for the normalized distributed capacitances of the strips are then

$$C_a = \eta Y_a \qquad C_b = \eta Y_b \qquad C_{ab} = \eta Y_{ab} \qquad \text{where}$$
$$Y_a = \frac{1}{Z_a} \qquad Y_b = \frac{1}{Z_b} \qquad Y_{ab} = \frac{1}{Z_{ab}} \tag{2.90}$$

and where η is the absolute intrinsic impedance of the dielectric medium. For the odd and even mode admittances, Y_{oea}, Y_{oeb}, Y_{ooa}, and Y_{oob}, there are four equations for the general case of lines of unequal width. These are

$$Y_{oea} = \frac{C_a}{\eta} = Y_a \qquad Y_{ooa} = \frac{2C_{ab} + C_a}{\eta} = 2Y_{ab} + Y_a$$
$$Y_{oeb} = \frac{C_b}{\eta} = Y_b \qquad Y_{oob} = \frac{2C_{ab} + C_b}{\eta} = 2Y_{ab} + Y_b \tag{2.91}$$

or transposing for the distributed capacitances, we obtain

$$C_a = \eta Y_{oea} \qquad C_b = \eta Y_{oeb} \qquad C_{ab} = \frac{\eta(Y_{ooa} - Y_{oea})}{2} \tag{2.92}$$

When the strips are of equal width (i.e., a symmetrical coupled-line section) we have

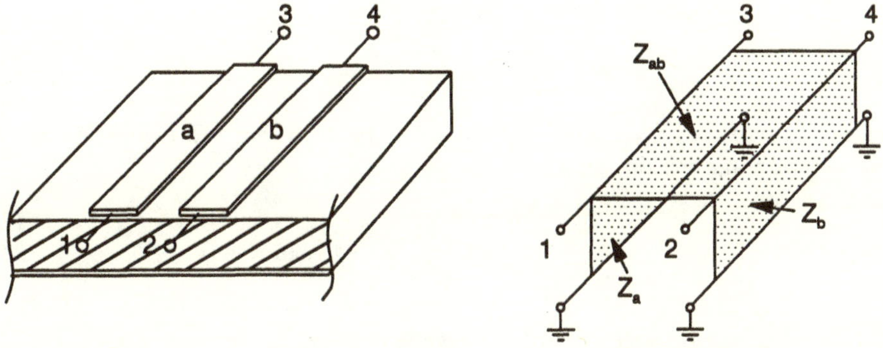

$$C_a = \eta Y_a \qquad Y_a = 1/Z_a$$

$$C_{ab} = \eta Y_{ab} \qquad Y_{ab} = 1/Z_{ab}$$

$$C_b = \eta Y_b \qquad Y_b = 1/Z_b$$

$$C_a = \eta Y_{oea} \qquad Y_{oea} = Y_a$$

$$C_{ab} = \eta (Y_{ooa} - Y_{oea})/2 \qquad Y_{ooa} = Y_a + 2Y_{ab}$$

$$C_b = \eta Y_{oeb} \qquad Y_{oeb} = Y_b$$

$$Y_{oob} = Y_b + 2Y_{ab}$$

For symmetrical lines:

$$Z_{oe} = Z_a = Z_b = \frac{1}{Y_{oea}} = \frac{1}{Y_{oeb}}$$

$$Z_{oo} = \frac{Z_{ab}Z_b}{Z_{ab} + 2Z_a} = \frac{1}{Y_{ooa}} = \frac{1}{Y_{oob}}$$

Figure 2.43 The ideal three-line equivalent circuit of a pair of coupled lines.

$$Y_{oea} = Y_{oeb} \quad \text{and} \quad Y_{ooa} = Y_{oob} \tag{2.93}$$

and for *only* this special case, the odd and even mode impedances of the strips are equal to the reciprocal of the odd and even mode admittances. Hence,

$$Z_{oea} = Z_{oeb} = Z_{oe} = Z_a = Z_b \qquad \text{and}$$
$$Z_{ooa} = Z_{oob} = Z_{oo} = \frac{Z_{ab}Z_a}{Z_{ab} + 2Z_a} \tag{2.94}$$

Using the odd and even mode impedances, we find that the expression for the voltage coupling coefficient is

$$c = 20\ \mathrm{Log}_{10}\left\{\frac{Z_{oe} - Z_{oo}}{Z_{oe} + Z_{oo}}\right\}\ \mathrm{dB} \tag{2.95}$$

and the expression for the coupler *section impedance*, Z_{sec}, is

$$Z_{sec} = \sqrt{Z_{oe}Z_{oo}} \tag{2.96}$$

This equivalent circuit is extremely useful if a simple analysis of a circuit containing a coupled-line section must be performed before the introduction of circuit parasitics. It also serves to define some of the most important design parameters for coupled-line sections.

Either distributed capacitances or odd and even mode immittances are the intermediate electrical parameters from which the physical dimensions of coupled lines are calculated. Most physical models in the literature give odd and even mode impedances in terms of physical dimensions, although some are able to work in reverse and give dimensions from impedances. When more than two strips are involved or the strips are of unequal width, it is more appropriate to work with distributed capacitances rather than odd and even mode impedances/admittances because this involves fewer parameters. References [28–35] are some of the most useful available for the physical modeling of coupled lines in the various transmission line media. They include treatments of edge-coupled strips in triplate stripline and microstrip as well as Lange couplers [28–35]. References [32,33] deal with asymmetrical microstrip couplers in particular. As a good practical rule of thumb, microstrip edge-coupled sections are realizable for coupling ratios between –5 and –25 dB. The gap between the strips in an edge-coupled section with a coupling ratio greater than –5 dB would normally be unacceptably small. Lange couplers are appropriate for tighter couplings in the range of –2 to –5 dB. Asymmetrical coupled lines should be avoided whenever possible, because an accurate calculation of their physical dimensions is generally much more difficult (i.e., the models for asymmetrical lines are significantly inferior to those for symmetrical lines).

To enable coupled-line sections to form part of a ladder network, they must be configured in such a way that they can be inserted into the ladder as two-port elements. This means that two of the open ends of the coupled lines must be nominated as the input and output, while the rest must be terminated with the appropriate reactance branches. It is essential that the terminations be reactive for consistency with the overall design procedure, which requires that both *S*-plane and *f*-plane prototype networks be reactance two-ports. The most simple terminations are open or short circuits applied directly to the "free" ends of the coupled lines. However, the terminations can generally be reactance branches of arbitrary shape and complexity. The more complicated terminations would be realized as commensurate line, multisection stubs terminated in either an open or short circuit. As well as reactance terminations, some coupled-line sections may employ conducting bridges between the ends of adjacent coupled lines.

The two-port equivalent circuit of a coupled-line section embedded in a ladder network depends on which of the two ends of the lines are selected as the input and output and on the exact combination of reactance terminations applied to the other ends. There are a large number of combinations and permutations that can be conceived, especially when more than two coupled lines are involved. However, as far as MICs and MMICs are concerned, there are relatively few that are really useful, and these have been assembled into the set coupled-line sections illustrated in Figure 2.44.

Examples 1 and 2 of Figure 2.44 are of primary significance because they enable the realization of a series open-circuit stub. As previously explained, this is something that was not possible using only directly coupled distributed elements. In the first example, input and output connections are made to opposite ends of alternate strips, while the remaining free ends are left open-circuit. As shown in the figure, the *f*-plane equivalent circuit is a pair of open-circuit stubs separated by a length of interconnecting line, all of the same electrical length. This is the equivalent in the *S*-plane of a pair of series capacitors separated by a unit element. One of the capacitors is redundant, and therefore when treated in isolation from the rest of the prototype network, the section has a degree of 2. The relevant transmission zeros are located at $S = (0, j0)$ and $S = (1, j0)$. In the second example, the input and output connections are assigned to the same ends of the two adjacent coupled lines, leaving the opposite ends open-circuit. The *f*-plane equivalent in this case is a single series stub positioned between a pair of shunt open-circuit stubs, which in the *S*-plane is equivalent to a series capacitor in between a pair of shunt capacitors. One of the shunt capacitors is redundant, which again makes the degree of the section 2. Its transmission zeros are located at $S = (0, j\infty)$ and $S = (0, j0)$. For both examples 1 and 2, coupled lines of equal width will produce symmetrical equivalent circuits containing no ideal transformers. However, by making the lines of unequal width, the equivalent circuits become asymmetrical and an ideal transformer is introduced. This can be used to good effect for making impedance level transformations either within the internal structure of a prototype or between the terminating resistors at its two ends. Both elements find widespread use in MICs and MMICs, offering the added advantage of an integral dc break.

The third and fourth coupled-line sections of Figure 2.44 realize shunt short-circuit stubs. Type 3 is effectively the dual of type 1, for which the open circuits on the ends of the two lines are replaced by short circuits. In the *f*-plane, the

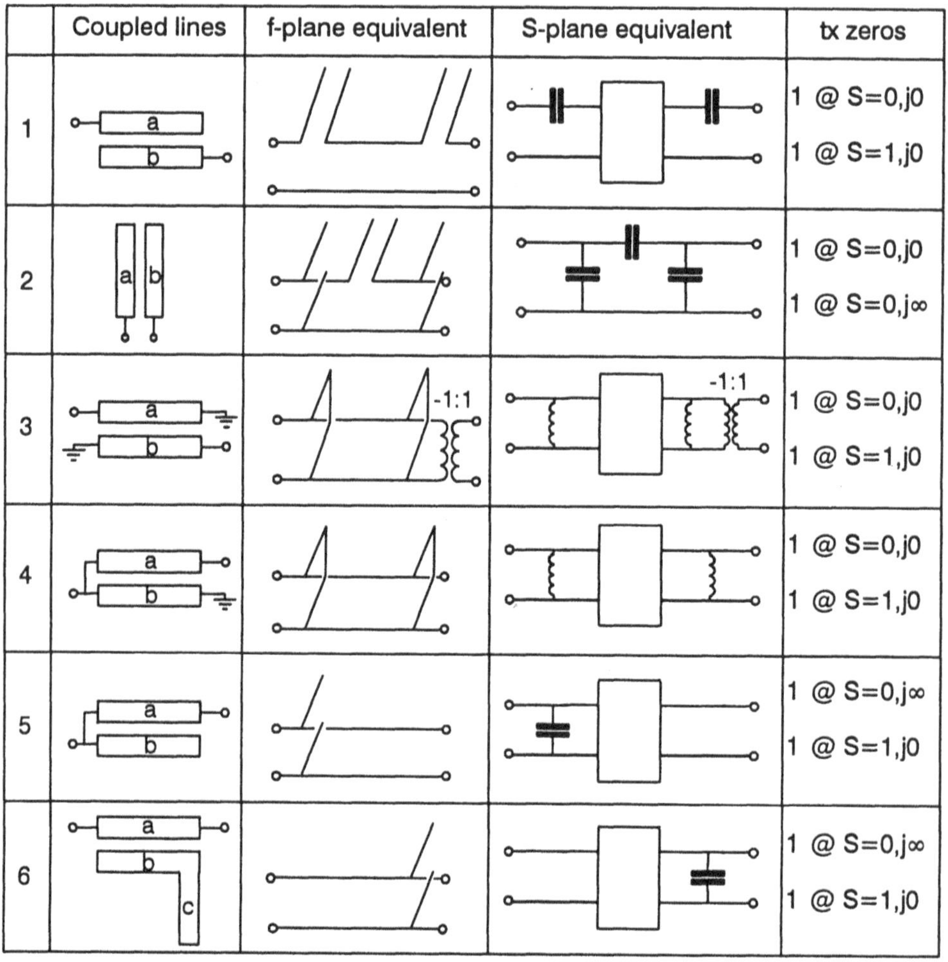

Figure 2.44 Some common coupled-line sections with their *f*-plane and *S*-plane equivalent circuits.

equivalent circuit is a pair of shunt short-circuit stubs separated by an interconnecting line, but there is an additional ideal transformer with a negative turns ratio. The negative turns ratio reverses the polarity of any signal passing through the section, which is equivalent to a frequency-independent phase shift of 180 degrees. This is an extremely useful property that can be exploited in a wide variety of MICs. It forms the basis of the wideband balun circuits that will be described in Chapter 4. A symmetrical pair of coupled lines will have a symmetrical equivalent circuit with a transformer turns ratio of −1:1. Any change in symmetry will change the magnitude of the turns ratio, which again can be used for impedance level transformations. In the *S*-plane, the equivalent circuit of the section is a pair of inductors separated by a unit element, plus the ideal transformer. The two relevant transmission zeros are located at $S = (1, j0)$ and $S = (0, j0)$.

Example 4 in Figure 2.44 has an identical equivalent circuit and the same transmission zeros as those of example 3, except for the absence of the transformer with the negative turns ratio. Instead of the short circuit to ground at one end of one of the lines, the section uses a conducting link between the ends of the lines nearest the input. This has the advantage of eliminating one ground connection, but it also has an effect on the symmetry of the equivalent circuit. In this section, the slot between the lines creates the shunt short-circuit stub to the right-hand side of the equivalent circuit, whereas in the previous section the slot created the interconnecting transmission line. Hence, a symmetrical coupled-line section of type 4 will not usually produce a symmetrical equivalent circuit. Its principle advantages are that it is a highly compact section requiring only one ground connection and that it can be used as a noninverting companion section to a section of type 3.

The last two examples of Figure 2.44 are alternative methods for realizing a through transmission line and a shunt open-circuit stub. Sometimes referred to as a *spurline* section, example 5 has an interconnecting link across the input end of the two lines with an open-circuit on the opposite end of one of the lines. In concept it is rather like a directly coupled shunt open-circuit stub and an interconnecting transmission line with the stub folded over through an angle of 90 degrees to be parallel with the line. It is therefore an extremely compact realization of the line and stub, but the proximity coupling between the line and the stub also allows the section to realize much higher values of stub impedance. Its *S*-plane equivalent circuit is a shunt capacitor in cascade with a unit element, and its transmission zeros are at $S = (0, j\infty)$ and $S = (1, j0)$. Example 6 uses a real open-circuit stub to create the ideal stub in the equivalent circuit. However, the stub is coupled to the through transmission line by the proximity coupling between the pair of coupled lines and not by any direct connection to the through line. This makes the coupling to the through line very weak, allowing the section to realize stubs with extremely high characteristic impedances, in excess of 1,000Ω. Coupled-line sections of this type are most useful in narrowband high-*Q* filter circuits.

For each of the coupled-line sections of Figure 2.44, there is a set of formulas that must be used for converting between the distributed capacitances of the physical lines and the element values of the equivalent circuit. The conversion is an essential part of the design procedure for any circuit containing coupled lines, and a collection of all six sets of equations can be found in Appendix B.

2.5.2.7 Dealing with Losses

In practical MICs and MMICs, ohmic losses cannot be avoided. Because of the predominance of transmission line elements in these circuits, ohmic losses are largely associated with the current flow in the surface metallization (i.e., the conductor losses) and with the displacement current flow in the dielectric substrate (i.e., dielectric losses). The balance between conductor and dielectric losses depends on the physical properties of the materials involved and the operating frequency. Conductor losses generally increase as the square root of frequency, while dielectric losses are directly proportional to frequency. To minimize conductor losses, it is always important to ensure that the thickness of the metallization is several times greater than the surface penetration of the current flow (i.e., the skin depth).

The effects of ohmic losses in a circuit are more far-reaching than just reducing the amplitude of a transmitted signal. They also distort the shape of a circuit's frequency responses. Losses tend to be most pronounced near the passband edges of a frequency response where the reactive elements of a network are close to resonance. In the center of the passband, insertion loss may show only a gradual rise with frequency, similar to that which would be expected of a simple length of transmission line. However, closer to the passband edges, the elements of a network carry much larger circulating currents, and it is these that are the primary cause of the higher losses. The increased losses are a strong function of the slope of the passband skirts. Steeper skirts give higher losses, making the passband shape become progressively more rounded. Hence, highly selective narrowband circuits suffer most from these distortions of passband shape. Conversely, most active circuits suffer relatively little because of their generally low selectivity.

As presented in this chapter, the exact synthesis design procedures create prototype networks that contain no dissipative loss. Section 2.3.6 defined the prototype networks as generalized ladder networks comprising only pure reactive elements. To extend the procedures for the synthesis of lossy prototypes would be possible but would increase the complexity of the mathematical procedures and have an adverse effect on topological flexibility. On balance, the extension to deal with lossy prototypes would not be worthwhile.

In the unlikely event that a performance specification calls for a passband response without any discernible rounding at the passband edges, there are two approaches that can be used that do not involve modifying the basic synthesis procedures. Both achieve the desired aim by effectively degrading the input reflection coefficient in the central region of the passband. First, during the generation of the $Z_1(S)$ polynomials, some of the reflection zeros can be moved away from the $j\omega$ axis to predistort the loss-free frequency response of the prototype in such a way that the lossy circuit will emerge with a flat response. This would usually involve some iteration. Secondly, the losses of the real network can be included as parasitic resistance elements in the f-plane equivalent circuit, after which the rounded frequency responses can be corrected by computer optimization. For certain special filter circuits, the predistortion approach during synthesis will be the better option, but for the majority of circuits, especially active circuits where a mixture of lumped and distributed elements might have been used, correction by optimization is preferred.

In conclusion, the omission of losses during prototype synthesis has *no* significant impact on the overall scope of the generalized design procedure.

2.5.2.8 Other Circuit Parasitics and Discontinuities

With reference to Figure 2.2, the last stages in the overall circuit design procedure involve substituting realistic physical models complete with parasitics in place of the ideal elements of the *f*-plane equivalent circuit. This more realistic equivalent circuit is then analyzed to check if the parasitics of the real elements disturb the frequency responses beyond the limits set by the design specification. If the result is acceptable, then the circuit design procedure can be considered complete, the circuit having been completely defined in terms of its physical dimensions and its predicted performance. However, as indicated by Figure 2.2, if the result is unacceptable, the design procedure continues with some carefully constrained optimization.

Some of the most common sources of parasitics to be found in MIC and MMICs are:

1. Transmission line T junctions;
2. The open ends of open-circuit stubs;
3. The grounded ends of short-circuit stubs;
4. Bends in transmission lines;
5. Abrupt steps in the width of transmission lines;
6. Unwanted proximity couplings.

These are all associated with transmission line elements and are illustrated by the simple sketches drawn in Figure 2.45. Most of the discontinuities add an excess susceptance and a path length error in the transmission line element to which they relate. They can be modeled by an equivalent circuit comprising an appropriate combination of lumped capacitors, resistors, ideal transformers, and transmission lines of positive or negative length.

For details of the modeling of these and other transmission line discontinuities, reference should be made to the literature. Some literature as early as [36] is still highly relevant for discontinuities in stripline and microstrip transmission lines, while [37–41] are more recent studies giving more rigorous models for a much wider range of circuit elements. Models of most of the physical elements relevant for MICs and MMICs will also be found in the databases and user manuals of commercially available CAD software.

Because of the incorporation of physical element models into modern CAD software, a realistic analysis of an integrated circuit complete with parasitics has become a relatively easy task. Little, if any, reference to the literature is required if such software is available. However, it is important to sound some notes of caution.

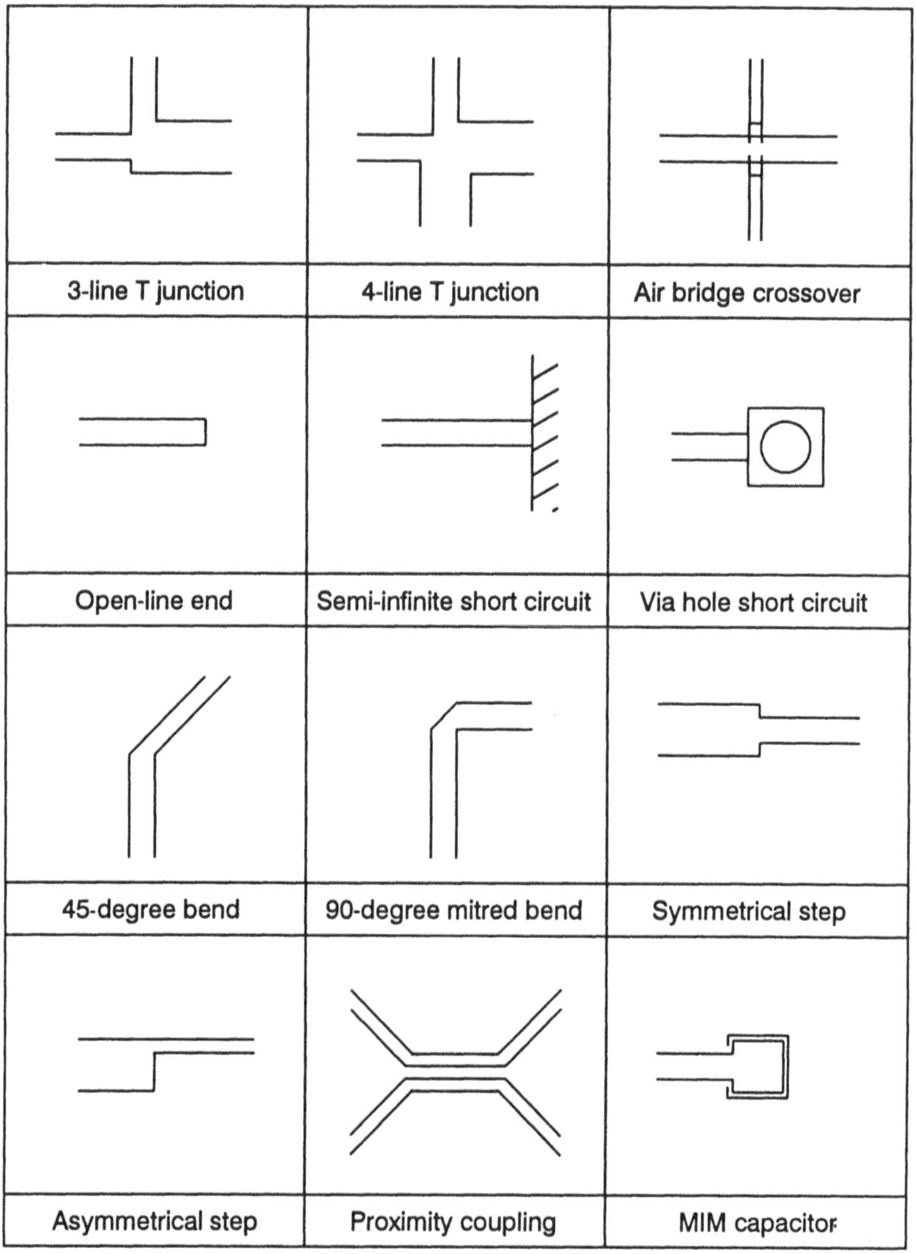

Figure 2.45 A set of common microstrip transmission line discontinuities found in MICs.

First, it is important to observe the ranges of validity for the individual circuit element models and, in particular, to ensure that any optimization does not take element dimensions outside the ranges of validity. Second, as good as the various models for physical elements may be, they are strictly only valid when the element in question is isolated from the rest of the circuit and when the modes of the entry and exit waves have been fully established. They are much less reliable when inserted into a circuit in which several discontinuities are in close proximity. Neighboring discontinuities disturb the purity of the basic microstrip mode to the point where, in very close proximity, the models of two adjacent discontinuities are completely useless. In real circuits, particularly MMICs on GaAs, circuit elements and discontinuities have to be in close proximity to conserve space, and therefore the use of the various equivalent circuit models is not always justified. As a good rule of thumb, model validity rapidly deteriorates as the separation of the relevant elements becomes less than the thickness of the substrate. Above this separation, the use of the models is usually to be recommended, but otherwise it is often better to omit the discontinuity models altogether.

If optimization is required toward the end of the design procedure, it is important to adjust only those elements that are thought to be the principle cause of any distortion in the frequency responses and to set realistic targets for the circuit performance. Identifying which elements need to be adjusted is a matter of experience and of applying the basic principles of design by synthesis that have been set out in this text. For example, errors in the frequency responses near the upper edge of the passband can be attributed to the low-pass element values in the network, while errors near the lower edge can be attributed to the high-pass elements. Any adjustment of redundant elements should be undertaken with special care to avoid the possibility of an indeterminate solution. Optimization targets should be set initially to the predicted performance of the ideal *f*-plane equivalent circuit, focusing primarily on the passband frequency range. Occasionally, however, it is also necessary to specify some target performances in the stopband. In any event, a well-designed circuit should only require superficial optimization. While the benefits of optimization are acknowledged, the reader should be aware that optimization is one of the best opportunities in the design procedure to destroy what is basically a good circuit design.

Although not shown in Figure 2.2, there is one more final stage in the design procedure that could be undertaken to give an even more accurate prediction of circuit performance. This would be a three-dimensional (3-D) field simulation of the whole physical circuit, which would take account of all unwanted parasitics and cross-couplings between elements. Three-dimensional field simulators do not suffer from the limitations of the previous segmented-model approach, where models become invalid in highly compact circuits. It is only in the last few years that simulators with sufficient geometrical versatility have become available for tackling MMICs. The more accurate simulations reduce design risk, which is especially important for MMICs because of the high cost of fabrication.

Several 3-D simulators are now becoming commercially available, but at present one of the most effective packages for dealing with MMICs is called LAYAN [42]. LAYAN is a derivative of FACET [43], a program written by researchers at Philips Research Laboratories which analyzes circuit behavior directly from physical geometry. It derives a lumped-element equivalent circuit of all the transmission line parts of the circuit and includes all the effects of parasitic capacitive and inductive cross-couplings, as well as all dielectric and ohmic losses. The models of the active elements of the circuit are then added and the whole equivalent circuit loaded into a conventional CAD circuit simulator to produce the predicted circuit performance. An example of the application of FACET to the repair of an MMIC distributed amplifier can be found in [44].

2.6 A PRACTICAL REVIEW OF THE DESIGN PROCEDURE

Having given a broad outline of the generalized exact synthesis procedure for microwave circuits in Section 2.1 and covered the technical details of each step of the whole procedure in Sections 2.2 through 2.5, we will in this section make one more pass through the procedure to reinforce the main principles and to highlight more of the practical considerations that must be made. General trends and the practical consequences of making a particular choice of prototype structure will be covered, together with some important rules of thumb. The section will concentrate on the synthesis aspects of the *S*-plane and *f*-plane prototype networks, which form the major part of the design procedure above the AA line in Figure 2.2. It will not attempt to re-cover the physical modeling aspects of the procedure that have already been presented in Section 2.5. It will also review only the design of distributed or mixed distributed/lumped microwave circuits; it will not specifically deal with the design of low-frequency lumped circuits. The aim is to lay down a set of practical design rules that will be relevant to the forthcoming design examples in Chapters 3, 4, and 5 and almost any other type of MIC that might be considered.

2.6.1 Partitioning Circuits Into Reactive Two-Ports

As proposed, the exact network synthesis technique is a highly systematic design approach that can be applied to almost any type of microwave circuit including active circuits such as amplifiers, mixers, oscillators and switches, despite the fact that the exact synthesis part of the procedure deals only with two-port reactance networks connected between a pair of resistive terminations. So long as a complete circuit can be partitioned into a set of two-port reactance ladder networks together with a set of active devices, it should be possible to apply the synthesis techniques. Partitioning, therefore, is the first step in the design procedure for any circuit comprising anything more than just a simple passive two-port. As will be seen in Chapter 5, which deals with active circuit design, the complex nature of the impedance seen looking into an active device does not conflict with the need to synthesize a prototype network terminated in a resistive load. Any capacitive or inductive component that forms part of the complex impedance of the device can be treated as part of the two-port matching network. The reactive component is extracted along with the rest of the matching network elements, leaving the resistive component of the active device to be treated as the load.

2.6.2 Contents of Circuit Library/Database

As indicated in Figure 2.2, the synthesis design procedure draws on the contents of a circuit library or database. Essentially, this database contains sets of component parts from which almost any kind of prototype circuit can be constructed. Some of the elementary components such as lines, stubs, and certain coupled-line sections are intrinsic to the synthesis procedure and effectively will have been installed in the database by the very adoption of the synthesis approach. However, other more complex combinations of circuit elements or even complete networks will be added during the course of using the design procedure in practice. In concept, this database pertains mostly to the element configurations of *f*-plane and *S*-plane prototype networks rather than the precise geometrical configurations of complete microwave circuits. The correspondence of *f*-plane and *S*-plane elements with the physical elements can also be considered part of the database, but the geometrical details of the physical elements would be more appropriately held in another database such as that of the chosen CAD circuit simulator.

At the lowest level, the database contains individual (i.e., directly coupled) circuit elements, which, for the *S*-plane prototype, comprise unit elements and all the various reactance branches listed in Figure 2.15. The unit elements correspond to transmission zeros at $S = (1, j0)$ and the reactance branches to the $j\omega$ axis zeros also listed in Figure 2.15. When transformed into the *f*-plane, the unit elements become interconnecting transmission lines and the reactance branches become shunt or series stubs (Figure 2.13). Individual *S*-plane inductors and capacitors correspond to single-section stubs, while any reactance branches of higher degree correspond to multiple-section stubs of the types described in Appendix A. As previously explained, although present in the database, the series reactance branches would not normally be considered available for use in a new *S*-plane prototype network, because of the difficulties of physical realization in microstrip and other similar transmission line media. The possible exception is the series capacitor, which can be realized using coupled-line sections and sometimes by lumped capacitors. The elements most commonly selected from this level of the database will be unit elements ($S = (1, j0)$), shunt capacitors ($S = (0, j\infty)$), and shunt inductors ($S = (0, j0)$).

At an intermediate level, the database contains special groups of *S*-plane circuit elements. Some of these may have been recorded by the designer as a particularly useful combination of elements for a given application. Most, however, are likely to involve the coupled-line sections already described. The most useful coupled-line sections were listed in Figure 2.44, together with their f-plane and S-plane equivalent circuits. Transmission zero distributions for the coupled-line sections or the special group of circuit elements should also be regarded as part of the database.

Contained in the top level of the database will be the prototypes of complete circuits that have previously been found to have attractive electrical or physical properties. These may be circuits with ultrawide or ultranarrow bandwidth capabilities, circuits that are highly effective impedance level transformers, circuits that are useful for matching into active devices, or simply circuits that are especially easy to fabricate. Whatever the attractive property, the S-plane equivalent circuit representation must be kept in the database along with the relevant transmission zero distribution. It is in entering these data into the database that the abilities to recognize redundant elements, to determine the relevant type of frequency response, and to understand the constraints that exist on the topology of a network are particularly important.

2.6.3 Choosing the Two-Port Network Configuration

When faced with designing a new MIC, choosing the composition of the relevant two-port networks to be synthesized is the first critical stage of the design process, and some general guidelines are given below. Because of the central role of the S-plane prototype, most of the guidelines will be given with reference to this network. For the purposes of this discussion, it has been assumed that the transmission line medium of the finished microwave circuit will be microstrip, but the guidelines are of general relevance for realization in any of the other transmission line media described in Section 2.5.

Always Start With Commensurate Line Networks

Whether the resultant microwave circuit is to be a pure distributed circuit or a mixture of lumped and distributed elements, the synthesis should always be of an S-plane prototype for which the f-plane equivalent is a commensurate line network. As previously explained in Section 2.3.5, the S-plane and f-plane networks are related by the Richards Transformation, which is only valid if all the transmission lines are of equal length. The length of the lines will be determined after the precise network frequency responses have been specified later in the procedure.

Keep the Network as Simple as Possible

One of the most important general guidelines for a circuit design is to keep the layout as simple as possible, spreading out the elements so that parasitic couplings are kept to a minimum. Space permitting, a circuit in which the elements are spread out is easy to model and more likely to function according to simulations after its fabrication. This means that wherever possible, directly coupled elements should be used in favor of coupled-line sections. Coupled-line sections are considerably more difficult to model and are also the cause of higher circuit losses. For the *S*-plane prototype, this means using roughly equal numbers of unit elements and reactive branches, where the prime function of the unit elements is to act as separators. Broadly speaking, the aim should be to create a ladder network comprising an alternating sequence of unit elements and reactance branches.

Use a Minimum Number of Elements

Most prototype networks tend to have a topology that is a repetitive sequence of circuit elements. At this early stage, it may not be apparent how many of the repeating groups of elements will be needed to achieve the required electrical performance. Hence, the most effective approach is to begin with the minimum number of elements that experience suggests will be necessary and then to increase the number of elements if, after the synthesis, predicted performance is found to be inadequate.

Using Unit Elements

Unit elements make a relatively weak contribution to circuit selectivity, so they would not normally be added to a network in greater numbers than are necessary to separate the reactance branches. However, they do have some other useful attributes, one of which is the ability to transform series inductors into shunt capacitors (see Kuroda Identities, Section 2.4.1). While providing enough unit elements to act as element separators, their numbers and distribution should also be chosen to ensure that all the low-pass elements of a network are present in the form of shunt capacitors.

Occasionally it may be necessary to build a network in which two or more unit elements are wired in cascade without a reactance branch at each junction. Certain types of wideband impedance-transforming networks have this sort of topology. However, networks based on cascaded unit elements tend to be inefficient in terms of their selectivity and physical size for a given number of elements. More importantly, they depend for their correct behavior on the differences between the impedances of the adjacent unit elements, which makes it impossible to apply any straightforward length/impedance trade-offs. There are no topological degrees of freedom for networks comprising only unit elements.

Avoid Certain Types of Element

The difficulty of realizing series elements has already been discussed (Section 2.5.2.5). Hence, all series reactive branches should be avoided with the possible exception of series capacitors. All the shunt reactance branches can, in principle, be realized as a cascade of one or more stubs. However, fabricating high-quality short circuits in most types of printed circuit transmission line media can be difficult, and this is often a good reason for minimizing the number of short-circuit stubs. Short circuits in MMICs are readily available in the form of via holes, but electrically these are both inductive and lossy. The same is generally true for hybrid microstrip circuits. In triplate stripline, it may be possible to make better quality short circuits, but these are extremely inconvenient to fabricate, requiring connections to both the upper and lower ground planes. In order to minimize short-circuit stubs in a microwave circuit, shunt inductors must be minimized in the *S*-plane prototype. This also applies to any of the higher order reactance branches of the second canonical form, illustrated in Figure 2.15.

Despite the disadvantages of short-circuit stubs, they will be employed in the majority of active circuits because of their value as impedance-transforming elements and their ability to provide easy access for dc supplies. Most *S*-plane prototypes for active circuits will therefore include at least one shunt inductor. If only one shunt inductor is present, the *S*-plane prototype must be of the second canonical form unless the inductor is preceded by a series capacitor at a point closer to the input port.

Bandwidth Implications

The fractional bandwidth requirement of a circuit has important implications for its element composition. Circuits with a bandwidth of more than an octave (i.e., >67% fractional bandwidth) are generally categorized as wideband and necessitate strong coupling between their circuit elements. Hence, wideband circuits will be most easily realized using directly coupled rather than capacitively coupled elements. Coupled-line sections are generally realizable in circuits with an octave bandwidth, but their physical dimensions rapidly become impracticable for bandwidths in excess of an octave. In the *S*-plane prototype, providing the strong coupling means giving preference to unit elements and shunt reactance branches. Series capacitors should only be used if they can be realized in lumped form. Because of the frequent absence of series capacitors, it is more common to select high-pass and band-pass prototypes of the second canonical form for bandwidths greater than an octave.

For narrowband circuits (i.e., <15% fractional bandwidth), the trends governing element composition are the reverse of those for wideband circuits. Coupling between elements is weak and structures based on directly coupled elements are usually difficult to realize. Most narrowband circuits will need to use coupled-line sections and be based on an *S*-plane prototype containing series capacitors. Hence, the first canonical form of the network will normally be most appropriate.

For networks that must have bandwidths above 15% and below 67%, the factors influencing the choice of elements are a mixture of those for the narrowband and wideband circuits. Further specific advice for this bandwidth range is difficult to offer. However, it is generally true that networks in this bandwidth range are more difficult to realize in printed circuit transmission line media than wideband or narrowband types, and unless a bandwidth in this range is absolutely necessary, the range should be avoided. For filter circuits, this can present a problem, because the electrical specification is almost certain to place a tight constraint on the passband width. Fortunately for active circuits, this is less likely to be a problem because (1) there is a greater opportunity to use lumped elements, and (2) the specification of an active circuit will usually only place a loose constraint on the passband width.

Networks Performing Impedance Transformation

The most fundamental rule for networks that must perform an impedance transformation between dissimilar source and load resistors is that they cannot have a low-pass frequency response. An impedance-transforming network must be either high-pass or band-pass. It is true that a low-pass prototype with a Tchebycheff equal-ripple response and an even number of elements has unequal source and load resistors, apparently contradicting the basic rule. However, strictly speaking, the prototype is not quite low-pass, since it does not have zero loss at zero frequency. It would be more accurately described as a band-pass prototype with a lower pass-band edge very close to zero frequency. Hence, although low-pass responses must be excluded for impedance-transforming networks, this does not mean that low-pass elements must be excluded. Indeed, a network containing only low-pass elements is a perfectly acceptable, if rather ineffective, composition for an impedance transformer.

High-pass elements are much more effective impedance-transforming elements than low-pass elements or unit elements. Generally, the more high-pass elements that are present, the greater the impedance ratio that can be achieved for a given number of elements and a given bandwidth. However, there is good reason to use only a single high-pass element in a transformer and make up the rest of the network with either low-pass elements or unit elements. This is because high-pass elements can be split into several redundant parts and transformed through the other types of element with Kuroda Identities. Extremely precise control of the impedance ratio can be achieved in this way. As previously stated, moving series capacitors to the left decreases impedances to the right, while moving them to the right increases impedances to the right. The reverse is true for shunt inductors.

If an impedance-transforming network contains only high-pass elements and unit elements, its frequency response can be either high-pass or band-pass. If low-pass elements are present, however, the response is forced to be band-pass. Networks containing all three types of elements will usually be the most effective for a given number of elements. It is often helpful to arrange for one or more capacitor T or π sections to be present in the network, since adjustment of the symmetry of such sections is a useful method of adjusting impedance ratio.

Complex-Impedance-Matching Networks

Design of matching networks to operate between complex terminations rather than pure resistors will often be necessary for active circuits. In these circumstances, an equivalent circuit must first be derived for each complex termination in the form of a single resistor plus a set of reactance elements. The reactance elements must be created in such a way that they can be treated as part of the matching network and the single resistor treated as the load. Most frequently, the equivalent circuit will be a resistor plus a single *S*-plane capacitor or inductor in either shunt or in parallel. Occasionally, two reactance elements will be required. The type of element must be taken into consideration when specifying the frequency response of the network. Low-pass elements (series inductors or shunt capacitors) force a low-pass or band-pass response, high-pass elements (series capacitors or shunt inductors) force a high-pass or band-pass response, and when both types are present, the response must be band-pass. More details of complex impedance matching will be covered in Chapter 5.

High-Selectivity Networks

Most circuits requiring high selectivity at the passband edges fall into the category of filters, and it is in filters that the need for second-order (or higher) reactance branches most frequently arises. It is possible to achieve high selectivity by simply using large numbers of high-pass or low-pass elements in a network. However, for a given number of elements, second-order reactance branches with their transmission zeros at finite nonzero frequencies will be much more effective. For high selectivity at the low-pass edge of the passband (i.e., the upper edge), one or more shunt second-order elements can be selected, and for each of these, at least one low-pass element must be present on the input side of the element. For high selectivity at the high-pass edge (i.e., the lower edge) of the band, one or more shunt second-order elements can again be selected, but for each of these, at least one high-pass element must be present on the input side. The low-pass or high-pass zero-shifting elements associated with the second-order elements can, if necessary, be moved to other positions in the network with Kuroda Identities and do not have to be immediately adjacent to the second-order elements. Details of the constraints associated with using these elements were given in Section 2.3.6.3

The Need for dc Breaks and dc Returns

Most matching networks for use in active circuits will need to include a dc break and/or a dc earth return. Each dc break will be provided with a series capacitor and each dc return by a shunt inductor. Both are high-pass elements, and if both are required, there must be at least two transmission zeros at $S = (0, j0)$. The positioning of the active device will determine whether the capacitor precedes the inductor or vice versa. However, since both elements are high-pass elements, their precise position in the network will also depend on any impedance level changes required. Putting the elements at opposite ends of the network will cause the largest impedance level change from input to output. Occasionally, the direction of the impedance level change will conflict with the needs of the dc supplies. In this case, the solution is to add another high-pass element of the appropriate type and readjust accordingly the position of all three high-pass elements in the network.

Designating Redundant Elements

The primary reason for introducing redundant elements is to ease any difficulties with the physical realization of a network. They make no contribution to a network's frequency selectivity. Their introduction allows changes to be made to either the types or the values of the prototype elements.

Redundant unit elements are most likely to be required in narrowband circuits where bandwidths are 20% or less. They are not often needed in wideband circuits. Without redundant unit elements, element values in narrowband circuits tend to be extreme, and by moving unit elements into the network from both ends, internal element values can be scaled up or down toward the relevant range of physical realizability. For this type of application, it is usual to introduce only two redundant unit elements, one from each end of the network. Only when a cluster of low-pass elements must be transformed into elements of the same type may there be a need for several redundant unit elements to be introduced at once.

Redundant reactance branches can help produce prototype networks with a near uniform topology. Typically, series capacitors may be split into several redundant parts and spread throughout the rest of the network. In this case, the series capacitor must be the only nonredundant high-pass element present, since it is not possible to transform high-pass elements through any other high-pass elements. Alternatively, a shunt capacitor may be selected for splitting and redistributing throughout a network, in which case the network cannot contain any other low-pass elements. Like unit elements, redundant reactance branches are also useful when either internal or external impedance levels need to be adjusted. Either the redundant elements themselves must be high-pass or the elements through which they are transformed must be high-pass in order to achieve the impedance level adjustment.

2.6.4 Choosing Transmission Zero Locations, Degree, and Network Frequency Responses

Selecting the Set of Transmission Zeros and Network Degree

Such is the relationship between the topology of the S-plane prototype and its set of transmission zeros that the exercise of choosing topology will also have fixed the location of most of the zeros (see Section 2.3.6). If we ignore redundant elements, each high-pass element will correspond to a zero at $S = (0, j0)$, each low-pass element to a zero at $S = (0, j\infty)$, and each unit element to a zero at $S = (1, j0)$. Any zeros corresponding to second or higher degree reactance branches must be on the $j\omega$ axis, and when present, it is essential that there be at least one zero at $S = (0, j0)$ or $S = (0, j\infty)$, depending on whether ω is above or below the passband. The number of zeros will be determined by the number of elements in the prototype. Numbers should be kept to a minimum at the start of a synthesis and then increased as necessary to achieve the required frequency responses. Any increases tend to apply to groups of zeros, so that the pro rata distribution of zeros remains unchanged. This extends the length of the prototype network without changing its basic topology.

As previously stated, band-pass S-plane prototype networks must be of an even degree. If after choosing the transmission zero distribution, the degree is not even, then it is essential that the appropriate change be made to the distribution to make the degree even. This will often entail adding or removing a zero at $S = (0, j0)$ or $S = (0, j\infty)$

Deciding on Low-Pass, High-Pass, or Band-Pass Frequency Responses

As in the case of the distribution of transmission zeros, choosing the network topology will have largely established whether the S-plane response is low-pass, high-pass, or band-pass. Different combinations of network elements will apply different constraints on the type of frequency response. Some of these leave a limited freedom of choice, while others leave no choice at all. By examining the combination of network elements, it is possible to conclude that:

1. If there are both high-pass and low-pass elements present in the network, then the response is forced to be band-pass.
2. If there are low-pass elements but no high-pass elements, then the response could be either low-pass or band-pass.

3. If there are high-pass elements but no low-pass elements, then the response could be either high-pass or band-pass.
4. If there are only unit elements in the network, then the response could be low-pass, high-pass, or band-pass.

Other factors influence the choice of frequency response. If a band-stop response is required in the *f*-plane, then the *S*-plane response should be low-pass. All the *f*-plane elements will be a quarter of a wavelength long at the commensurate frequency in the center of the stopband. Choosing a low-pass *S*-plane frequency response also offers the opportunity to use lumped elements to realize the reactive branches, since this does not significantly disturb the lower *f*-plane passband and normally just extends the upper end of the stopband. Low-pass prototypes should not be chosen if any impedance transformation is required.

A high-pass *S*-plane prototype should be chosen for a band-pass response in the *f*-plane if the width of the upper stopband is not an important requirement. All the *f*-plane elements of the network will be a quarter of a wavelength long at the commensurate frequency, which must be in the center of the passband. The width of the passband will determine the width of all the stopbands. *It is not possible* to use any lumped elements in such networks without destroying half of the passband response. Networks based on high-pass *S*-plane prototypes require relatively long but relatively few transmission line elements for a given selectivity. They are extremely effective impedance-transforming structures but their use is mostly confined to completely passive circuits. Most distributed narrowband band-pass filters will be based on using high-pass *S*-plane prototypes.

A band-pass *S*-plane prototype should be chosen for a band-pass response in the *f*-plane when a wide upper stopband is an important performance requirement. All the *f*-plane elements of the network will be a quarter of a wavelength long at the commensurate frequency, which must now be in the center of the stopband. This allows the width of the stopband not only to be wide, but also to be controlled independently of the width of the passband. Band-pass *S*-plane prototypes produce extremely compact *f*-plane networks because the comparatively high commensurate frequency reduces the length of the transmission line elements. Furthermore, being electrically short in the region of the passband, some of the transmission line stubs can be replaced by lumped elements without significantly disturbing the passband. This further reduces the size of circuits and provides the important opportunity to use lumped series capacitors. Most matching networks in active microwave circuits will be based on band-pass *S*-plane prototypes.

Choosing Commensurate Frequency

For band-stop *f*-plane networks with a low-pass *S*-plane response, the commensurate frequency will determine the center of the stopband. However, since the width of the stopband is determined by both the passband width and the commensurate frequency, almost any commensurate frequency could be chosen to satisfy a particular minimum stopband width. If a precise stopband width is not essential, the commensurate frequency should be set as high as possible to give the shortest possible transmission lines with realizable impedance values.

For band-pass *f*-plane networks with a high-pass *S*-plane response, there is usually no freedom to change the commensurate frequency, since this must be in the center of the passband. For band-pass *f*-plane networks with a band-pass *S*-plane response, the commensurate frequency and the passband width together determine the width of the stopband. As in the case of band-stop networks, commensurate frequency must be in the center of the stopband if the ideal stopband characteristics are to be preserved. However, if some change to the stopband characteristics can be tolerated, then the commensurate frequency can be set to whatever value gives the best physical realization. The higher the commensurate frequency, the shorter the transmission line elements. For general-purpose matching networks, the recommended beginning is to set the commensurate frequency to between two and three times the center frequency of the passband.

Passband Specifications and Precise Location of Transmission Zeros

Once the type of *S*-plane response has been established, the details of the passband characteristics remain to be specified. In most cases this means choosing between a Tchebycheff or Butterworth transfer characteristic, specifying the relevant ripple and/or band edge loss parameters and specifying the locations of the edges of the passband. When special passband characteristics such as those with a positive or negative slope or a linear displacement are required, the specifications are more complex, and for details on these, the reader should refer to Sections 2.3.4 and 2.3.7. The comments in the following paragraphs are made with special reference to a conventional Tchebycheff equal-ripple passband characteristic, but are of general relevance to any of the other types available.

The frequencies of the passband edges in the *S*-plane are calculated from those of the *f*-plane response using the Richards Transformation (see Section 2.3.5). If the specification allows, the passband should be made significantly wider than the minimum acceptable range, since this (1) gives a worthwhile margin for error in the physical realization, (2) lowers the losses in the central region of the passband, and (3) lowers the sensitivity of the network to variations in element values. Wideband networks in general have lower losses and are more tolerant to errors in element values than narrowband networks. However, wideband networks tend to require more elements than narrowband networks for the same frequency selectivity and for the same impedance-transforming capability. Hence, an optimum balance must be struck between the passband width and the number of network elements.

In networks specifically designed as impedance transformers, bandwidth is one of the most important parameters for adjusting impedance ratio. It is also a useful means of adjusting the element values in a network, so long as the specified electrical performance can still be met. As a general rule, *changes to the position of the upper passband edge will have a most pronounced effect on the values of the low-pass elements in the network, whereas changes to the position of the lower edge will have the most pronounced effect on the high-pass elements.*

It is rare for the electrical specification of a network to call for a precise passband insertion loss ripple. More frequently, a specification will call for a passband ripple that is equal to or less than some maximum value. This allows the passband ripple to become another adjustable parameter with which load/source impedance ratios or network element values can be modified. In comparison with bandwidth, the ripple parameter tends to have a slightly more delicate influence on impedance ratio or element values. Like bandwidth, it also affects the passband selectivity for a fixed number of network elements. Typical values of ripple are in the range of 0.01 to 1.0 dB. When passband ripple is specified in terms of peak reflection coefficient, a simple conversion into an equivalent insertion loss ripple is usually made for the purposes of the synthesis (see Section 2.3.2). As an example, a −17-dB peak reflection coefficient specification would convert to an insertion loss ripple specification of 0.1 dB (or −0.1 dB in term of transmission coefficient).

Frequency responses in the stopband are strongly affected by the precise location of finite, nonzero transmission zeros on the $j\omega$ axis. These would be associated with any multielement reactance branches that might be present in the prototype. The zeros can be moved to adjust the magnitude of the insertion loss ripples in the stopband and the slope of the passband skirts. By careful positioning, virtually any shape and depth of stopband floor can be achieved, including, for example, an equal-ripple characteristic. As the zeros are placed closer to the edges of the passband, the passband skirts become progressively steeper, but this causes an increase in the magnitude of the stopband ripples. There is always, therefore, a trade-off between the slope of the passband skirts and the depth of the stopband floor. The optimum position of the stopband zeros is best determined by iteration.

There are another two consequences of placing the transmission zeros close to the passband edges that need to be taken into consideration. The first is that as the zeros move closer to the passband edge, the sensitivity of the circuit to variations in element values is increased. The second is that while finite, nonzero transmission zeros placed very close to the passband edge will have no effect on the magnitude of the insertion loss ripples in the passband of an ideal prototype, they will cause increased passband insertion loss near the passband edges of a real circuit containing ohmic losses. Both effects are important, and just how close to the passband edges the zeros can be placed will depend on the planned fabrication process and the target electrical specification.

With reference to the flow diagram of Figure 2.2, there is an important reentry point into the design path immediately above the point at which passband and stopband characteristics are specified in detail. Changes in the passband and/or stopband characteristics allow fine adjustments to be made to either electrical performance or the element values of a network. Entering at this point brings about less radical modifications than entering higher up in the procedure, where the numbers of transmission zeros or the types of transmission zeros would be changed.

2.6.5 Singly vs. Doubly Terminated Networks and First vs. Second Canonical Forms

The choice between synthesizing a singly or doubly terminated prototype will depend on the practical application in question, but it will be more common to choose a doubly terminated network. This is because in most cases, a filter or matching network will have to operate between source and load impedances that have finite internal resistance. Very few real microwave power sources have sufficiently high or low internal resistances for them to approximate ideal current or voltage sources. However, there are two notable cases in which a singly rather than a doubly terminated network would be preferred. The first is in the field of diplexer design, where the input ends of a pair of filters with contiguous passbands have to be connected in parallel at a common input port. There are several conditions that need to be met for correct operation of such a diplexer, but one of the most important is that the two filters must be singly terminated. Only when the two filters are singly terminated does the necessary mutual cancellation of input susceptances take place. Under these circumstances, the input to the diplexer takes on the same input reflection characteristics as those of a single two-port filter, whose passband extends over the combined range of the two diplexer passbands.

The second situation in which a singly terminated network would be preferred is in the design of solid-state power amplifiers, where an output matching network must extract the maximum available power from an active device (usually an FET) for delivery to the external load. It is relatively easy to extract the maximum available power at a single frequency, but it is impossible to create any network, singly or doubly terminated, that would be capable of extracting maximum available power over an instantaneous band of frequencies. However, certain types of networks are able to extract *most* of the available power over a given bandwidth, and for this purpose, singly terminated networks are more effective. This is because they are able to present the output terminal of the active device with a near constant value of load resistance, whereas a doubly terminated network would attempt to present a constant match.

In choosing between the first or second canonical forms of a network (i.e., first or second duals), it must be remembered that the input impedance of a network of the first type will always tend toward an open circuit in the stopband, while that of the second type will tend toward a short circuit. This is relevant whenever a pair of networks needs to be connected together in either series or parallel. For a series connection, both networks must be of the second canonical form, while for a parallel connection, such as the case of the diplexer mentioned in the previous paragraph, they must be of the first canonical form.

When designing matching networks, it will often be the type of the first element adjacent to the input port that will determine the choice of network dual. In the absence of redundant unit elements, a series reactance branch at the input corresponds to a first dual, and a shunt reactance branch to a second. Reference should be made to Section 2.3.6.3 and Figure 2.16 for a reminder of the constraints placed on reactance branch combinations in generalized ladder networks.

2.6.6 Tips on Prototype Synthesis

As previously explained and illustrated in Figure 2.2, synthesizing the *S*-plane prototype network is a two-stage process, the first dealing with the generation of an intermediate set of polynomials from which frequency responses can be determined, and the second dealing with the extraction of the elements of the network. It is helpful to treat the two stages of the process separately, since this allows the problem of generating the desired frequency responses to be solved without the complication of having to extract the elements of the network. Iteration within the two-stage process is much more efficient and versatile than iterating around both stages together.

The precise details of how the synthesis would proceed will depend on the software being used. Names of at least two and a reference to several other commercially available software packages for prototype synthesis will be given in Section 2.7. Some of these split the synthesis process into two parts, enabling an early prediction of the likely frequency responses of the network to be obtained. For those that do not, it may be necessary to determine the frequency responses using a separate circuit simulator. If this is the case, the elements of the *S*-plane prototype will first have to be transformed into the equivalent *f*-plane distributed network for analysis.

For the element extraction part of the process, it is advisable to extract the required *S*-plane prototype in its simplest form, having temporarily eliminated as many redundant elements as possible. It is also good practice to extract a network that is as nearly symmetrical as possible, irrespective of whether the required final form of the network is symmetrical or not. Both these actions help with the numerical precision in the software and avoid extremes in the values of certain elements that can occur in highly asymmetrical networks. When the extraction is complete, redundant elements can be reintroduced and the network manipulated back into the most appropriate form for physical realization.

2.6.7 Tips on Network Transformation

To move individual elements within an *S*-plane prototype either for the purposes of adjusting topology or for adjusting impedance levels, Kuroda Identities (Section

2.4.1) are most appropriate. When only reactance branches are involved, some of the transformations listed in Appendix A will also be helpful. Redundant reactance branches can be created by splitting any of the existing internal branches of the network. When redundant unit elements are required, they can be introduced from either the load or the source end of the network, although there is a restriction that applies for singly terminated networks, which prevents unit elements from being introduced from the source end. The same restriction applies if the phase of the S_{11} frequency response must be preserved. If more than one network element must be moved simultaneously, the method of admittance matrix scaling (see Section 2.4.2) is more appropriate than using Kuroda Identities. The method is particularly useful for distributing several parts of a redundant reactive element throughout a network and making them all have the same value.

When topological adjustments to the S-plane prototype are complete, the prototype must be transformed into the f-plane. So long as the transmission lines remain of equal length and equal to a quarter of a wavelength at the commensurate frequency, the transformation will be exact. For high-pass S-plane prototypes corresponding to a band-pass prototype in the f-plane, there is unlikely to be any possibility of applying length/impedance trade-offs to the transmission line elements. Unless the topology or impedance levels of this network are unacceptable, the network should be ready for passing on to the physical realization part of the design procedure without any need for the fine-tuning optimization loop (Figure 2.2). For the f-plane networks derived from low-pass or band-pass S-plane prototypes, length/impedance trade-offs should be applied to the transmission lines as necessary to bring characteristic impedances into the most desirable range for physical realization. Any open-circuit stubs should also be converted to lumped capacitors as required. In this pseudoideal form, the f-plane network should be analyzed and optimized to correct any of the slight disturbances that may have been caused to the frequency responses by these approximate transformations of elements.

Assuming frequency responses, topology, and element values are all acceptable, the circuit design procedure above the AA line in Figure 2.2 can now be considered complete, unless any subsequent difficulties with physical realization force a return.

2.7 AVAILABILITY OF SOFTWARE

Despite the potential and versatility of the exact synthesis approach to microwave circuit design, its adoption by the microwave circuit design fraternity does depend on the availability of synthesis software. Fortunately, there is a ready supply of suitable software on a commercial basis from at least 10 different companies throughout the world. Many of these companies are included in a report published by MSN [45], which shows the results of a comprehensive survey of suppliers of CAD tools.

Two of the major players in the microwave CAD software market are HP-EEsof and Compact Software. Both companies can provide powerful synthesis tools. In the case of HP-EEsof, the synthesis package is called E-Syn [46], which at the time of writing runs stand-alone, but can export data in the appropriate format to the LIBRA Series IV Microwave and RF circuit design suite. In the case of Compact Software, the synthesis tool used to be known by the name of FILSYN, but is now believed to be available only as a module integrated inside their SuperCompact [47] CAD suite.

In the event that a commercial network synthesis package is unobtainable, then given enough time and effort, it is well within the capabilities of most circuit designers to write a suitable program. All the necessary theory has been presented in the earlier parts of this chapter. The author has written such a program, called NETSYN, with which all the practical circuits about to be described in Chapters 3, 4, and 5 were synthesized. An overview of NETSYN and its data file format will be given in the following section. Unfortunately, this program is unavailable to third-party users except by special arrangement, but in describing the data format, it is hoped that the fundamentals of the generalized synthesis design procedure will be reinforced.

In anticipation that the reader will have access to the E-Syn software from HP-EEsof, each of the practical examples in Chapters 3, 4, and 5 that were synthesized by NETSYN will be accompanied by a description of how the synthesis could be repeated using E-Syn. The relevant numerical data for each network will be provided in a table and the procedure described in the text. To conclude this chapter, a general overview and introduction to the E-Syn software will be given in Section 2.7.2 and a design example given in Section 2.7.3.

2.7.1 Designer Software: The NETSYN Program

NETSYN comprises approximately 4,000 lines of FORTRAN code that has been compiled to run on a personal computer (PC), on an IBM mainframe, or on any of the latest Unix-based workstations. Its speed of execution depends on the computer hardware involved, but typical run times for a fast PC with a math coprocessor are in the region of 1 or 2 seconds. NETSYN uses a very simple user interface based on reading data from either a data file or direct from the keyboard. When reading direct from the keyboard, the program runs in a conversational mode, prompting the user for all the data required. Three output files are produced after a run: the first is the main results file containing the types and values of the elements of the synthesized network, the second is an intermediate file containing the $Z_1(S)$ polynomials, and the third is an optional plotting file containing the network frequency responses.

An example of an input data file for the NETSYN program has been listed in Figure 2.46. To give this example some relevance, these are the same data that were used to generate the prototype network and frequency responses previously shown in Figures 2.27 and 2.28 and described at the end of Section 2.3.8. As will

```
***************************************************************************
*  Input data for the NETSYN program corresponding to the network and    *
*  frequency responses given in Figures 27 and 28  of Chapter 2           *
***************************************************************************
*Give the key word POLGEN to trigger polynomial generation
POLGEN
*Choose a polynomial generator.  Specify 1 for equi-ripple responses.
1
*Specify a LP, BP or HP S-plane response (1=LP or BP, 2=HP).
1
*Give S-plane passband specifications
* ω1           ω2          tp(dB)       to(dB)      tslope(+/- dB)
0.3640      0.8391        0.1          0.0             0.0
*Specify number of tx zeros
5
*Give list of locations of tx zeros (i.e. 2 numbers (σ and ω))
*Use 0  99999 for zeros at (0,j∞)
1  0
0  99999
0  0.2679
0  0
1  0
*Give S-plane plotting range  (start, step & stop values of omega)
0.0  0.01  2.0
*Specify doubly or singly terminated (1=doubly  2=singly)
1
*Specify 1st or 2nd canonical forms
1
*End of polynomial generation data
***************************************************************************
*Give key word MAKENET to trigger network extraction
MAKENET
*Now specify the number of each tx zero to be created.  For zeros at
*S=(0,j0) or S=(0,j∞), the number must be followed by either 1 for
*partial creation of the zero or 2 for complete creation of the zero.
1
1  2
1
1  2
1
*Extract another network? (111=yes   222=no)
222
*End of network extraction data
***************************************************************************
*Give the key word GENPLT to trigger the production of a plotting file
GENPLT
*Give commensurate frequency
18
*End of plotting data
***************************************************************************
```

Figure 2.46 A sample data file for the NETSYN synthesis program.

be apparent from the listing, most of the file is just text in the form of comment lines beginning with an asterisk (*). These comments help remind the user of the proper sequence for the numerical data and are ignored during program execution. They are not essential for the running of the program. The rest of the relevant data are a collection of numbers that act as switches in determining the various options for the synthesis, or alternatively specify the details of the network's frequency characteristics.

In accordance with the synthesis procedures described in Sections 2.2.4 and 2.2.5, NETSYN embodies the two main processes of polynomial generation and network extraction, as well as a third concerned with producing graphical results. These processes correspond to the three key words POLGEN, MAKENET, and GENPLT, which are optionally present in the data file. When NETSYN encounters one of these key words in the data file, it starts the appropriate process which then proceeds to read the appropriate numerical data. For POLGEN, these numerical data specify the type of polynomial generator, the frequency characteristic required, and the rest of the passband parameters. They also include the number and types of the transmission zeros, the plotting frequency range, the choice of a doubly or singly terminated network, and the choice of first or second canonical form. When POLGEN has finished, the $Z_1(S)$ polynomials are stored in an external file. For MAKENET, the numerical data specify the order in which the transmission zeros are to be created. For those zeros at $S = (0, j0)$ or $S = (0, j\infty)$, the data also determine whether this is to involve a partial or complete element extraction, and if partial, what the value of the element should be. MAKENET writes the details of the extracted network to the main results file. For GENPLT, only the commensurate frequency needs to be specified to allow the S-plane frequency responses to be transformed into the f-plane. GENPLT produces an ASCII file of (x, y) data representing the input return loss and insertion loss responses of the generic network.

It should be emphasized that NETSYN synthesizes s-plane or S-plane prototype networks and the user must make the appropriate transformations into and out of the f-plane. Passband specifications and transmission zero locations must be given in terms of ω, the imaginary part of the complex variable S or s. For S-plane prototypes, the transformation is $\omega = \tan\{(\pi/2) \times (f/f_c)\}$, whereas for s-plane prototypes it is $\omega = 2\pi f$. The network element values for an S-plane prototype will represent the characteristic impedances/admittances of the transmission lines of a distributed f-plane network, while those of an s-plane prototype will represent the inductance/capacitance values of a lumped f-plane network.

2.7.2 An Overview of the E-Syn Software Available from HP-EEsof

According to the support team at HP-EEsof, E-Syn was originally written as an aid to matching network design and was not created especially for filter design. Therefore, in its present form, it has particularly good functionality in the area of matching-network design, but is not particularly well developed for advanced filters. Its most obvious weakness with regard to filter design is its inability to synthesize elliptic or pseudoelliptic function distributed filters (i.e., distributed filters with finite nonzero transmission zeros). The version of the program under discussion is the version released under Series IV, version 5 of the LIBRA suite of software. As previously mentioned, this runs stand-alone using its own window manager. The following overview of program functionality and user interaction has been compiled by the author after working with the software over a relatively short period of time. The observations will not, therefore, necessarily be the same as those that might be made by a more experienced user of E-Syn, nor can they in any way be attributed to the suppliers of the software (HP-EEsof).

E-Syn's user interface comprises six windows in an arrangement indicated by the drawing in Figure 2.47. The arrows indicate the accessibility of the windows from their neighbors. The two most important windows are the Specifications Window and the Synthesized Network Window, which correspond to the two main parts of the synthesis procedure (i.e., the solution of the approximation problem and the network extraction). Hence, E-Syn, like NETSYN, separates the procedure into two parts, the required frequency characteristics and basic network composition being declared in the Specifications Window and the precise network topology being declared in the Synthesized Network Window. Additionally, as indicated in Figure 2.47, there is a Network Analysis Window accessible from the Synthesized Network Window, which allows the frequency responses of the extracted network to be calculated. It also allows some numerical optimization of the ideal circuit, should this be found to be necessary after making any approximate transformations of the network at an earlier stage.

The Transformed Network Window shown in Figure 2.47 is only accessible for lumped band-pass networks whose elements have been extracted in a manual mode. In this window, E-Syn will perform an automatic search for alternative sequences of elements that meet more closely the specified values of source and

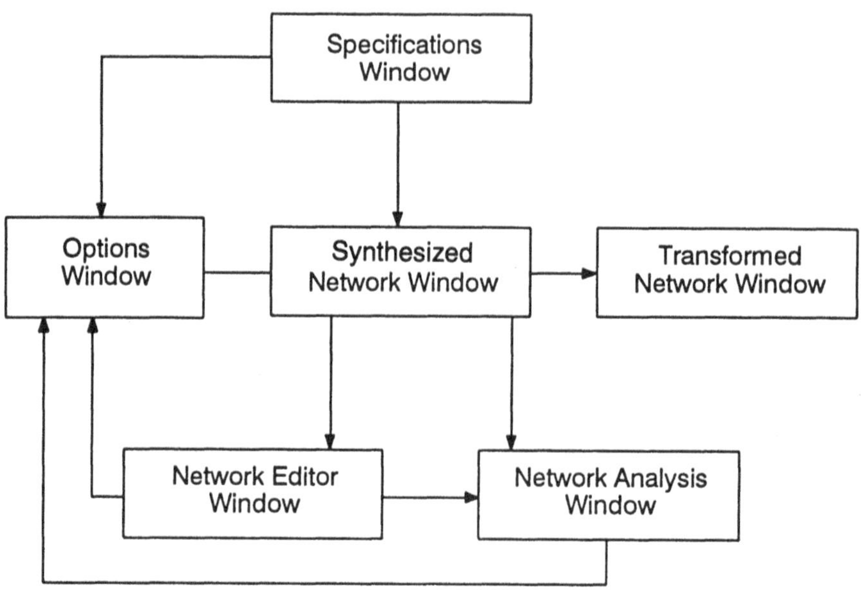

Figure 2.47 Window organization in E-Syn.

load resistances. It is not applicable for singly terminated networks. The Network Editor Window provides the opportunity to convert one or all of the elements of a network from lumped to distributed types or vice versa. The conversions are necessarily approximations, and in the case of a distributed result, the transmission lines will not all be of the same length. Hence, the result of a conversion in the Network Editor Window must always be analyzed and optimized to restore the required frequency responses. Finally, the Options Window and all its subordinates allow a variety of options to be specified for program and unit defaults, synthesis, and analysis.

In terms of its synthesis capabilities, E-Syn is able to synthesize the following kinds of networks:

1. Lumped low-pass, high-pass, band-pass, and band-stop networks with Tchebycheff or Butterworth passband characteristics. Elliptic function versions are also available for both equal-ripple and maximally flat passbands. The passband is specified in terms of band edge frequencies, a ripple magnitude (decibels), an offset value (decibels), and in the case of nonelliptic band-pass networks, a slope parameter (decibels).
2. Distributed low-pass, high-pass, band-pass, quarter-wave (QW)-band-pass, and QW-band-stop networks with Tchebycheff or Butterworth passband characteristics. As previously explained in Sections 2.3.4 and 2.3.5, it is impossible for any commensurate line distributed network to have either a true low-pass or a true high-pass response. Hence, the low-pass and high-pass options that are offered by E-Syn are not what their names imply. They are in fact the same as the QW-band-pass and QW-band-stop options, respectively, except that the passband and stopband edge frequencies are specified in a different way. For the low-pass and QW-band-stop options, the transmission lines are a quarter of a wavelength long at the center of the stopband, and for the high-pass and QW-band-pass options, they are quarter of a wavelength long at the center of the passband. The one response that is distinct from the rest is the band-pass response, where the transmission line elements are electrically short in the region of the passband and the number of elements employed is constrained to be even. It is transformed into a band-pass response in the *S*-plane while the others are transformed into either low-pass or high-pass *S*-plane responses.

   There are no elliptic responses available for distributed networks. Passband ripple and offset parameters must be specified for all response types, and an additional slope parameter which can be either positive or negative (decibels) must be specified for a band-pass response.
3. Optimum matching networks for lumped complex loads using any of the above lumped network types, but excluding any with a band-stop response. This mode is triggered as soon as either the input or output termination is specified complex. E-Syn can also synthesize impedance-transforming net-

works for use between purely resistive terminations. This option is activated by selecting *transformer* as the required type of frequency response. It is available for either lumped or distributed networks and is equivalent to the synthesis of a band-pass network, except that high-pass elements are not allowed and the program does a little more to help in calculating the passband ripple value required for a given impedance ratio.

4. Doubly and singly terminated versions of networks of the above types, with the exception of singly terminated impedance-matching transformers and the QW-band-stop types.

E-Syn is designed primarily for interactive use wherein input data is read from the console after the appropriate prompt from the program. Output in the form of network element values or frequency response data is normally displayed on the screen, but can also be stored permanently in named files when required. A finished network design can also be saved in the format required for direct importation into the LIBRA circuit design software, where it can be further developed into the final, physical circuit design.

Before dealing with a specific design example, it may be helpful to state the most noticeable differences between using E-Syn and a program like NETSYN.

First, E-Syn makes absolutely no reference to the complex frequency plane (either to the *s*-plane or to the *S*-plane), and all the frequency information requested in the Specifications Window is directly relevant to the *f*-plane. Similarly, all the elements extracted in the Synthesized Network Window are *f*-plane elements. In other words, the inductors and capacitors of the lumped networks are given values in terms of henries and farads, respectively, plus the appropriate exponent, and the transmission lines and stubs of the distributed networks are *all* expressed in terms of their characteristic impedances. Obviously, this avoids the need to understand the concepts of prototypes defined in the complex plane, which may be considered an advantage by the inexperienced user. However, it is a little tedious for the experienced user, and overall leads to a less clear understanding of what is actually happening during the synthesis. It could be argued that it is also responsible for the reduced capability of E-Syn compared to NETSYN in synthesizing distributed networks.

Second, E-Syn offers two modes of operation for the network extraction process. By selecting a manual mode in the Specifications Window and then pressing function key F8 (SYN), E-Syn will compute the necessary intermediate polynomials and stop without performing any network extraction. At this point the Synthesized Network Window is entered and then the user may extract the elements one by one until the network is complete. Alternatively, by selecting an automatic mode and pressing F8, E-Syn will compute the intermediate polynomials and then go on to extract the elements of the network itself. It applies some built-in artificial intelligence to the task and attempts to extract one or more valid networks that satisfy the specified values of terminating resistance. Unfortunately, there is no guarantee of success and in the author's experience, the automatic mode is not particularly useful for *distributed* networks. It should also be noted that there is no mention of transmission zeros during element extraction. Elements available for extraction are listed by their type, not by the transmission zero to which they relate.

As far as source and load impedance specifications are concerned, E-Syn requests the value of both terminations. However, only the source termination is forced to have the specified value. The load resistance is determined by the element extraction process, which may or may not produce a value the same as that requested. Its value is most relevant as a target for network extraction when working in an automatic mode. Singly terminated networks are specified by setting the output termination to a value of either ZERO or INF. In marked contrast to NETSYN, it is not permitted to set the input termination to zero or infinity. This causes a certain amount of difficulty in establishing the real part of the impedance/admittance looking into the network from the short- or open-circuit load, which is relevant for the design of certain special types of matching networks.

Selection of the first or second canonical forms of a network is made by the action of extracting the first network element. If a series element is extracted first, the network will be of the first canonical form, whereas if a shunt element is extracted first, it will be of the second canonical form. If transmission lines are present, one of which is to be extracted first, it will be possible to determine which of the two duals will be created by examining the types of the other elements listed. E-Syn offers the option of partial element extraction when extracting in a manual mode.

2.7.3 A Design Example Based on Using E-Syn

A design example will now be described based on using E-Syn for the prototype network synthesis. To make this as general a case as possible, it was decided to attempt a new band-pass filter design for realization in microstrip or triplate stripline, whose topology has not been seen before in the literature. To avoid any serious difficulties with circuit parasitics, the passband of 4 to 8 GHz was chosen to be in the low microwave frequency range and the commensurate frequency was set to 15 GHz. Hence, the center of the stopband will be at 15 GHz and the next higher passband will not be encountered until a frequency of 22 GHz. The response of the corresponding *S*-plane prototype must therefore be band-pass.

In choosing topology, the only guiding principle was that the numbers of low-pass, high-pass, and all-pass elements (i.e., transmission lines) would be roughly equal. The equal numbers of low-pass and high-pass elements would make the skirts of the passband have roughly equal slopes, and a similar number of transmission lines would ensure that the circuit elements were adequately separated. Hence, with no particular specification to meet in terms of passband selectivity, three low-pass elements, three high-pass elements, and four transmission lines were chosen.

This makes a total of ten nonredundant elements and gives the network the necessary even degree.

To begin the prototype synthesis with E-Syn, the program opens with an introductory window which then clears to reveal the Specifications Window. By using the cursor keys and function key F3, a DISTRIBUTED, TCHEBYCHEFF, BAND-PASS network was selected. The passband edge frequencies were then set to 4 and 8 GHz, respectively, and the electrical length of the transmission lines was set to 48 degrees at 8 GHz. The 48 degrees is equivalent to an electrical length of 90 degrees at 15 GHz. Input and output terminations can be set to the external system impedance of 50Ω if required, but this is not really recommended. To facilitate subsequent network transformations, it is better to set both to a value of 1Ω. On the next line down in the window, E-Syn invites the user to select the basic types of element from which the network will be constructed. Selecting four transmission lines, three low-pass stubs, and four high-pass stubs is equivalent to specifying four S-plane transmission zeros at $S = (1, j0)$, three at $S = (0, j\infty)$, and three at $S = (0, j0)$, respectively. The program will complain if the total is not even. Finally, the window requires passband ripple, offset, and slope parameters to be specified. In this case the passband ripple (T_p) was set to 0.01 dB, while the others were set to zero.

Before actually starting the synthesis, it is possible to gain some idea of the likely frequency selectivity of the network by moving into the RESPONSE window and setting the frequencies F_1 and F_2 to arbitrary values. Once these are set, the program will automatically compute the insertion losses at the two frequencies whenever the parameters in the Specifications Window are changed. Reference should be made to the E-Syn User Guide [46] for information about the ROOTS subsidiary window, but for the present this window can be ignored.

On returning to the Specifications Window, the network can now be synthesized and by default the synthesis mode will be set to AUTO. However, if a synthesis is attempted in this mode by pressing the function key F8, E-Syn will fail to extract any useful network. Hence, the mode must be toggled to MAN by pressing F0, and then by pressing F8, E-Syn will successfully compute the intermediate polynomials and move the user into the Synthesized Network Window. Here the network elements can be extracted successively using function key F6. Each time F6 is pressed, the program lists the various types of elements that can be extracted at that point. At the start of extraction, the number of options may be large, but this decreases gradually toward the end of the process.

In this example, when F6 is pressed for the first time, there are seven elements listed that could be extracted, two of which are nonredundant transmission lines (TLIN). Selecting the higher of the two characteristic impedances ensures that the network will be of the first canonical form. The rest of the elements were extracted using F6 in such a way that the resultant network had a good chance of successful physical realization. The resulting network is shown in Figure 2.48(a) after translation into an equivalent nonredundant S-plane prototype. Converting to an $\hat{S}$-plane

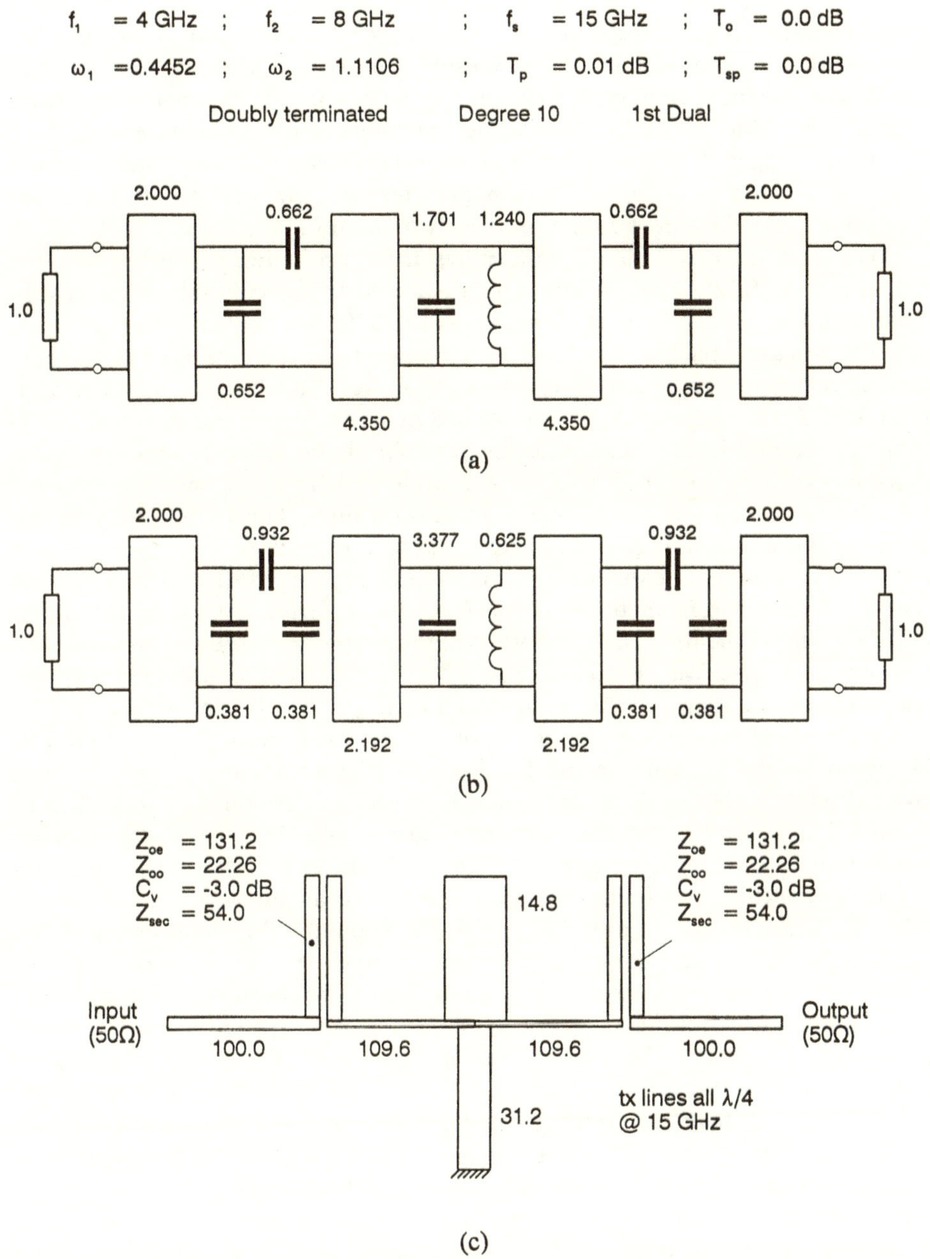

Figure 2.48 A 4- to 8-GHz band-pass filter synthesized by E-Syn: (a) nonredundant *S*-plane; (b) *S*-plane prototype after transformation; (c) stripline/microstrip realization.

prototype not only helps with subsequent network transformation, but also makes the network consistent with the other *S*-plane prototype networks described in this text. It is important to remember to invert the impedances of the open-circuit stubs to obtain the values of the corresponding capacitors in the *S*-plane network.

By pressing shift F8 in the Synthesized Network Window, E-Syn will store the details of the extracted network in a named text file. By pressing F9, E-Syn will store the extracted network as a netlist ready for importation into LIBRA.

On examination of the element values of the nonredundant prototype, it is clear that the middle pair of unit elements would be unrealizable as microstrip transmission lines after scaling impedances for 50Ω terminations. Furthermore, the series *S*-plane capacitors would not be physically realizable without transforming them into lumped capacitors. To overcome these difficulties, an extra pair of redundant *S*-plane capacitors are introduced by transforming the capacitor L sections into symmetrical π sections, as indicated in Figure 2.48(b). This lowers the impedances in the middle of the network while making the capacitor π sections realizable using capacitively coupled transmission lines. The transformations are made using the formulas given in Section A.5 of Appendix A. It should be noted that E-Syn is capable of extracting the network of Figure 2.48(b) directly without resorting to the transformations. This is done by performing a partial extraction of the capacitors of the π sections using the function key shift F8 (NxtEV). The difficulty is in specifying the correct value of the element that would result in symmetrical π sections, but the solution can be found by trial and error.

Figure 2.48(c) shows a possible layout of the filter circuit for realization in microstrip or triplate stripline and for use with 50Ω terminations. The elements have not been drawn to scale and some may prove to be difficult to realize in practice. The central shunt stubs are rather low in impedance, and the coupled-line sections require a tight coupling ratio of –3 dB, which without the addition of a lumped capacitor would necessitate the use of an interdigitated structure. Reference should be made to the Figure B.2 of Appendix B for conversion of the capacitor π sections into the coupled-line parameters. Despite these reservations, however, the circuit does have some potential and deserves further investigation.

Without the addition of losses or parasitics, E-Syn predicts the frequency responses of the filter to be those shown in Figure 2.49. These were obtained by pressing F0 in the Synthesized Network Window and exporting an *S*-parameter data file (shift F0) from the Network Analysis Window. The data format is suitable for reading directly into LIBRA. As anticipated, the slopes of the passband skirts are roughly equal as a result of choosing equal numbers of low-pass and high-pass circuit elements. The return loss over the 4- to 8-GHz passband ripples above 26 dB, which is consistent with the 0.01-dB insertion loss specification.

Work on this design example will not be extended any further, since it has served its purpose as a demonstration vehicle for synthesis by E-Syn.

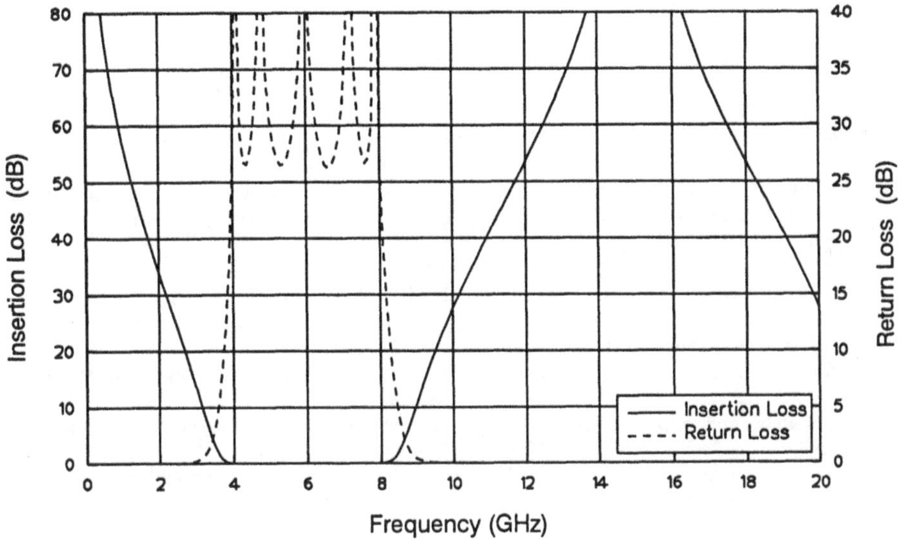

Figure 2.49 Frequency responses of a 4- to 8-GHz filter synthesized by E-Syn.

References

[1] Richards, P. I., "Resistor-Transmission Line Circuits," *Proc. IRE*, Vol. 36, February 1948, pp. 217–220.

[2] Brune, O., "Synthesis of a Two Terminal Network Whose Driving-Point Impedance Is a Prescribed Function of Frequency," *J. Mathematics and Physics*, Vol. 10, No. 3, October 1931, pp. 191–236.

[3] Bode, H. W., Network Analysis and Feedback Amplifier Design, New York: D. Van Nostrand Co., 1945.

[4] Wenzel, R. J., "Application of Exact Synthesis Methods to Multi-Channel Filter Design," *IEEE Trans. Microwave Theory and Techniques*, Vol. MTT-13, January 1965, pp. 5–15.

[5] Temes, G. C., and S. K. Mitra, *Modern Filter Theory and Design*, Wiley & Sons, 1973.

[6] Kuroda, K., "Derivation Methods of Distributed Constant Filters From Lumped Constant Filters," Text for lectures at Joint Meeting of Konsoi Branch of Elec. Commun., of Elec., and Illumin. Engineers of Japan, October 1952, pp. 32.

[7] Levy, R., "A General Equivalent Circuit Transformation for Distributed Networks," *IEEE Trans. Circuit Theory [Correspondence]*, Vol. CT-12, September 1965, pp. 457–458.

[8] Grover, F. W., Chap. 16 in *Inductance Calculations: Working Formulas and Tables*, Dover Publications, 1962, pp. 142–162.

[9] Howe, H., Jr., *Stripline Circuit Design*, Dedham, MA: Artech House, 1974.

[10] Gupta, K. C., R. Garg, and I. J. Bahl, *Microstrip Lines and Slotlines*, Dedham, MA: Artech House, 1979.

[11] Edwards, T. C., *Foundations for Microstrip Circuit Design*, Wiley & Sons, February 1983.

[12] Hoffmann, R. K., *Handbook of Microwave Integrated Circuits*, Dedham, MA: Artech House, 1987.

[13] Cohn, S. B., "Characteristic Impedance of the Shielded-Strip Transmission Line," *IRE Trans. Microwave Theory and Techniques*, Vol. MTT-2, No. 2, July 1954, pp. 52–57.

[14] Cohn, S. B., "Problems in Strip Transmission Lines," *IRE Trans. Microwave Theory and Techniques*, Vol. MTT-3, March 1955, pp. 119–126.

[15] Cohn, S. B., "Thickness Correction for Capacitive Obstacles and Strip Conductors," *IRE Trans. Microwave Theory and Techniques*, Vol. MTT-8, November 1960, pp. 638–644.

[16] Bahl, I. J., and R. Garg, "A Designer's Guide to Stripline Circuits," *Microwaves*, January 1978, pp. 90–96.
[17] Weil, C., "The Characteristic Impedance of Rectangular Transmission Lines With a Thin Centre Conductor and Air Dielectric," *IEEE Trans. Microwave Theory and Techniques*, Vol. MTT-26, 1978, pp. 238–242.
[18] Smith, J. I., "The Even and Odd Mode Capacitance Parameters for Coupled Lines in Suspended Substrate," *IEEE Trans. Microwave Theory and Techniques*, Vol. MTT-19, No. 5, pp. 424–431.
[19] Bryant, T. G., and J. A. Weiss, "Parameters of Microstrip Transmission Lines and of Coupled Pairs of Microstrip Lines," *IEEE Trans. Microwave Theory and Techniques*, Vol. MTT-16, No. 12, December 1968, pp. 1021–1027.
[20] Wheeler, H. A., "Transmission-Line Properties of a Strip on a Dielectric Sheet on a Plane," *IEEE Trans. Microwave Theory and Techniques*, Vol. MTT-25, No. 8, August 1977, pp. 631–647.
[21] Hammerstad, E., "Equations for Microstrip Circuit Design," *Conf. Proc. 5th European Microwave Conference*, September 1973, pp. 268–272.
[22] Hammerstad, E., and O. Jensen, "Accurate Models for Microstrip Computer-Aided Design," IEEE MTT-S Symposium Digest, June 1980, pp. 407–409.
[23] Bates, R. N., and M. D. Coleman, "Millimetre Wave Finline Balanced Mixers," *Proc. Ninth European Microwave Conference*, September 1979, pp. 721–725.
[24] Wen, C. P., "Coplanar Waveguide: A Surface Strip Transmission Line Suitable for Non-Reciprocal Gyromagnetic Device Applications," *IEEE Trans. Microwave Theory and Techniques*, Vol. MTT-17, 1969, pp. 1087–1090.
[25] Hanna, V. F., "Finite Boundary Corrections to Coplanar Stripline Analysis," *Electronic Letters*, Vol. 16, No. 15, July 1980, pp. 604–606.
[26] Rowe, D. A., and Y. L. Binneg, "Numerical Analysis of Shielded Coplanar Waveguides," *IEEE Trans. Microwave Theory and Techniques*, Vol. MTT-31, No. 11, November 1983, pp. 911–915.
[27] Ghione, G., and C. Naldi, "Analytical Formulas for Coplanar Lines in Hybrid and Monolithic MICs," *Electronic Letters*, Vol. 20, No. 4, February 1984.
[28] Lange, J., "Interdigitated Strip-Line Quadrature Hybrid," *IEEE Trans. Microwave Theory and Techniques*, Vol. MTT-17, December 1969, pp. 1150–1151.
[29] Cohn, S. B., "Shielded Coupled-Strip Transmission Lines," *IRE Trans. Microwave Theory and Techniques*, Vol. MTT-3, October 1955, pp. 29–38.
[30] Perlow, S. M., "Analysis of Edge-Coupled Shielded Strip and Slabline Structures," *IEEE Trans. Microwave Theory and Techniques*, Vol. MTT-35, No. 5, May 1987, pp. 522–529.
[31] Reisch, R. E., "Designing Coupled Lines With a Pocket Calculator," *Microwaves*, June 1978, pp. 88–95.
[32] Kirschning, M., and R. H. Jansen, "Accurate Wide-Range Design Equations for the Frequency-Dependent Characteristics of Parallel Coupled Microstrip Lines," *IEEE Trans. Microwave Theory and Techniques*, Vol. MTT-32, No. 1, January 1984, pp. 83–90.
[33] Tripathi, V. K., "Asymmetric Coupled Transmission Lines in an Inhomogeneous Medium," *IEEE Trans. Microwave Theory and Techniques*, Vol. MTT-23, September 1975.
[34] Tripathi, V. K., and Y. K. Chin, "Analysis of the General Non-Symmetrical Directional Coupler With Arbitrary Terminations," *Proc. IEEE*, Vol. 129, December 1982, pp. 360.
[35] Osmani, R. M., "Synthesis of Lange Couplers," *IEEE Trans. Microwave Theory and Techniques*, Vol. MTT-29, No. 2, February 1981, pp. 168–170.
[36] Matthaei, G. L., L. Young, and E. M. T. Jones, *Microwave Filters, Impedance-Matching Networks and Coupling Structures*, New York: McGraw-Hill 1964.
[37] Gupta, K. C., R. Garg, and R. Ghadha, *Computer-Aided Design of Microwave Circuits*, Dedham, MA: Artech House, 1981, pp. 197–199.
[38] Kirschning, M., R. H. Jansen, and N. H. L. Koster, "Accurate Model for Open-End Effect of Microstrip Lines," *Electronic Letters*, Vol. 17, No. 3, February 1981, pp. 123–125.
[39] Kerchning, M., R. H. Jansen, and N. H. L. Koster, "Measurement and Computer-Aided Modelling of Microstrip Discontinuities by an Improved Resonator Method," *1983 MTT-S International Microwave Symposium Digest*, May 1983, pp. 495–497.

[40] Koster, N. H. L., and R. H. Jansen, "The Microstrip Step Discontinuity: A Revised Description," *IEEE Trans. Microwave Theory and Techniques*, Vol. MTT-34, No. 2, February 1986, pp. 213–223.

[41] Pettenpaul, E., et al., "CAD Models of Lumped Elements on GaAs up to 18 GHz," *IEEE Trans. Microwave Theory and Techniques*, Vol. MTT-36, No. 2, February 1988, pp. 294–303.

[42] LAYAN: *Electromagnetic Layout Simulator*, Number One Systems Ltd., Huntingdon, U.K., 1995.

[43] Milsom, R. F., "FACET—A CAE System for RF Analogue Simulation Including Layout," 26th ACM/IEEE Design Automation Conf., 1989, pp. 622–625.

[44] Healy, M., and B. J. Minnis "Electromagnetic Layout Analysis Improves MMIC Design," *Microwave Engineering Europe*, February 1993, pp. 23–27.

[45] Editorial, "MSN CAE Survey," *Microwave Systems News*, May 1989, pp. 62–63.

[46] HP-EEsof, *E-Syn (Series IV) User's Guide—Microwave and RF Design Synthesis*, Hewlett-Packard Ltd., Bracknell, Berkshire, U.K., February 1994.

[47] *SuperCompact RF/Microwave Software*, Compact Software, Paterson, NJ.

CHAPTER 3

Synthesis of High-Selectivity Printed Circuit Bandpass Filters

The next three chapters will present a selection of practical examples of microwave circuits designed by the exact synthesis procedure. In this chapter, high-selectivity printed circuit filters will be the focus of attention. Filters are a good choice with which to begin because they tend to place the widest possible demands on the capability of the procedure. Almost every important part of the design procedure is fully exercised by the synthesis of filters. They are also a good choice because being purely passive structures, the synthesis techniques can be applied directly without incurring the complications associated with the modeling of active devices. Since any type of microwave circuit can be split into a collection of two-port networks, the passive ones of which are essentially filters, it can be argued that an ability to design filters amounts to an ability to design almost anything else.

The chapter is split into two main sections. Section 3.1 deals with filters derived from high-pass *S*-plane prototypes. These filters have the advantage of simplicity, requiring comparatively few elements for a given frequency selectivity. Their disadvantages are that their elements can be rather long and their stopbands are not particularly wide. In contrast, Section 3.2 deals with band-pass filters derived from band-pass *S*-plane prototypes. These filters use relatively large numbers of elements, but the elements are short by virtue of being a quarter of a wavelength long in the center of the stopband. They have the added advantages that stopband widths can be specified semiarbitrarily and the filters can employ a mixture of lumped and distributed elements. In both cases, the filters to be described are printed circuit structures suitable for applications where reproducibility and low cost are important considerations.

Most of the work relating to the filters of Sections 3.1 and 3.2 has been previously published by the author in [1–3]. The discussion in Sections 3.1 and

3.2, therefore, will concentrate mainly on the synthesis aspects of the design procedure and the practical results. It should also be noted that the practical filters to be described were constructed for laboratory demonstration purposes only. They were developed as engineering prototypes and require some refinement to bring their performance and physical construction up to the standard required for commercial production.

3.1 COMMENSURATE LINE FILTERS FROM HIGH-PASS S-PLANE PROTOTYPES

Of the many types of commensurate line band-pass filters being used today, perhaps the most well known are the so-called *line-and-stub* and *coupled-line* varieties. Coincidentally, these two types of filters have prototypes that are network duals with identical distributions of transmission zeros. The relevant frequency transformation for filters of this class has been described in Chapter 2, Section 2.3.5, and illustrated in Figure 2.12. As previously stated, the transformation forces all the transmission lines of the microwave network to be of the same electrical length at the center frequency of the passband f_0 (i.e., $f_0 = f_s$). It also forces the center frequency of the next higher passband to be located at $3f_0$ and the center of the stopband to be at $2f_0$. The stopband width, therefore, is determined by the passband width and is not an independent control parameter.

The line-and-stub filter type is illustrated in Figure 3.1, accompanied by its f-plane and S-plane equivalent circuits. In the example shown, the stripline structure uses five pairs of shunt short-circuit stubs and four interconnecting transmission lines. This corresponds to an f-plane equivalent circuit containing five stubs and four transmission lines, or an S-plane equivalent circuit containing five inductors and four unit elements. All except one of the shunt inductors are redundant and these consequently produce only a single transmission zero at $S = (0, j0)$. The four unit elements each contribute a single transmission zero at $S = (1, j0)$. It is unusual for filters of this type to use redundant unit elements.

Being a directly coupled structure, the line-and-stub filter is the most appropriate for applications in which a wide passband width is required. Its element values are most easily realized for passbands of an octave or more. When realized as a printed circuit, it is helpful if all the stubs are of the same physical length, and for a dispersive medium like microstrip, the consequence is that the stubs are also required to be of the same characteristic impedance. This is easily accomplished using an admittance matrix transformation of the prototype network.

In view of the similarity between the synthesis procedure for the line-and-stub and coupled-line filters, a design example for a line-and-stub filter will not be given here. As a helpful exercise, a design example will now be given for a coupled-line filter and the reader is encouraged to repeat the procedure for the line-and-stub filter. The only modification required will be the specification of a prototype of the second rather than the first canonical form.

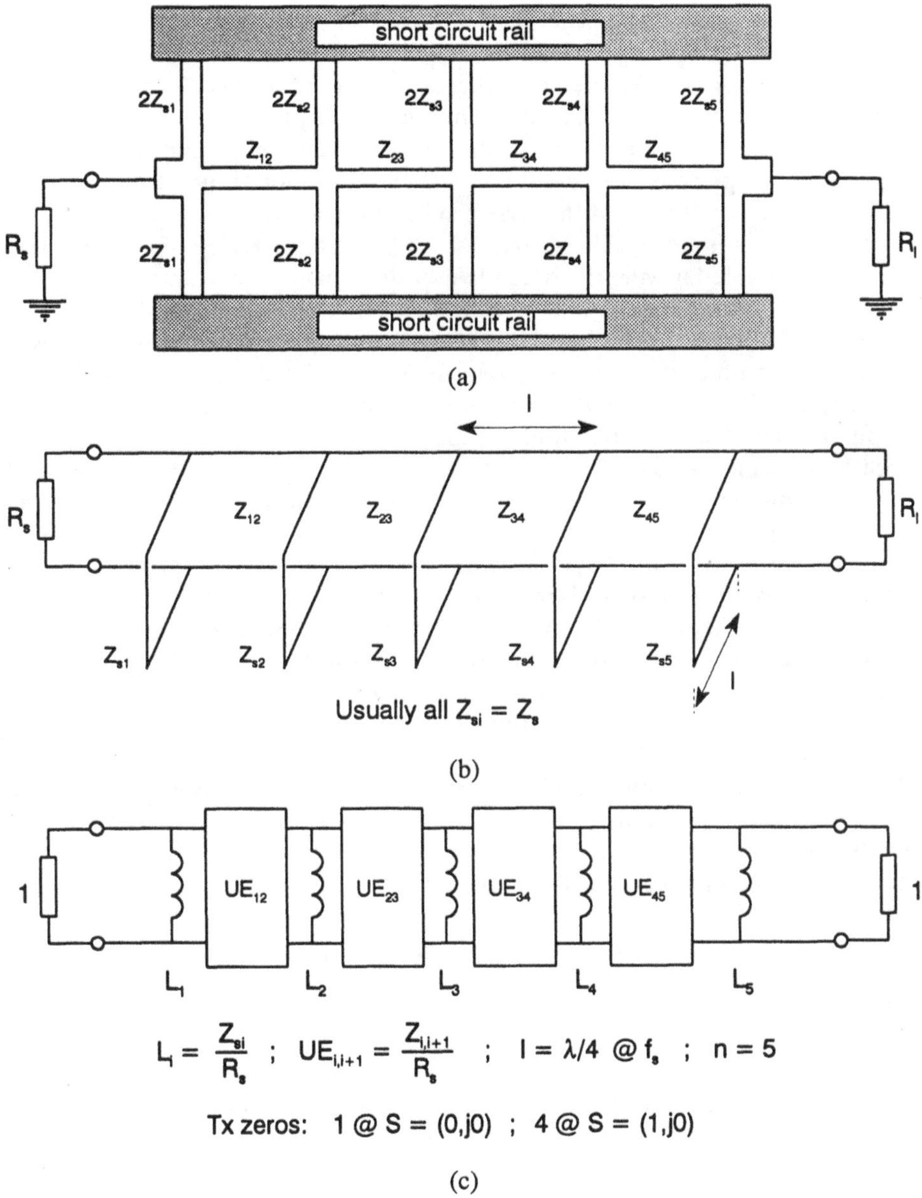

Figure 3.1 The classical line-and-stub band-pass filter: (a) stripline filter realization; (b) *f*-plane equivalent circuit; (c) *S*-plane equivalent circuit.

A sketch of the classical coupled-line filter structure is given in Figure 3.2(a), together with its *f*-plane and *S*-plane equivalent circuits in Figure 3.2(b,c). As previously mentioned, for the same degree and passband specifications, this is the exact dual of the line-and-stub filter. Hence, in the illustration, the *f*-plane prototype comprises five series open-circuit stubs and four interconnecting lines, while the *S*-plane prototype comprises five series capacitors and four unit elements. The five capacitors contribute to only a single transmission zero at $S = (0, j0)$.

Because it is a capacitively coupled structure, the coupled-line filter is inherently less suitable for wide fractional bandwidths than the line-and-stub filter, and historically the filter has tended to be used for relatively narrow bandwidths of 15% or less [4]. However, its bandwidth capability is strongly dependent on the presence of redundant unit elements. For example, in Figure 3.2(c), the network can be synthesized with four nonredundant unit elements, giving it an overall degree of 5. In this situation, the bandwidth capability is in excess of an octave, as will shortly be demonstrated. Alternatively, the network can be synthesized with only two nonredundant unit elements, the other two redundant unit elements having been introduced from the source and load. In this case, the degree of the network would only be 3 and the bandwidth capability would be typically less than 15%. The reason for the difference is that the redundant unit elements cause the internal elements of the network to be transformed to a lower impedance, making the gaps between the coupled lines realizable for only narrow bandwidths. It was in [1] that the author first described the effect of leaving out the redundant elements and demonstrated the physical realization of coupled-line filters with bandwidths as high as 100%.

Two practical coupled-line filters were described in [1]. Both were originally designed by an approximate design procedure, but will now be redesigned by exact synthesis. The objective is not to achieve better results, but rather to give a contrasting demonstration of the universal design procedure.

3.1.1 Two Simple Printed Circuit Coupled-Line Filters

The two practical coupled-line filters have passbands of 4 to 8 GHz and 3 to 9 GHz, respectively. Photographs of the filters illustrating their physical construction and the topology of their printed circuits are shown in Figures 3.3 and 3.4, respectively. Both filters were fabricated in triplate stripline using a copper-clad, glass-reinforced plastic material (RT Duroid 5870 by Rogers Corporation) of dielectric constant 2.32 and sheet thickness of 1.52 mm. To save costs, no metal clamping plates were used, and the pairs of dielectric sheets were glued together with a thermoplastic bonding film. The bonding film has low losses at microwave frequencies and helps to fill any air gaps between the coupled-line sections as well as serving as an adhesive. External connections to the filters were made via self-clamping coaxial receptacles that have thin gold-plated tabs projecting into the stripline to contact the metal tracks.

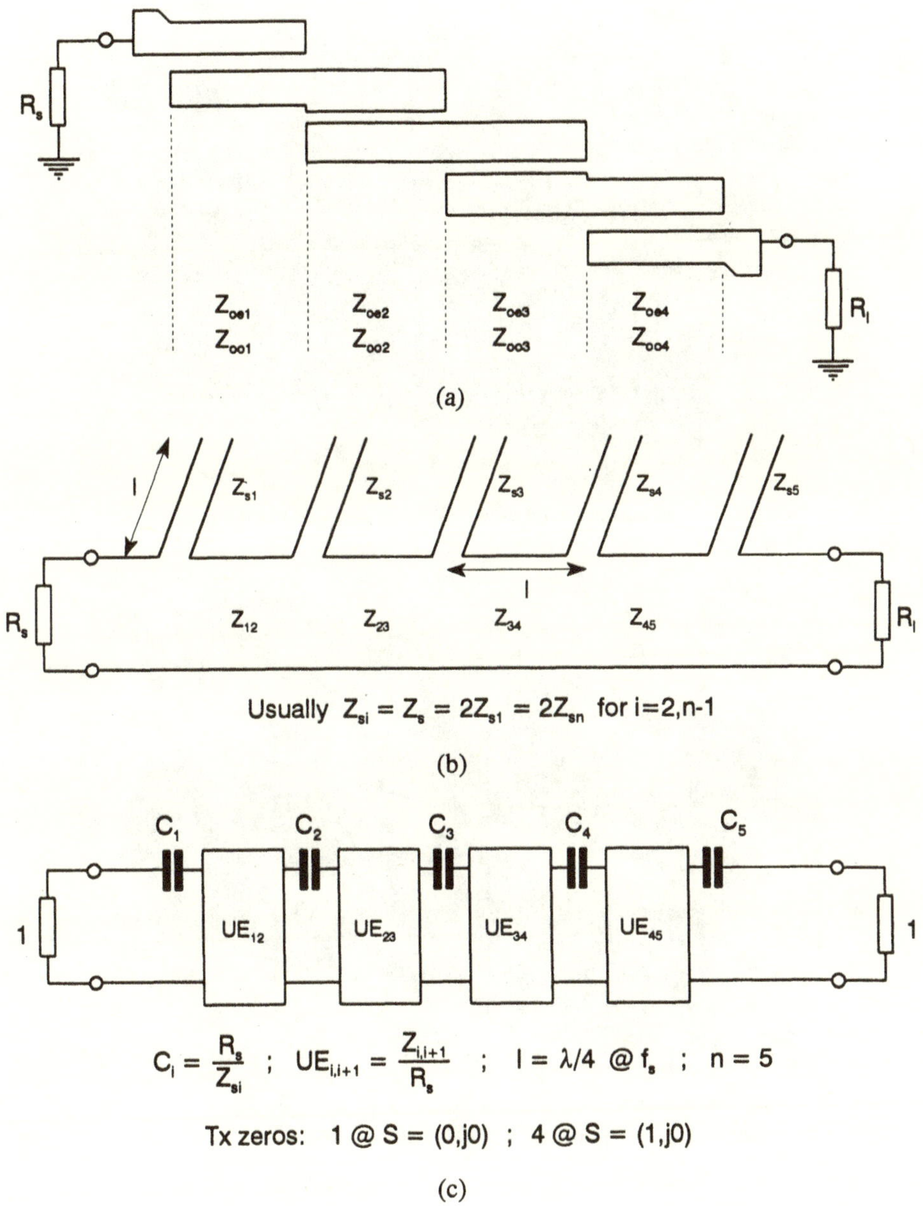

Figure 3.2 The classical coupled-line band-pass filter: (a) stripline filter realization; (b) *f*-plane equivalent circuit; (c) *S*-plane equivalent circuit.

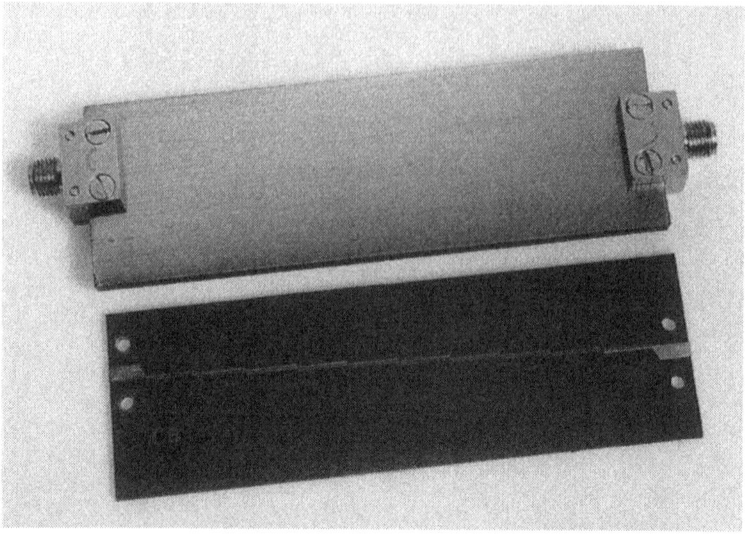

Figure 3.3 Photograph of an experimental 4- to 8-GHz coupled-line filter.

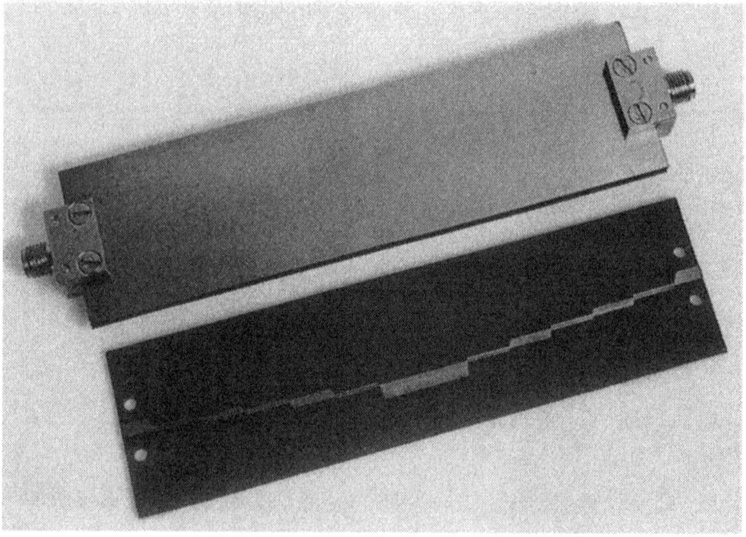

Figure 3.4 Photograph of an experimental 3- to 9-GHz coupled-line filter.

3.1.1.1 4- to 8-GHz Filter

The 4- to 8-GHz filter was required to provide more than 60 dB of attenuation in its stopbands, with a cutoff rate such that the 60 dB was reached at a frequency displacement of 0.8 GHz from the passband edges. The displacement corresponds to 10% of the upper passband edge frequency. Passband losses were required to be generally <1.5 dB, with a maximum permitted loss of 3 dB close to the band edges. Return losses at the terminals of the filter had to be better than 15 dB throughout the passband.

With reference to the universal design procedure and the flow diagram of Figure 2.2, the first step in this and any other component design is to choose the structure of the basic circuit. For this filter, the structure is well known and there is no real need to be creative in terms of circuit topology. The filter is based on a high-pass *S*-plane prototype of the type already described with one transmission zero at $S = (0, \mathrm{j}0)$ and several at $S = (1, \mathrm{j}0)$. The passband edges in the *f*-plane are chosen to be at 4 and 8 GHz, respectively. In general, it is wise to synthesize a passband wider than the minimum performance requirement to allow for modeling and fabrication tolerances. However, the 4- and 8-GHz frequencies are convenient round figures for the prototype demonstrator. The commensurate frequency is at 6 GHz, the center of the passband.

For this filter, there are relatively few parameters that need to be assigned in the *S*-plane. With the location of all the transmission zeros already known, the only uncertainty is the number required at $S = (1, \mathrm{j}0)$ (i.e., unit elements) to give the necessary rate of cutoff. This is best resolved by a little trial and error: a suitably small number is chosen at first, the frequency responses are computed during the first part of the synthesis procedure, and then the number is increased as required. As previously mentioned, this ability to optimize the frequency responses without having to extract a network is an attractive feature of design by synthesis. The passband is specified as high-pass with a cutoff frequency of $\omega_1 = 1.7321$. The cutoff frequency is calculated by taking the lower passband edge in the *f*-plane (4 GHz) and applying the Richards Transformation ($\omega = \tan((\pi/2) \times (f/f_s))$. In order to achieve a return loss comfortably better than 15 dB, the insertion loss ripple is specified to be 0.1 dB. There is no need for any insertion loss offset or slope in the passband, and hence the two parameters T_o and T_{sp} can be set to zero. This is all the information required to perform the first part of the synthesis.

Using NETSYN, the result of the first part of the synthesis is that 10 transmission zeros at $S = (1, \mathrm{j}0)$ are required to produce the required selectivity, making the overall degree of the network 11. All the unit elements are nonredundant, and in view of the bandwidth, none need to be added later. After specifying the network to be doubly terminated and of the first canonical form, the first part of the synthesis will also have generated the necessary $Z_1(S)$ polynomials in readiness for the network extraction. Extraction can then proceed and this is executed by specifying the order in which the transmission zeros are to be realized. The precise way this is done depends on the software being used (see Chapter 2, Section 2.7). In contrast to NETSYN, E-Syn [5] prompts the user to make a choice of element type at each successive extraction rather than choose a transmission zero.

Because of the level of redundancy in the resultant prototype network, it is highly desirable to extract the network in its nonredundant, symmetrical form, as shown in Figure 3.5(a). Here, five unit elements (i.e., zeros at $S = (1, \mathrm{j}0)$) have

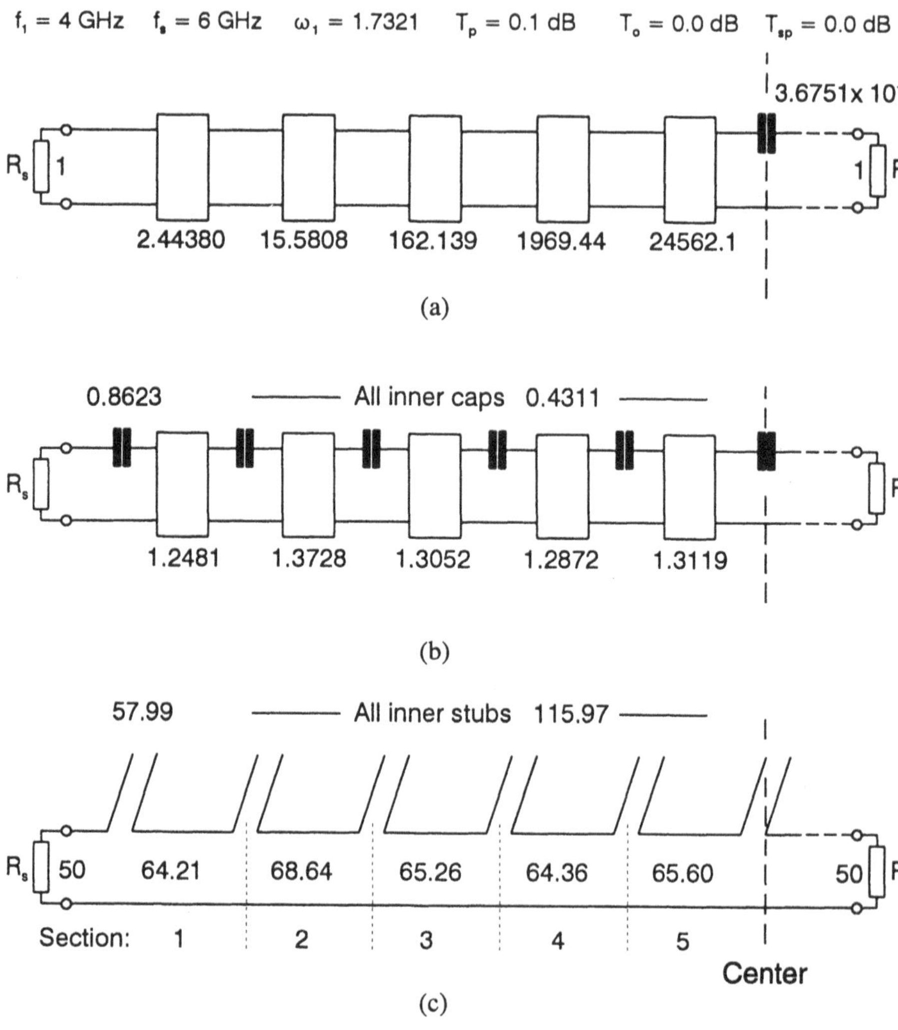

Figure 3.5 *S*-plane and *f*-plane equivalent circuits of the 4- to 8-GHz coupled-line filter: (a) *S*-plane equivalent circuit after extraction; (b) *S*-plane equivalent circuit after transformation; (c) *f*-plane equivalent circuit.

been extracted in succession, followed by a single series capacitor (i.e., zero at $S = (0, j0)$), followed by the remaining five unit elements. For the sake of convenience, only one half of the network is shown in the figure.

It is interesting to note that in the nonredundant network of Figure 3.5(a), there is a gradual rise in impedance level toward the center of the network. This is caused by the presence of the single capacitor in the middle. Moving the capacitor about the network with a Kuroda Identity (Chapter 2, Section 2.4.1) tends to make the high point of the impedance follow the capacitor. Similarly, if fractional parts of the capacitor are moved about the network, the distribution of the impedance level can be adjusted at will. This is what must be done to create the redundant form of the prototype for the coupled-line filter shown in Figure 3.5(b). An admittance matrix transformation (see Chapter 2, Sections 2.4.2 and 2.7) has produced the required network in which all the innermost capacitors are equal to half the value of the outermost capacitors. This allows all the corresponding coupled-line sections to be symmetrical. After some impedance scaling for 50Ω terminations, the *S*-plane network is then transformed into the *f*-plane equivalent circuit of Figure 3.5(c). It is in this form that the network can be analyzed readily by a conventional circuit simulator.

Another point worth noting is that the network of Figure 3.5(c) is an exact solution, meaning that if the network were to be analyzed, its frequency response would be exactly the same as that predicted by the synthesis software. In its derivation, no approximations have been made such as the conversion of lumped elements into distributed elements or vice versa, and consequently there is no need for any circuit optimization prior to the introduction of physical models (see the optimization loop in Figure 2.2). The network is ready for direct translation into the physical dimensions of the coupled-line filter.

The nonredundant network of Figure 3.5(a) can be synthesized using E-Syn. The necessary specifications for the synthesis are summarized in Table 3.1. The term MIL signifies minimum insertion loss and is equivalent to the previously defined parameter T_o. The term Nextel is an abbreviation for "extract next element" and is the name of one of the function keys.

Conversion into physical dimensions is accomplished in two stages. The first involves the subdivision of the *S*-plane or *f*-plane equivalent circuits into sections of the form shown in Figure 3.6. Using the formulas given, the element values within each of these sections are then converted into either the distributed capacitances or the odd and even mode impedances of pairs of coupled lines. When the sections are symmetrical, it is most convenient to work with the odd and even mode imped-

Table 3.1
Synthesis of the 4- to 8-GHz Coupled-Line Filter by E-Syn

Network type	Distributed	Tchebycheff	QW-band-pass	
Passband	f_1 = 4 GHz	f_2 = 8 GHz	T_p = 0.1 dB	MIL = 0 dB
Terminations	R_s = 1	R_l = 1	(only relevant for auto mode)	
Elements	Ten transmission lines	One stub		
Mode (F0)	Manual extraction			
Nextel (F6)	Use successively to extract next element from choice given			
Analyze (F0)	Analysis after extraction			

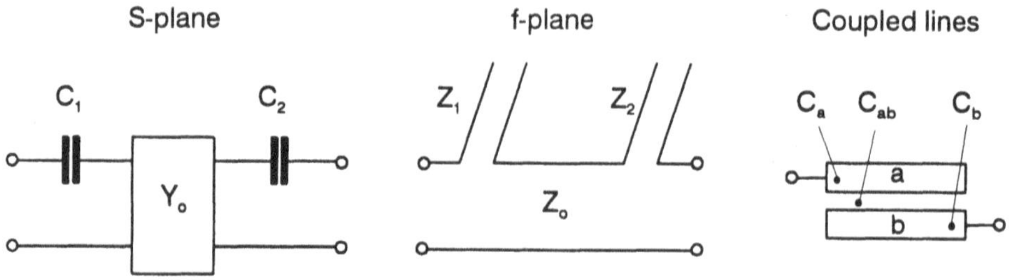

$$C_a = \frac{aC_1Y_0}{Y_0 + C_1 + C_2} \qquad C_b = \frac{C_aC_2}{C_1} \qquad C_{ab} = \frac{aC_1C_2}{Y_0 + C_1 + C_2} \qquad a = \frac{377}{\sqrt{\varepsilon_r}}$$

$$Z_1 = \frac{1}{C_1} \qquad Z_2 = \frac{1}{C_2} \qquad Z_0 = \frac{1}{Y_0}$$

For a symmetrical section:

$$Z_1 = Z_2 \qquad Z_{oo} = Z_1 \qquad Z_{oe} = Z_1 + 2Z_0$$

Figure 3.6 Conversion of *S*-plane and *f*-plane elements into coupled-line design parameters.

ances. The second stage then involves the conversion of the chosen electrical coupled-line parameters into the dimensions and the separation of the lines. For narrow striplines of the type needed for the 4- to 8-GHz filter, the exact equations published by Cohn [6] are the most appropriate. These ignore strip thickness but remain valid for narrow strips where the fringing fields at the edges of the strips are in very close proximity.

Table 3.2 gives the electrical parameters and the corresponding physical dimensions of all the 10 coupled sections of the 4- to 8-GHz filter derived directly from the *f*-plane prototype network of Figure 3.5(c). Included in the table are the coupling coefficients (C_v) and the geometrical mean impedance levels (Z_{sec}) of the sections. The filter is almost uniform along its length, there being relatively little variation in the physical dimensions of the sections. A coupling coefficient of –5.5 dB is close to the limit of what is achievable with edge-coupled striplines in RT Duroid 5870 with a ground plane spacing of 3.043 mm.

Measured and theoretical (zero ohmic losses) insertion loss responses for the 4- to 8-GHz filter are plotted in Figure 3.7. As indicated, there is excellent agreement between the theoretical response and the response of the real filter as measured using a Hewlett Packard (HP) 8409 network analyzer. Passband losses are under 1 dB in the central region, rising to a maximum loss of approximately 3 dB at the upper passband edge. Stopband rejection reaches 60 dB at 8.8 GHz (10%

Table 3.2
Electrical and Physical Parameters for Sections of the 4- to 8-GHz Filter

Parameter	*Section*				
	Coupled Lines 1 & 10	*Coupled Lines 2 & 9*	*Coupled Lines 3 & 8*	*Coupled Lines 4 & 7*	*Coupled Lines 5 & 6*
Z_{oo} (Ω)	57.99	57.99	57.99	57.99	57.99
Z_{oe} (Ω)	186.41	195.27	188.51	186.71	189.19
C_v (dB)	–5.59	–5.32	–5.52	–5.58	–5.50
Z_{sec} (Ω)	103.97	106.41	104.55	104.05	104.74
Width (mm)	0.319	0.284	0.310	0.318	0.308
Separation (mm)	0.106	0.094	0.103	0.106	0.102

RT Duroid 5870:

Dielectric constant	= 2.32
Conductor thickness	= 0.034 mm
Ground plane separation	= 3.043 mm
A quarter wavelength at 6 GHz	= 8.207 mm

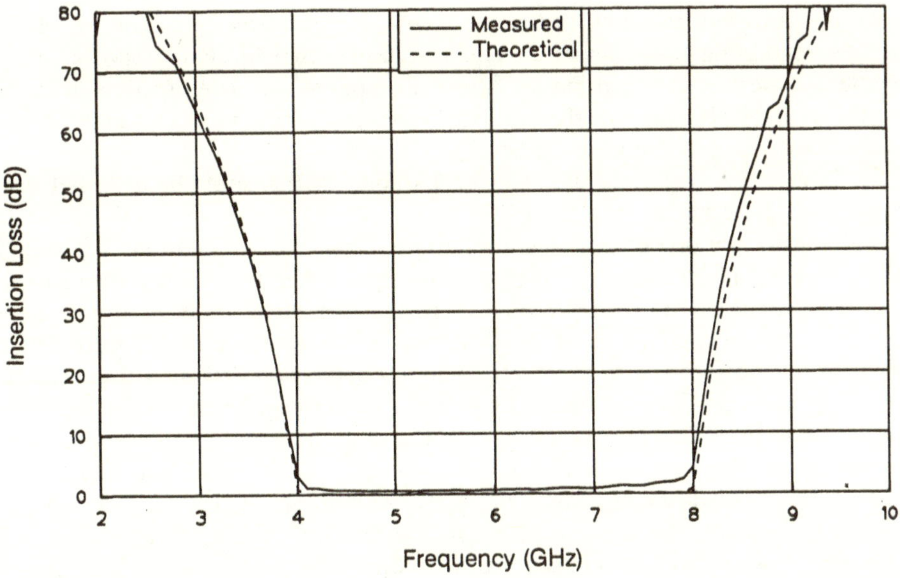

Figure 3.7 Measured and predicted insertion loss responses of the 4- to 8-GHz coupled-line filter.

displacement from the passband edge) and exceeds 80 dB in places. The filter therefore meets the original performance specification. Furthermore, such deep stopband rejection and the absence of spurious modes of propagation are a credit to the fundamental behavior of the triplate stripline transmission line medium. A microstrip circuit would not have achieved better than 50 dB of stopband rejection due to leakage into surface modes.

3.1.1.2 3- to 9-GHz Filter

The performance requirement for the 3- to 9-GHz filter was almost exactly the same as that for the 4- to 8-GHz filter except for the change in passband width. Once again, 60 dB of stopband attenuation had to be reached at a 10% frequency displacement (0.9 GHz) from the upper passband edge. Passband specifications in the *f*-plane were f_1 = 3 GHz, f_2 = 9 GHz, T_p = 0.1 dB, T_o = 0 dB (offset), and T_{sp} = 0 dB (slope). In the *S*-plane, the response was specified high-pass with a cutoff frequency of ω_1 = 1.000. There must be a single transmission zero at $S = (0, j0)$ for the series capacitor of the prototype and an as yet unknown number at $S = (1, j0)$ for the unit elements. The network is doubly terminated and of the first canonical form.

Iteration around the first stage of the synthesis using NETSYN establishes that 12 unit elements are needed to give the required selectivity. Hence, the degree of the network will be 13. Extracting the nonredundant form of the *S*-plane prototype involves taking out the first six unit elements, then a series capacitor, and finally the six remaining unit elements. The resultant nonredundant network and its element values are given in Figure 3.8(a). Compared with the previous prototype for the 4- to 8-GHz filter, the dynamic range of the element values is much reduced as a consequence of the wider passband.

Table 3.3 below gives the relevant parameters that must be assigned if the synthesis is performed using E-Syn.

Transformation of the nonredundant prototype into the redundant prototype of Figure 3.8(b) is performed using the same type of admittance matrix transformation used for the 4- to 8-GHz filter except that the distribution of the capacitor values is differently constrained. As illustrated in the photograph of Figure 3.4, the physical structure of the 3- to 9-GHz filter has been modified such that only half of the sections are coupled lines, the sections in between being simple lengths of transmission line. This has the effect of increasing the coupling between adjacent sections and allows the increased bandwidth to be achieved without making the gaps in the coupled-line sections completely unrealizable. With reference to Figure 3.8(b), this means that the capacitors in the interior of the network will not be split and assigned to two adjacent coupled-line sections as was the case for the 4- to 8-GHz filter. Instead, each pair of capacitors and the unit element in between will be assigned entirely to a single coupled-line section. Hence, there must be an

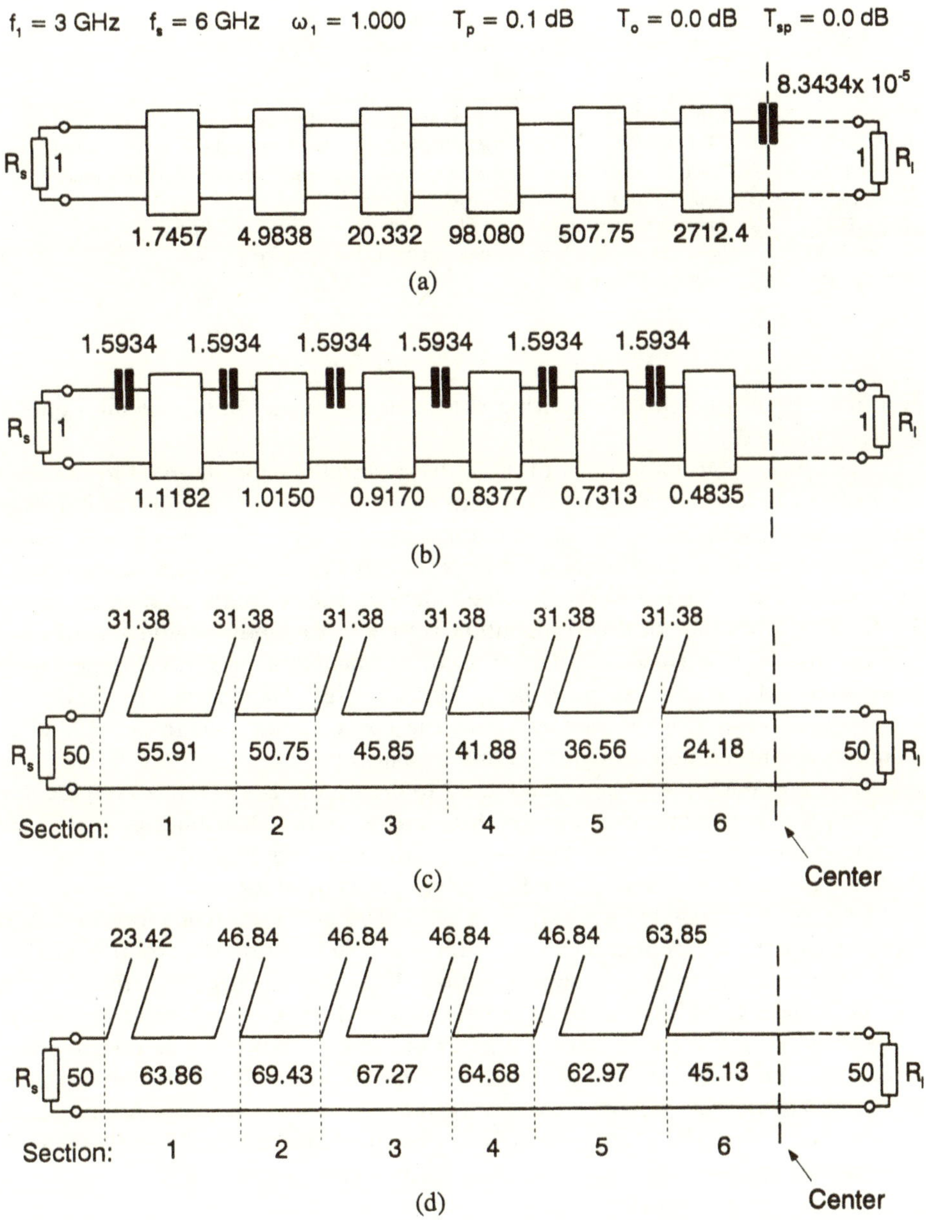

Figure 3.8 *S*-plane and *f*-plane equivalent circuits of the 3- to 9-GHz coupled-line filter: (a) *S*-plane equivalent circuit after extraction; (b) *S*-plane equivalent circuit after transformation; (c) *f*-plane equivalent circuit; (d) alternative *f*-plane equivalent circuit.

Table 3.3
Synthesis of the 3- to 9-GHz Coupled-Line Filter by E-Syn

Network type	Distributed	Tchebycheff	QW-band-pass	
Passband	$f_1 = 3$ GHz	$f_2 = 9$ GHz	$T_p = 0.1$ dB	MIL = 0 dB
Terminations	$R_s = 1$	$R_l = 1$	(only relevant for auto mode)	
Elements	Twelve transmission lines	One stub		
Mode (F0)	Manual extraction			
Nextel (F6)	Use successively to extract next element from choice given			
Analyze (F0)	Analysis after extraction			

even number of capacitors, requiring the capacitor in the middle of the network to be eliminated.

The most logical approach to the redistribution of capacitance is to first split the center capacitor of Figure 3.8(a) into two halves and move the two halves to the left and right through the adjacent unit elements with a Kuroda Identity. Then, in each half network, an admittance matrix transformation is applied which creates six capacitors of equal value. The resultant element values are those given in Figure 3.8(b). The advantage of this distribution is that all the coupled-line sections will be symmetrical, which facilitates a more accurate calculation of physical dimensions. The disadvantage is that, as Figure 3.8(c) shows, the impedance levels in the equivalent f-plane network are rather low, with the consequence that the coupled-line sections are unrealizable in conventional triplate stripline.

The approach that was actually used to create the 3- to 9-GHz coupled-line filter was to take the network of Figure 3.8(a) and begin by redistributing capacitance in exactly the same way as for the 4- to 8-GHz filter. That is, all the 11 internal capacitors were constrained to be equal to half the value of the outer two capacitors. Then the center capacitor was split into two halves and each half moved left and right through the adjacent unit element with a Kuroda Identity. After transformation into the f-plane, the result is shown in Figure 3.8(d). As indicated, impedance levels have been significantly increased by this alternative distribution of capacitance, and the realization of the coupled lines is now not quite so difficult. The disadvantage is that two of the coupled-line sections in each half of the filter (sections 1 and 5 in Figure 3.8(d)) have become asymmetrical.

In each half of the filter there are three coupled-line sections. For the symmetrical section (section 3 in Figure 3.8(d)), Cohn's exact equations [6] for strips of zero thickness are applied as before. However, for the two asymmetrical sections, there are no known exact formulas for calculating their physical dimensions, and therefore a method is used that involves applying Cohn's equations twice in each section. In order to do this, two pairs of odd and even mode impedances have to be calculated from the f-plane equivalent circuit. They are calculated by using the relations given in Figure 3.6 after making the assumption that each of the two different stub values belongs to a separate symmetrical section. Strictly speaking, these are not, therefore, the true odd and even mode impedances of an asymmetrical coupled-line section, but they do facilitate the use of this realization procedure. Using Cohn's equations twice then gives directly the widths of the two coupled strips, while their separation is obtained by taking the mean of the two computed values of separation. This method may not be exact, but it does give good practical results.

All the electrical and physical parameters of the lines and coupled lines of the 3- to 9-GHz filter are listed in Table 3.4. The widths of the simple lines were calculated using Cohn [7]. With the exception of the first and twelfth sections, the physical dimensions are all readily realizable in triplate stripline. The difficulty with the first and twelfth sections is the separation between the strips, which is only 0.028 mm. In triplate stripline and without special fabrication processes, such a gap is not realizable, and consequently, in the filter that was actually fabricated, this gap was increased to 0.055 mm. At the same time, the width of both strips in these sections was made equal to 0.400 mm. There were no other compensatory changes made to the circuit dimensions.

The uncompensated increase in the gap between the strips of the outer coupled-line sections has the effect of decreasing the coupling coefficient. In terms

Table 3.4
Electrical and Physical Parameters for Sections of the 3- to 9-GHz Filter (Asymmetrical Coupled Sections)

Parameter	*Section*					
	Coupled Lines 1 & 12	*Simple Lines 2 & 11*	*Coupled Lines 3 & 10*	*Simple Lines 4 & 9*	*Coupled Lines 5 & 8*	*Simple Lines 6 & 7*
Z_{ooa} (Ω)	23.42	—	46.84	—	46.84	—
Z_{oob} (Ω)	46.84	—	—	—	63.85	—
Z_{oea} (Ω)	151.14	—	181.38	—	172.76	—
Z_{oeb} (Ω)	174.56	—	—	—	189.78	—
C_{va} (dB)	−2.71	—	−4.59	—	−4.83	—
C_{vb} (dB)	−4.78	—	—	—	−6.08	—
Z_{seca} (Ω)	59.49	69.43	92.17	64.68	89.96	45.13
Z_{secb} (Ω)	90.42	—	—	—	110.08	—
W_a (mm)	0.585	1.282	0.368	1.479	0.412	2.724
W_b (mm)	0.402	—	—	—	0.290	—
S (mm)	0.028	—	0.051	—	0.098	—

RT Duroid 5870:	Dielectric constant	= 2.32
	Conductor thickness	= 0.034 mm
	Ground plane separation	= 3.043 mm
	Quarter wavelength at 6 GHz	= 8.207 mm

of the equivalent circuit of Figure 3.8(d), this increases the value of the outer series stub from 23.42Ω to the 46Ω of its neighbor. While undesirable, this has a surprisingly modest effect on the frequency responses of the filter, as shown in Figure 3.9. The plot of the theoretical insertion loss response (zero ohmic losses) shows an increase in the ripple near the passband edges but virtually no change in the width of the passband. When measured on an HP8409 network analyzer, the practical result also shows the increased ripple superimposed on rising ohmic losses toward the upper end of the passband. While this characteristic is not entirely desirable, the filter may still be a useful component. Its limitations are associated with the physical realization problem of edge-coupled striplines and not with the synthesis design procedure. It is a first example of the sort of topological variations that can be considered when using exact synthesis techniques.

3.1.2 Two Enhanced Printed Circuit Coupled-Line Filters

Simple printed circuit coupled-line filters as described in the previous section can provide a frequency selectivity that is sufficient for a wide range of system applications. Although they are long, thin structures, they do not occupy a large area or volume, and if their length becomes inconvenient, it is possible to perform some folding of the coupled-line sections without serious adverse effects on performance. However, when the selectivity requirement becomes extreme as, for example, when

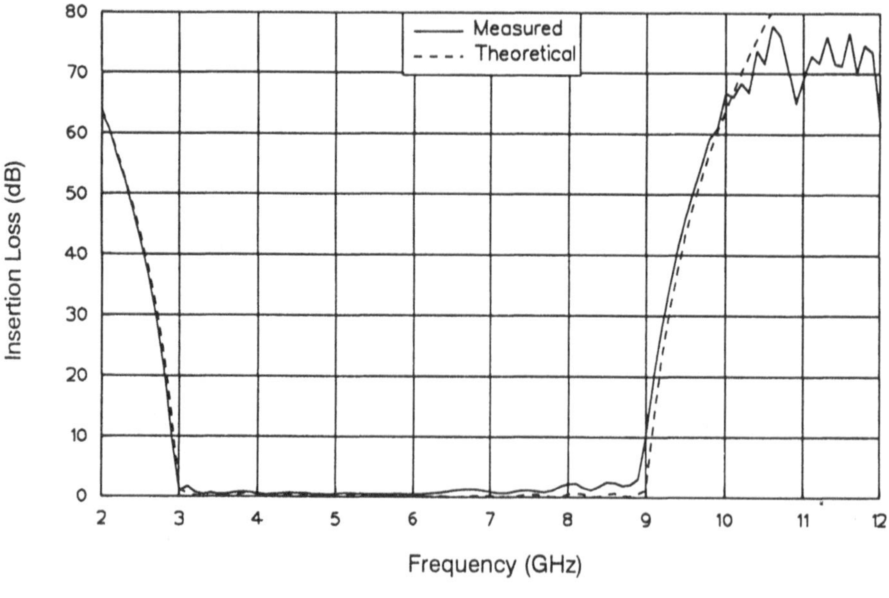

Figure 3.9 Measured and predicted insertion loss responses of the 3- to 9-GHz coupled-line filter.

an attenuation of more than 60 dB is required at a frequency displacement of less than 10% of the upper passband edge, the number of elements that would be required becomes impracticable. It is in this situation that the enhanced coupled-line filter becomes a more attractive option.

Enhanced coupled-line filters [2] are simple coupled-line filters in which double-length, open-circuit shunt stubs are added at the junction between each of the coupled-line sections. These stubs provide extra poles of attenuation at frequencies close to the passband edge, forcing the attenuation to reach very high levels very rapidly. In terms of the *S*-plane equivalent circuit, they take on the basic form shown in Figure 3.10, which is an alternate cascade of a unit element flanked by two series capacitors and a second-order shunt foster reactance branch. Each unit element and capacitor pair corresponds to a symmetrical coupled-line section according to the equivalence given in Figure 3.6. Each second-order reactance branch of the *S*-plane network corresponds to a double-length shunt stub according to the equivalences given in Figure 3.11. As indicated, the stub can be a simple pair of lines of characteristic impedances Z_1 and Z_2 connected in cascade, or alternatively a coupled-line section followed by a single-length stub of characteristic admittance Y_s. The choice between the two depends strongly on the location of the required transmission zero in the *S*-plane. For low-frequency zeros ($j\omega \leq j1$), which tend to occur for wide fractional bandwidths, the simple stub will be most appropriate, and is, in view of its simplicity, generally to be preferred. However, when the zero occurs at a relatively high frequency, as in the case of narrowband filters, the value of C in the *S*-plane equivalent circuit will be too small to be realizable with the simple stub and will only be realizable with the help of the coupled-line section. All the relevant formulas for transforming between the *S*-plane elements and the two types of stubs are given in Figure 3.11. The capacitances C_a, C_b, and C_{ab} are the normalized distributed capacitances of the coupled lines, which can be translated into odd and even mode impedances if required (see Chapter 2, Section 2.5.2.6).

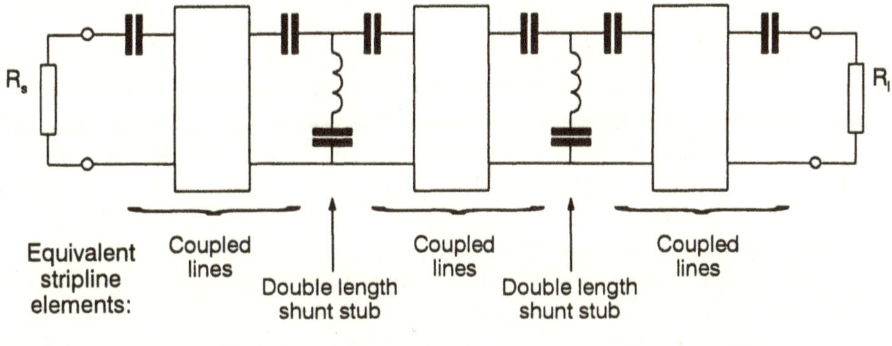

Figure 3.10 General form of the *S*-plane prototype for an enhanced coupled-line filter.

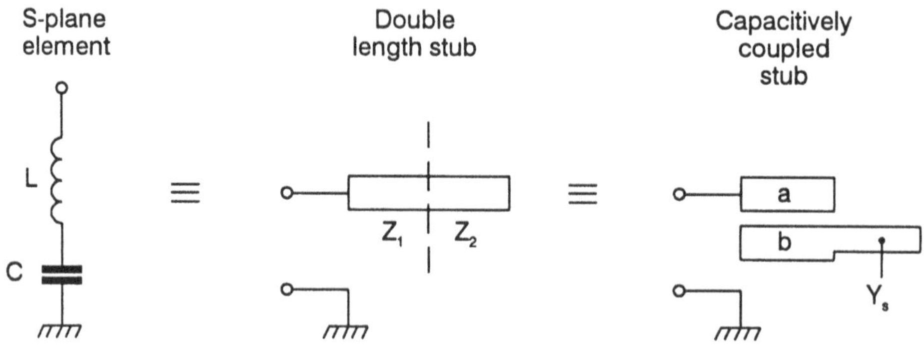

$$Z_1 = \left(L + \frac{1}{C}\right) \qquad Z_2 = \frac{Z_1}{LC}$$

$$a = \frac{377}{\sqrt{\varepsilon_r}}$$

$$C_{ab} = \frac{a\sqrt{Y_s\left(Y_s + \frac{1}{Z_1}\right)\left(\frac{1}{Z_1} + \frac{1}{Z_2}\right)}}{\left(Y_s - \frac{1}{Z_2}\right)\sqrt{Z_2}}$$

$$C_a = C_b = \frac{a\,Y_s\left(\frac{1}{Z_1} + \frac{1}{Z_2}\right)}{\left(Y_s - \frac{1}{Z_2}\right)} - C_{ab}$$

Figure 3.11 Second-order shunt stubs and their *S*-plane equivalent circuit.

Derivation of the equivalent circuits for enhanced coupled-line filters would be difficult by approximate procedures and would, in any case, require access to tables of lumped-element prototypes, which themselves would have been generated by synthesis. Exact synthesis offers the only really satisfactory approach. With reference to Figure 2.2 in Chapter 2, the enhancement of the basic coupled-line filter amounts to taking an existing circuit configuration from the top level of the circuit design library/database and adding to its *S*-plane equivalent one or more of the elementary types of elements (i.e., the second-order reactance branches) to be found in the first level. In this case, the need to add the second-order reactance branches to increase selectivity is relatively obvious. Where the skill of the circuit designer really counts is in ensuring that the new combination of circuit elements and the corresponding set of transmission zeros will be realizable and consistent with the fundamental rules governing synthesis (see Chapter 2, Section 2.3.6.3).

Two practical examples of enhanced coupled-line filters will now be described. The first was designed for a passband extending from approximately 2 to 4 GHz while the second was designed for an even wider passband of 2 to 6 GHz. Photographs of both filters are given in Figures 3.12 and 3.13, respectively. They are

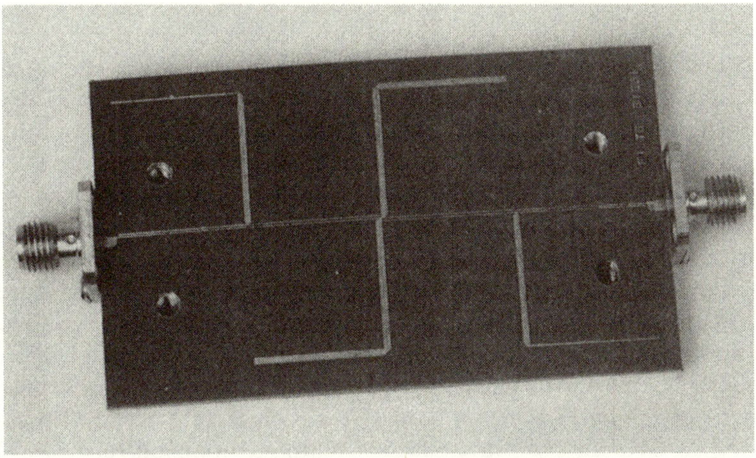

Figure 3.12 Photograph of the 2- to 4-GHz enhanced coupled-line filter.

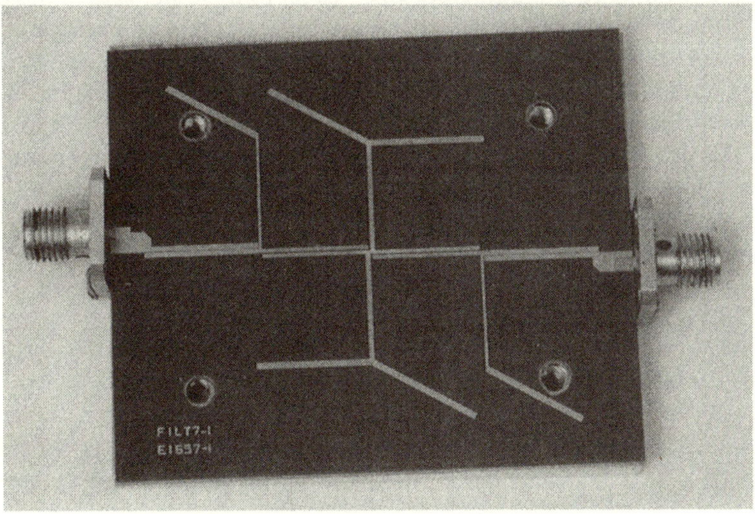

Figure 3.13 Photograph of the 2- to 6-GHz enhanced coupled-line filter.

shown with the top half of their triplate structures removed to reveal the layout of the printed circuits. Both were fabricated on the same RT Duroid 5870 material used for the previous coupled-line filters, but the sheet thickness for the 2- to 4-GHz filter was 0.76 mm, while that of the 2- to 6-GHz filter was 1.52 mm.

3.1.2.1 2- to 4-GHz Filter

In the photograph of Figure 3.12, the 2- to 4-GHz filter is seen comprising four coupled-line sections and four double-length stubs. The two stubs in the center are identical and wired in parallel, making the effective number of shunt stubs in the filter actually three rather than four. Each of the stubs has been folded at the midpoint through an angle of 90 deg to reduce circuit width. By requiring only four coupled-line sections, the overall length of the filter is reduced to less than half that of a simple coupled-line filter of roughly comparable performance. Overall dimensions are 70 × 40 mm. The size of the filter could be reduced still further if necessary by some more extensive folding of the circuit elements, but being a laboratory prototype, it was considered more important to keep the circuit parasitics to an absolute minimum rather than achieve the smallest possible physical size.

Although the passband of this filter is nominally 2 to 4 GHz, it was necessary for reasons associated with a system requirement to specify precise passband edge frequencies of f_1 = 2.033 and f_2 = 4.067 GHz. This is an octave bandwidth with a center at 3.05 GHz. In order to achieve a required stopband rejection of >70 dB at 4.3 GHz, two transmission zeros were placed at 4.321 GHz, which, by reflection around the commensurate frequency (3.05 GHz) to the other side of the passband, are also present at 1.779 GHz. Another transmission zero was placed at 4.575 GHz (reflecting to 1.525 GHz below the passband) to ensure that the floor of the stopband would be in excess of 80 dB. The precise position of these zeros and their effect on the stopband characteristic was determined by one or two iterations around the first stage of the synthesis process. To summarize the f-plane data for the filter, we then have:

f_1 = 2.033 GHz;
f_2 = 4.067 GHz;
f_s = 3.05 GHz;
f_{z1} = 1.779 GHz;
f_{z2} = 1.525 GHz.

When these f-plane specifications are translated into S-plane specifications using the Richards Transformation, the data required for synthesis become:

Response type: high-pass
S-plane specifications:

$\omega_1 = 1.7321$; $\omega_{z1} = 1.3032$; $\omega_{z2} = 1$ $\qquad$ T_p = 0.1 dB; T_o = 0 dB; T_{sp} = 0.0 dB.

Transmission zero specifications:

1 transmission zero at $S = 0, j0$ (series capacitor)
4 transmission zeros at $S = 1, j0$ (unit elements)
2 transmission zeros at $S = 0, j1.3032$ (second-order reactance branches)
1 transmission zero at $S = 0, j1.0000$ (second-order reactance branches)

Network degree: 11
Network type: doubly terminated, first dual

The list of transmission zeros is governed by the form of the basic *S*-plane equivalent circuit shown in Figure 3.10. The need for series capacitors forces the network to be a first dual, and since all the series capacitors except one are redundant, there can only ever be one transmission zero at $S = (0, j0)$, whatever the overall network degree. The number of second-order reactance branches is the dominant influence on frequency selectivity. One branch would be a perfectly acceptable minimum while four is the maximum number ever likely to be needed in practice. As previously stated, a total of three were chosen for the 2- to 4-GHz filter to achieve the required stopband characteristics. Setting two of the transmission zeros to the same frequency allows the network to be made symmetrical. The number of unit elements required is four, since this must always be one greater than the number of reactance branches to achieve the necessary separation of elements. If the number of second-order branches is p, the degree of the network for this type of filter will always be $(3p + 2)$. Having set $p = 3$, the degree of the 2- to 4-GHz filter network must be 11.

The initial *S*-plane prototype for the 2- to 4-GHz filter was generated from the above specifications by NETSYN in the form shown in Figure 3.14(a). In this form it contains a minimum number of redundant series capacitors. It should be noted that E-Syn is not capable of creating *S*-plane networks of this type with finite, nonzero transmission zeros, and if the reader wants to create such a network, then some alternative software will have to be obtained. The extraction proceeds by first extracting a unit element (transmission zero at $S = (1, j0)$) followed by a second-order reactance branch (transmission zero at $S = (0, j\omega_{z1})$). This sequence is repeated twice, making sure that the second-order branch with the transmission zero at $S = (0, j1)$ is extracted in the middle. Finally, the last series capacitor is extracted, followed by the last unit element. Each of the capacitors preceding a shunt reactance branch is a zero-shifting element (see Chapter 2, Section 2.3.8), whose extraction is performed automatically by the software as part of the extraction of the shunt branch. Only the series capacitor nearest the output load is extracted explicitly in respect of the transmission zero at $S = (0, j0)$. Because of this unequal distribution of capacitance, the network is not yet symmetrical, and there is a large impedance ratio between source and load. However, making the network symmetrical and equalizing the terminating resistors is a simple matter of moving an appropriate part of each capacitor to the left through the adjacent unit element. The necessary relations based on Kuroda Identities are given in Figure 2.26 of Chapter 2. Figure 3.14(b) shows the form of the symmetrical *S*-plane prototype after the redistribution of capacitors.

For conversion into the *f*-plane, all the unit elements become transmission lines, the series capacitors become series stubs, and the shunt reactance branches become double-length stubs according to the equivalence given in Figure 3.11. After scaling impedances for 50Ω terminations, the characteristic impedances of all the *f*-plane circuit elements are those given in Figure 3.14(c). As illustrated, the transmission line sections are all the same length, equal to a quarter of a wavelength

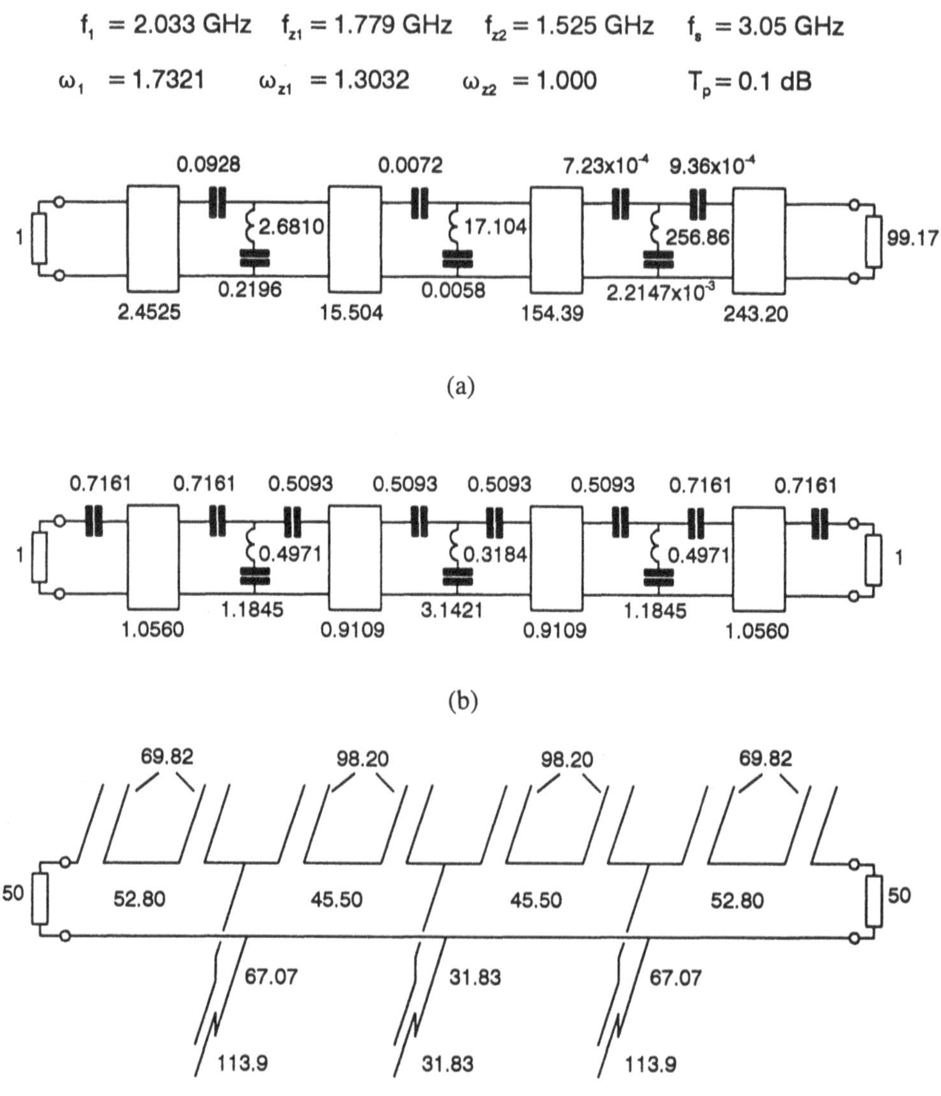

Figure 3.14 *S*-plane and *f*-plane prototypes for the 2- to 4-GHz enhanced coupled-line filter: (a) prototype after synthesis; (b) prototype after transformation; (c) *f*-plane equivalent circuit.

at a frequency of 3.05 GHz. Each of the sections comprising a pair of series stubs and an interconnecting transmission line is converted into the odd and even mode impedances of a symmetrical coupled-line section, and the physical dimensions are calculated using Cohn [6]. On this occasion, corrections are applied to the width and gap dimensions using Cohn [8] to take account of conductor thickness. As indicated in Table 3.5, all four coupled-line sections are readily realizable in the triplate stripline. The widths of the double-length stubs are calculated using Cohn [7]. Both of the outer double-length stubs are readily realizable, but the low impedance of the center double-length stub requires its realization as a pair of double-length stubs of twice the impedance wired in parallel (see Figure 3.12). It is interesting to note that the two lines of the center double-length stub are of equal impedance. This occurs in any double-length stub when the transmission zero is located at $S = (0, j\omega) = (0, j1)$.

After construction of the prototype filter, insertion loss measurements were made using an HP8510 network analyzer. By using video averaging, the dynamic range of the analyzer is on the order of 90 dB at frequencies below 6 GHz, which gives a good opportunity to examine the shape of the stopband floor of the filter. The measured insertion loss response is plotted in Figure 3.15, together with the theoretical response for the purposes of comparison. The figure indicates that the measured passband of the filter is very slightly higher and wider compared with the theoretical response, but is as accurately positioned as could be expected for a filter without benefit of tuning. Measured losses in the center of the band are below 1 dB, rising to approximately 3 dB at the band edges. Below the passband there is an extremely good agreement between the measured and theoretical responses, indicating that the transmission zeros associated with the shunt stubs have been

Table 3.5

Electrical and Physical Parameters for Sections of the 2- to 4-GHz Enhanced Coupled-Line Filter

Parameter	*Section*			
	Couple Lines 1 & 7	*2-Section Stub 2 & 6*	*Coupled Lines 3 & 5*	*2-Section Stub 4(2 in ‖)*
Z_{oo} (Ω)	69.82	—	98.20	—
Z_{oe} (Ω)	174.82	—	189.20	—
C_v (dB)	−7.35	—	−10.0	—
Z_{sec} (Ω)	110.5	67.07, 113.9	136.3	63.66, 63.66
Width (mm)	0.140	0.691, 0.189	0.081	0.765, 0.765
Separation (mm)	0.146	—	0.252	—

RT Duroid 5870:

Dielectric constant	= 2.32
Conductor thickness	= 0.017 mm
Ground plane separation	= 1.5215 mm
A quarter wavelength at 3.05 GHz	= 16.14 mm

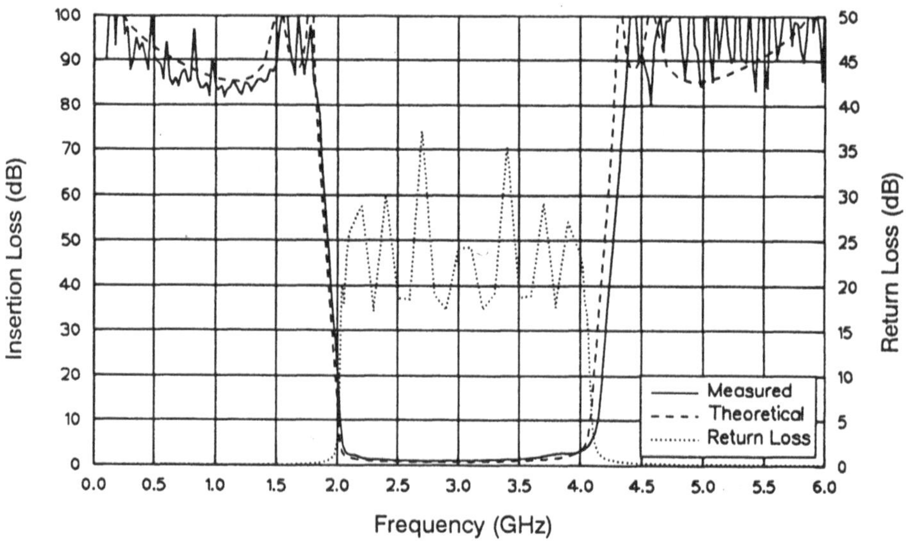

Figure 3.15 Measured and theoretical frequency responses of the 2- to 4-GHz enhanced coupled-line filter.

realized at the correct frequencies. Above the passband there appears to be a very slight error in the location of the nearest transmission zero, causing the first stop-band ripple to be a little higher than predicted and the second ripple to be a little lower than predicted. This behavior, in which the zeros are correctly placed below the passband but slightly in error above the passband, is caused by small differences in the effective electrical lengths of the coupled-line sections and the sections of the shunt stubs. In any event, this filter has excellent performance. Providing as it does 70 dB of insertion loss at 4.3 GHz, which is only a 5% fractional displacement from the upper band edge, the filter is considerably more selective than a conventional coupled-line filter similarly constructed in triplate stripline.

For the sake of clarity, only theoretical values of return loss are plotted in Figure 3.15. However, there is generally good agreement between the measured results and theory. As the response shows, the return loss ripples in the passband have minima at approximately 17 dB, which corresponds to the theoretical value of insertion loss ripple used for the synthesis of 0.1 dB.

3.1.2.2 2- to 6-GHz Filter

As shown by Figure 3.13, the 2- to 6-GHz filter has an almost identical circuit configuration to that of the 2- to 4-GHz filter. Once again, four coupled-line sections and effectively three shunt reactance branches are involved. The overall physical dimensions of the circuit are 60 × 50 mm. In realizing this filter, it was necessary to use a substrate thickness of 1.52 mm, which is twice the thickness used for the 2- to 4-GHz filter. The increased thickness was necessary primarily to ease the realization of the outermost coupled-line sections where the gap between the strips is extremely small. A thicker substrate, however, also decreases circuit losses.

The passband edges of this filter were specified at precisely 2.0 and 6.0 GHz, respectively. Commensurate frequency f_s was therefore 4.0 GHz. The higher commensurate frequency makes the elements of the filter shorter than those of the 2- to 4-GHz filter, and the ultrawide bandwidth causes a tighter degree of coupling in the coupled-line sections. Transmission zeros in the f-plane were placed at f_{z1} = 1.8 GHz and f_{z2} = 1.5 GHz, which reflect to the frequencies of 6.2 and 6.5 GHz above the passband. The S-plane specifications and distribution of transmission zeros are then identical to those of the 2- to 4-GHz filter, except that after applying the Richards Transformation we obtain:

$$\omega_1 = 1{,}000;\ \omega_{z1} = 0.8541 \text{ and } \omega_{z2} = 0.6682$$

Synthesis by NETSYN gives the semiredundant form of the S-plane prototype shown in Figure 3.16(a). Again, the asymmetrical distribution of capacitance causes an impedance level difference between source and load, but the ratio is less than that of the 2- to 4-GHz filter due to the wider bandwidth. Redistribution of capacitance for a symmetrical network gives the result shown in Figure 3.16(b), and the corresponding f-plane equivalent circuit is given in Figure 3.16(c). In the middle of the f-plane network, the line and stub values correspond to a coupling coefficient in the coupled-line sections of –6.0 dB, which is easily realizable. However, as in the case of the simple 3- to 9-GHz coupled-line filter, the coupling coefficient in the end sections is considerably higher, making realization difficult without overlapping striplines. To give the required coupling coefficient of –3.75 dB, the gap between the strips in the end sections must be as small as 40 μm. In the 34-μm thickness of the copper metallization, this is just beyond the capability of conventional wet etching techniques, but a gap of approximately the required dimension was eventually achieved using a combination of wet etching and laser trimming.

All three double-length shunt stubs of the f-plane network are easily realizable. The outer stubs are realizable without modification, but the center stub, owing to the low section impedances, has to be realized as a pair of identical double-length stubs in parallel, the second sections of which are split yet again into another pair of stubs connected in parallel (see Figure 3.13). Dimensions of all the stripline elements of the filter are given in Table 3.6.

Measured and theoretical frequency responses of the 2- to 6-GHz filter are plotted in Figure 3.17. As far as the position of the passband is concerned, there appears to be a very close agreement between measurement and theory, suggesting that the various sections of the filter have been fabricated with the correct length for a commensurate frequency of 4 GHz. Inside the passband, insertion losses at

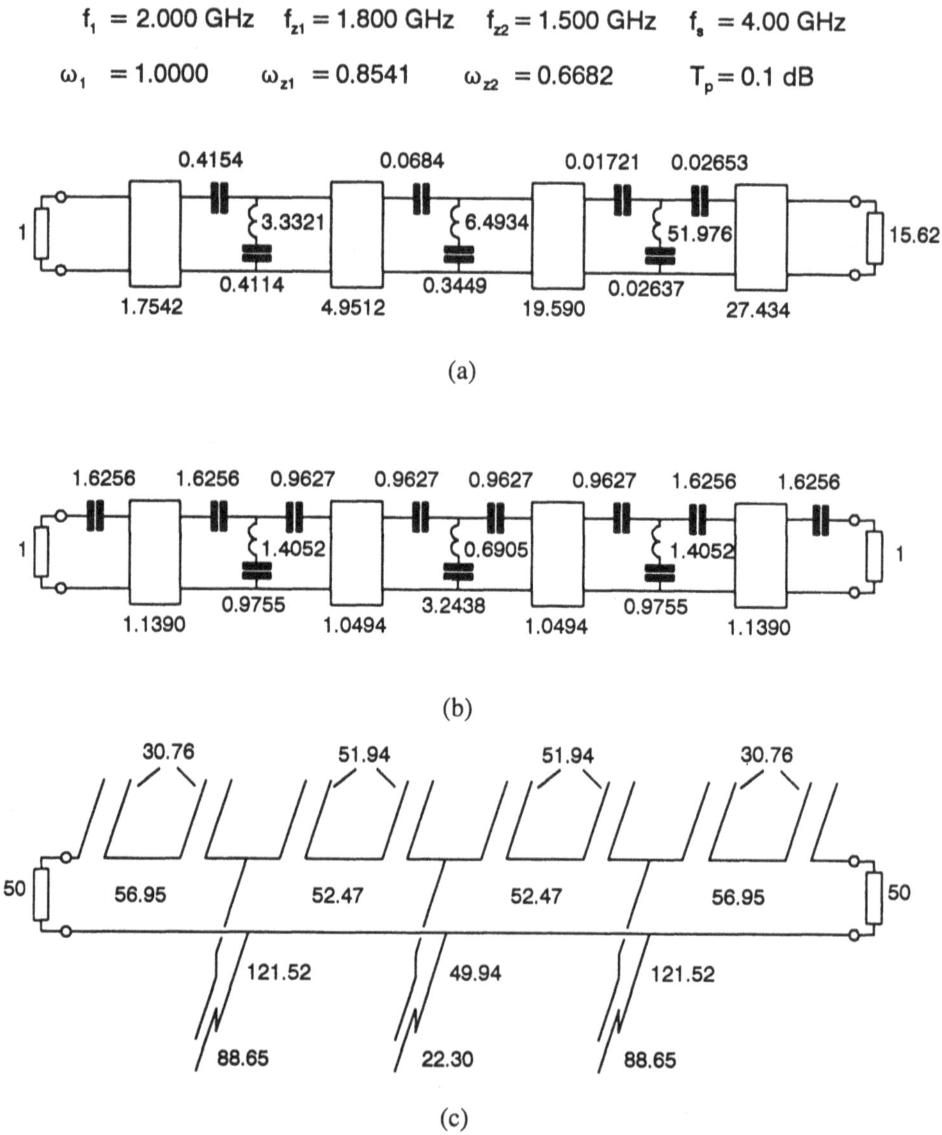

Figure 3.16 *S*-plane and *f*-plane prototypes for the 2- to 6-GHz enhanced coupled-line filter: (a) prototype after synthesis; (b) prototype after transformation; (c) *f*-plane equivalent circuit.

Table 3.6
Electrical and Physical Parameters for Sections of the 2- to 6-GHz Enhanced Coupled-Line Filter

Parameter	*Section*			
	Coupled Lines 1 & 7	*2-Section Stub 2 & 6*	*Coupled Lines 3 & 5*	*2-Section Stub (2 in ‖) & 4*
Z_{oo} (Ω)	30.76	—	519.4	—
Z_{oe} (Ω)	144.66	—	156.88	—
C_v (dB)	−3.75	—	−5.98	—
Z_{sec} (Ω)	66.71	121.52, 88.65	90.27	99.88, 89.20
Width (mm)	0.600	0.302, 0.734	0.440	0.545, 0.724
Separation (mm)	0.040	—	0.156	—

RT Duroid 5870:
- Dielectric constant = 2.32
- Conductor thickness = 0.034 mm
- Ground plane separation = 3.043 mm
- A quarter wavelength at 4 GHz = 12.31 mm

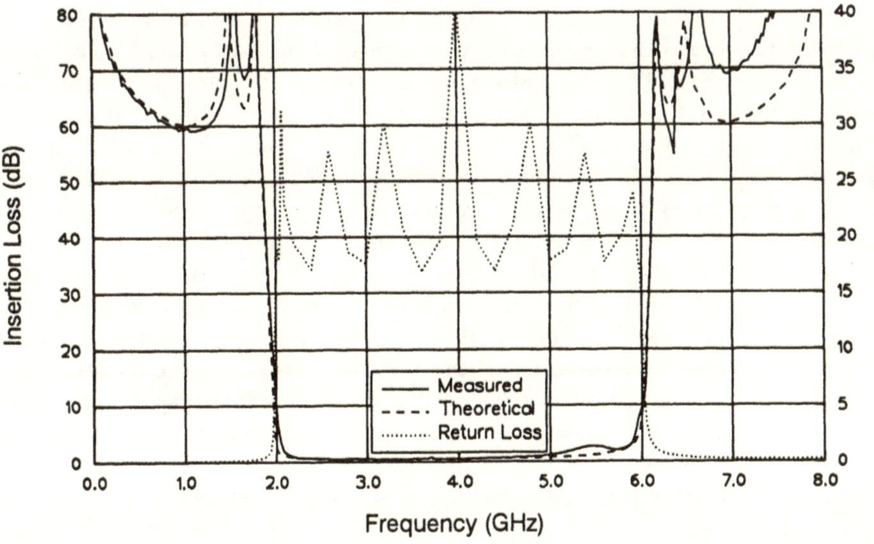

Figure 3.17 Measured and theoretical frequency responses of the 2- to 6-GHz enhanced coupled-line filter.

the center are a commendably low 0.5 dB, rising gradually toward the band edges before making a more abrupt rise close to the band edges. Some excess ripple can be seen in the measured response above 5 GHz. This is due to inadequate coupling in the outer coupled-line sections, despite the success in creating the 40-μm strip separation. One explanation may be that the coupling has been weakened by the presence of air instead of dielectric in the extremely narrow gap.

In the stopband below the passband, the ripples of the measured insertion loss are in extremely good agreement with theory, suggesting the transmission zeros of the shunt stubs are correctly placed. Above the passband, there has been some slight disturbance to the measured ripples caused by a very small displacement of the zeros. However, the effect is hardly of any consequence, and such is the selectivity of this filter that some misplacement of the zeros was to be expected.

The insertion loss immediately above the passband rises to a figure of over 70 dB at 6.175 GHz, where 6.175 GHz is only a 3% frequency displacement from the upper passband edge. A conventional coupled-line filter could not possibly have achieved this selectivity. Some modification, particularly in relation to the realization of the end-coupled sections is required to eliminate the raised passband ripple, but otherwise the filter has considerable potential.

3.2 NONCOMMENSURATE LINE FILTERS FROM BAND-PASS *S*-PLANE PROTOTYPES

As previously stated, several new classes of noncommensurate line filters were reported in [3] capable of providing both wide passbands and exceptionally wide stopbands. These new classes of filter were derived from band-pass *S*-plane prototypes, which, by the Richards Transformation, become band-pass *f*-plane prototypes whose transmission line elements are a quarter of a wavelength long at the center frequency of the stopband. The effect of the transformation on the frequency responses of the networks was illustrated in Figure 2.12(c) of Chapter 2. So long as the elements of the *f*-plane network remain of the same length and the commensurate frequency f_s is in the center of the stopband, the transformation is exact. In these circumstances, the main advantages of using a band-pass *S*-plane prototype are that the lengths of the transmission line elements are reduced for a given passband frequency range and there is almost complete freedom to specify independently the widths of the passband and the stopband. However, the use of a band-pass *S*-plane prototype also offers perhaps the most important practical advantage, which is the opportunity trade the length against the impedance of individual transmission line elements, and in some cases replace distributed elements with lumped elements. In so doing, a noncommensurate line network is created whose frequency responses can be held to within a close approximation of the exact responses. The basis for this was discussed in Section 2.4.3 of Chapter 2. In the filters to be described, some of the series transmission line elements have been realized as lumped series capacitors.

In the search for these new classes of filters, the general technical objective was to find structures that would be compact, inexpensive to manufacture, reproducible without the need for manual tuning, and versatile enough to be realizable for any fractional bandwidth in the range of 10% to 100%. Another important requirement was to be able to specify, semiarbitrarily, the width of the stopband. The relevant frequency range under consideration was 2 to 20 GHz. Within this general objective, special emphasis was to be given to filters with very wide passbands (e.g., $\delta f >$ octave), very wide stopbands, and extremely high rates of cutoff at the passband edges.

After consultation of the literature, no evidence was found of an existing filter structure that could meet satisfactorily this set of objectives. Of the many known filter structures, combline filters [9] were perhaps the most relevant, but these have been invariably realized in the form of 3-D cavity structures, which conflicted with the need for an inexpensive method of manufacture. This was another situation, therefore, in which design by synthesis provided the only likely design approach. It was necessary to invent and then synthesize some completely new structures, starting on the basis of employing only the most basic set of elementary circuit elements to be found in level 1 of the design database (see Chapter 2, Figure 2.2).

There were four new classes of filters introduced by [3], only one of which, the Class A type, will be described here. An explanation will be given of the general rationale behind the choosing of its topology and the synthesis of its *S*-plane prototype. A practical example will be described that has a 2- to 6-GHz passband, very steep passband skirts, and a passband extending to beyond 20 GHz.

3.2.1 Design Rationale

In view of the need for a low-cost manufacture, it was almost inevitable that the new classes of filters would be printed circuit structures. Triplate stripline was chosen as the preferred transmission line medium because it is relatively low-loss and supports a pure TEM mode of propagation. Suspended stripline could have been chosen if low losses had been an absolute first priority, but suspended stripline filters are significantly more expensive to manufacture in large quantities.

Printed circuit filters undoubtedly offer good reproducibility and low cost in volume production. Once the optimum layout has been determined and the photolithographic masks are available, large numbers of nearly identical devices can be fabricated. However, because of the relatively high initial costs of making the masks, it is important to achieve a final design with a minimum of iteration. It is, therefore, equally important to keep the circuit simple by striving to:

1. Avoid complicated sections consisting of multiple capacitively coupled strips;
2. Separate shunt or series elements with lengths of transmission line (unit elements in *S*-plane) so that each element corresponds to preferably one or no more than two transmission zeros.

This helps in the association of an error in performance with a particular circuit element and in the confident determination of the necessary modifications that must be made. It also ensures the most reliable stripline models are employed. In addition to keeping the circuit simple, short circuits should be avoided if at all possible, since good-quality short circuits are difficult to make and cannot be moved easily for the purposes of circuit tuning.

The above practical considerations have a strong influence on the choice of transmission zero locations. First, to provide the separation of shunt or series elements, approximately equal numbers of zeros at $S = (1, j0)$ (i.e., unit elements) and zeros on the $j\omega$ axis (i.e., reactance branches) should be specified. This also helps to keep the dynamic range of element values small. Second, to avoid short circuits and their corresponding shunt inductors in the S-plane network, only a single transmission zero may be specified at $S = (0, j0)$. Specifying a prototype of the first canonical form whose input impedance, $Z_1(S)$, tends to infinity at $S = (0, j0)$ will ensure that the only nonredundant high-pass element in the prototype is a series capacitor. There may be more than one series capacitor in the network, but these will all have been derived from the same nonredundant element. There would be no shunt inductors.

Applying these basic guiding principles, many different filter topologies were considered, most of which were physically undesirable. However, of those that were found to be practicable, perhaps the most attractive was the one called Class A, which has an S-plane prototype of the form shown in Figure 3.18. It consists of an alternating cascade of the two basic sections shown in Figure 3.18(b,c): the first, a

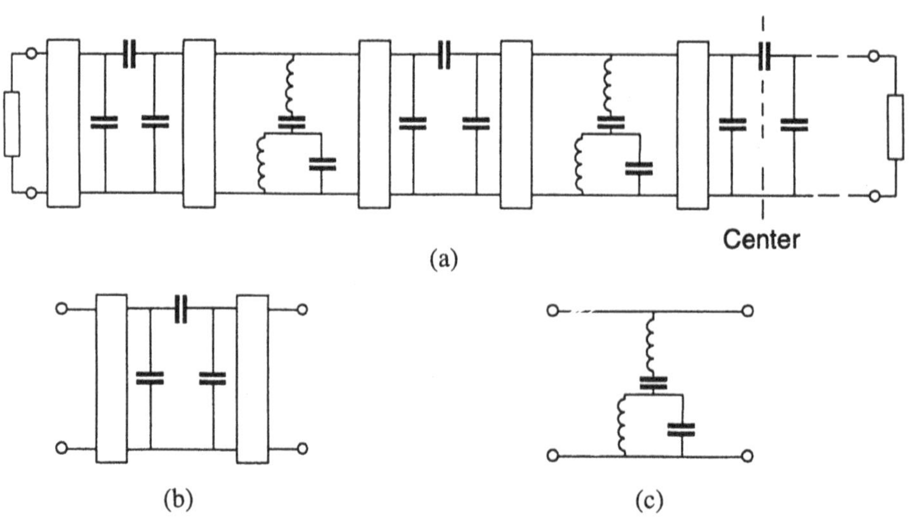

Figure 3.18 General form of the S-plane prototype for a Class A type noncommensurate line filter: (a) the prototype; (b) band-pass section; (c) fourth-order section.

band-pass section using two unit elements to provide two half-order transmission zeros at $S = (1, j0)$ plus a π section of capacitors contributing to single zeros of transmission at $S = (0, j0)$ and $S = (0, j\infty)$; the second, a fourth order section providing a pair of first-order $j\omega$ axis zeros, one at each side of the passband. In terms of transmission zeros, the prototype is specified thus:

$a = 1$	number of zeros at $S = (0, j0)$
$b = 1$	number of zeros at $S = (0, j\infty)$
$c = (p + 1) \times 2$	number of zeros at $S = (1, j0)$
$d = p$	number of zeros at $S = (0, j\omega_{z1})$
$e = p$	number of zeros at $S = (0, j\omega_{z2})$

where p is the number of fourth-order elements. This distribution gives the network a degree of $2 \times (3p + 2)$. As indicated by the set of transmission zeros, there are only single zeros at $S = (0, j0)$ and $S = (0, j\infty)$, which means that the prototype can contain only a single nonredundant high-pass element and a single nonredundant low-pass element. In the network of Figure 3.18, which is of the first canonical form, both these elements have been split into several different parts. The high-pass element, a series capacitor, has been split to form all the series capacitors of the capacitor π sections, and the low-pass element, a series inductor, has been split and transformed into all the shunt capacitors of the capacitor π sections.

Though not essential, it is strongly advisable to locate all the pairs of finite, nonzero transmission zeros (loss poles) at the same frequencies on both sides of the passband, since this leads to a smaller dynamic range of elements values. It also gives a nearly uniform distribution of element values throughout all except the end sections of the network. The precise location of the zeros should be chosen to be as close to the passband edges as is necessary to give the required skirt selectivity. Their number (i.e., the number of fourth-order sections) should be chosen to give the required depth to the stopband floor. Choosing the zeros with an equal displacement from the passband edges tends to assist with the realization of the fourth-order elements. However, the choice should also be made in the knowledge that the realization of the fourth-order elements becomes more difficult as the zeros in the S-plane move away from $S = (0, j1)$. Their realization becomes impossible for zeros specified below $S = (0, j0.2)$ due to the very large dynamic range of impedance values.

For the physical realization of a Class A type of filter, the unit elements of the S-plane network are transformed directly into lengths of transmission line a quarter of a wavelength long at the commensurate frequency. The capacitive π sections are transformed into coupled stripline elements of the form shown in Figure 3.19(a). When tight coupling is required, the total series coupling capacitance (C_2) is shared between the edges of the strips (i.e., the distributed fraction C_{ab}) and a lumped capacitor (C_s) mounted across the center of the section. The sharing of the coupling must be chosen to give an optimum combination of gap and capacitor dimensions. This fortuitous mixing of lumped and distributed elements is only

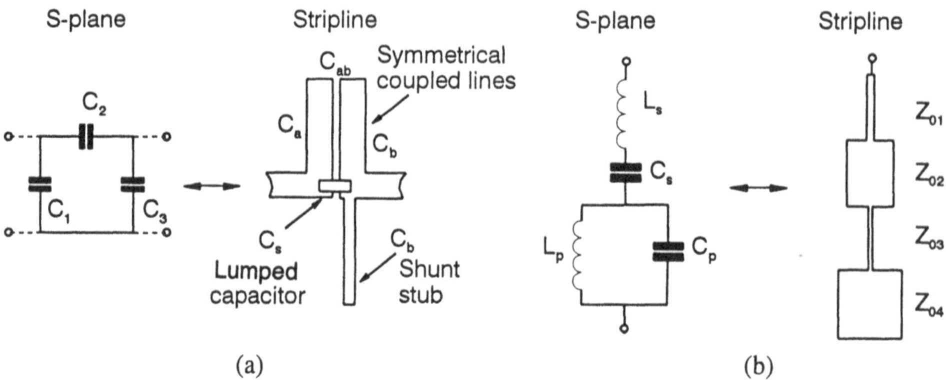

Figure 3.19 Stripline equivalences of a π section and fourth-order section: (a) a π section; (b) a fourth-order section.

possible because of the lumped character of the distributed elements, which are much shorter than a quarter of a wavelength at frequencies in the vicinity of the passband. The extra shunt stub in the stripline section is optional, allowing an asymmetrical capacitor π section to be realized without having to use an asymmetrical pair of coupled lines. As previously seen in Section 3.1.1.2, symmetrical pairs of coupled lines are to be preferred because of the superior physical models available.

Each fourth-order element can be realized using a stripline section of the form shown in Figure 3.19(b). This is an open-circuit shunt stub comprising a cascade of four transmission line sections of the same length but of different characteristic impedance. The formulas required to perform the conversion between the *S*-plane element and the stripline section can be found in Section A.2 of Appendix A. This realization of the fourth-order element is the one most likely to be used in practice. However, there is an alternative realization comprising a pair of open-circuit stubs in parallel, each one of which has only two transmission line sections instead of four. As seen in Section A.3 of Appendix A, the fourth-order *S*-plane element can be transformed into two second-order elements in parallel, which are in turn realizable as double-length open-circuit stubs (Section A.1, Appendix A). It is the separation of the relevant pairs of transmission zeros that determines the most suitable form of stub. For fractional separations of greater than 50% (i.e., passbands wider than 50%), the four-section stubs are more appropriate, especially if one of the zeros is located at a frequency close to the limit of $S = (0, j0.2)$. The double-section stubs are more appropriate for bandwidths less than 50%. It is also important to be aware that the commensurate frequency will influence the position and separation of the transmission zeros in the *S*-plane.

Broadly speaking, filters of this class are realizable for fractional bandwidths in the 50% to 100% range and for stopbands specified up to seven times the center frequency of the passband. However, generally the realization problem is eased as the specified stopband width decreases, and it may be possible to realize devices for bandwidths outside the above range if a small stopband width is acceptable. It should be noted that choosing a stopband width such that the second passband is as low as even three times the center frequency of the first could still be worthwhile, since the physical realization may be smaller or more desirable than that of a conventional filter design with the same performance.

3.2.2 A 2- to 6-GHz Band-Pass Filter with a 6- to 20-GHz Stopband

At the outset, the target electrical specification of the 2- to 6-GHz demonstrator filter was:

Insertion loss	<1 dB over the band 2 to 6 GHz;
Insertion loss	>65 dB within the bands 0 to 1.8 and 6.2 to 20 GHz;
Passband ripple	0.1 dB for >15-dB input return loss.

This is a particularly difficult requirement, since it calls for a low insertion loss of 1 dB at the passband edges as well as an extremely steep rate of cutoff. In fact, the specification is now known to be impossible to meet with a conventional dielectric filled triplate structure in which the unloaded Q of a resonator is only 300. Notwithstanding the difficulty, the design was attempted and the practical result is the filter whose stripline circuit has been photographed and shown in Figure 3.20.

As can be seen from the figure, the filter contains four of the four-section shunt stubs and five of the band-pass capacitively coupled sections. At the input on the left, the first of the capacitively coupled sections is a modified version of the section shown in Figure 3.19(a). It is an asymmetrical section in which one of the coupled lines has been completely eliminated (i.e., $C_a = C_{ab} = 0$) and all the series coupling is provided by the lumped capacitor C_s. The remaining half is then simply a pair of stubs of the same characteristic impedance wired in parallel (i.e., $C_b = C_b'$). The lumped capacitor and those capacitors throughout the rest of the filter are made from a gold foil cut to the appropriate size, bonded at one side of the gap between the striplines and insulated from the strip on the other side with an 8-μm-thick film of Kapton (a polyimide plastic with $\epsilon_r = 3$).

All of the inner capacitively coupled sections are symmetrical and use a combination of lumped capacitive coupling and distributed coupling between the strips. To ease realization, each of the coupled sections is actually a pair of coupled sections wired back to back, with the lumped capacitor and the external connections positioned along the center line. All of the shunt fourth-order stubs are identical. Each is an alternating cascade of high then low impedance transmission lines, the last line of which being of such low impedance that it has been necessary to use three transmission lines wired in parallel in the shape of a cross. This makes it

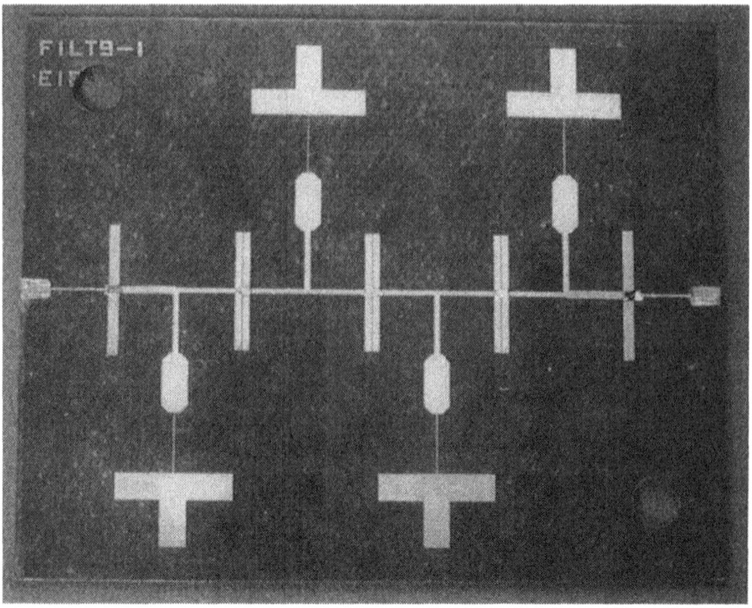

Figure 3.20 Printed circuit of a compact, noncommensurate line 2- to 6-GHz filter.

easier to correct for the discontinuities at the ends of the lines and at the junction with the high-impedance section. It can be seen that the low-impedance second section of the shunt branches has had the corners removed to compensate for the excess capacitance associated with the wide transverse edges of the section.

With four shunt reactance branches in the filter, the degree of the S-plane prototype must be 28 (i.e., $n = 2 \times (3p + 2)$, where $p = 4$). The passband, stopband, and transmission zero locations in both the f-plane and S-plane were specified as follows:

f-plane:

$f_1 = 1.950$ GHz; $f_2 = 6.050$ GHz; $f_s = 13.2$ GHz
$f_{z1} = 1.700$ GHz; $f_{z2} = 6.300$ GHz

S-plane (response-type band-pass):

$\omega_1 = 0.2363$; $\omega_2 = 0.8770$
$\omega_{z1} = 0.2051$; $\omega_{z2} = 0.9310$
$T_p = 0.1$ dB; $T_o = 0.0$ dB; $T_{sp} = 0.0$ dB

Transmission zero specifications:

1 transmission zero at $S = 0, j0$ (series capacitor)
1 transmission zero at $S = 0, j\infty$ (shunt capacitor)
10 transmission zeros at $S = 1, j0$ (unit elements)
4 transmission zeros at $S = 0, j0.2051$ (fourth-order reactance branches)
4 transmission zeros at $S = 0, j0.9310$ (fourth-order reactance branches)

Network degree: 28
Network type: doubly terminated, first dual

Synthesis by NETSYN of this degree 28 network directly from these specifications is possible in principle but prone to numerical problems in practice. A more reliable approach is to exploit the nearly uniform distribution of element values throughout most of the filter, synthesize a network of a much lower degree, and then replicate a suitable number of the inner sections. This is what was done to generate the symmetrical *S*-plane prototype of the 2- to 6-GHz filter, half of which is shown in Figure 3.21. The network was actually synthesized first with a degree of 16 (i.e., two shunt reactance branches) before doubling the number of band-pass and fourth-order stubs in the middle. There is very little perceptible error in the simulated frequency responses as a result of taking this approach.

Unfortunately, as was the case for the enhanced coupled-line filters of Section 3.1.2, it is not possible to synthesize the *S*-plane network for this type of filter using E-Syn.

The prototype shown in Figure 3.21 is not in the form that would usually be generated immediately after synthesis. For each of the fourth-order sections, NETSYN or similar software will extract two zero-shifting elements, a series capacitor and a series inductor, which would normally be extracted automatically just prior to the extraction of the fourth-order sections. To create the network of Figure 3.21, these elements must be moved to the left through the adjacent unit element with

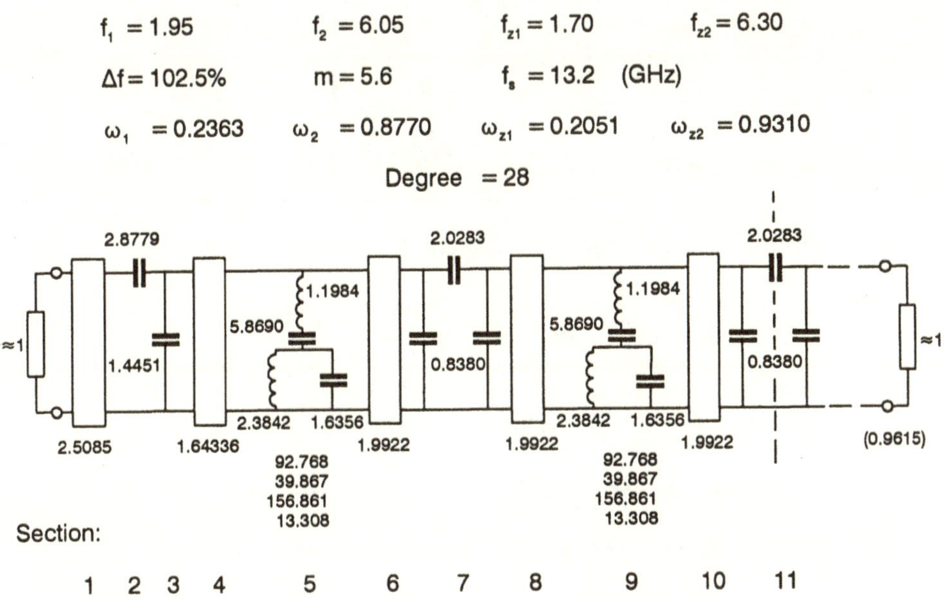

Figure 3.21 *S*-plane prototype of the noncommensurate line, 2- to 6-GHz filter.

Kuroda identities, where they become either the asymmetrical L section of capacitors at the input of the network or the symmetrical π sections of capacitors elsewhere. Hence, during the actual synthesis, the sequence of extraction is to remove repeatedly two unit elements followed by a fourth-order element (i.e., a pair of zeros at $S = (0, j0.2051)$ and $S = (0, j0.9310)$) until only single zeros at $S = (0, j0)$, $S = (0, j\infty)$, and $S = (1, j0)$ remain. Removing the zeros at $S = (0, j0)$ and $S = (0, j\infty)$ then creates the last L section of capacitors and the zero at $S = (1, j0)$ creates the last unit element. It should also be noted that the values of the elements of the network in Figure 3.21 have been scaled down slightly from 1 to 0.9615 to help with the physical realization of some of the high-impedance transmission lines.

Figure 3.22 contains details of all the f-plane and physical circuit elements of the 2- to 6-GHz filter, as derived from the S-plane network, after scaling for 50Ω terminations. As indicated, sections 1, 3, 4, 6, 8, and 10 (See both Figures 3.21 and 3.22) are essentially simple lengths of transmission line, the widths of the striplines

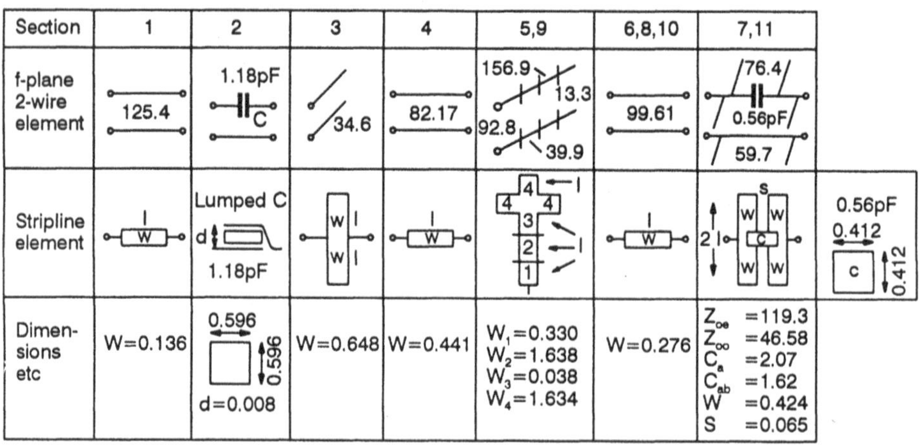

Commensurate length l = 3.730mm (ε_r = 2.32)

For transform into lumped C:- $C = \dfrac{Y_o \tan\left(\frac{\pi}{2}\frac{f_o}{f_s}\right)}{2\pi f_o}$

Material: RT Duroid 5870, thickness 0.793mm, 1/2 oz copper

t = 0.017 mm b = 1.522 mm 50 Ω line width = 1.165 mm

Kapton for capacitors: ε_r = 3.0 thickness (d) = 0.008 mm

Figure 3.22 f-plane and stripline elements of the noncommensurate line, 2- to 6-GHz filter.

having been calculated using Cohn [7]. Section 2 has been realized as a pure lumped capacitor with a capacitance value derived from the value of the corresponding *S*-plane capacitor using the simple relation given at the bottom (Note: Y_0 is the value of the *S*-plane capacitor ÷ 50). Sections 7 and 11 are coupled-line sections in which the required series reactance is shared between the lumped capacitor and the gap capacitance between the strips. Finally, sections 5 and 9 have been converted from the fourth-order elements of the *S*-plane prototype into the four-section shunt stub seen in the figure. Information concerning the material used for the construction of the 2- to 6-GHz filter is given at the bottom of Figure 3.22.

Measurements of the performance of the 2- to 6-GHz filter have been made using an HP8510 network analyzer. In order to be able to observe the width and depth of the stopband above 6 GHz, the analyzer was set to measure 801 frequencies over the 0.1- to 20-GHz frequency range. Plots of the measured and theoretical insertion loss responses are given in Figure 3.23. As indicated, the passband skirts of the filter are in precisely the correct position and the stopband ripples track the theoretical curve with very little deviation. This is good evidence of the correct placement of the transmission zeros on the two sides of the passband. In the center of the passband, losses are below 1 dB, while at 2 and 6 GHz, the losses have risen to approximately 3 dB. This is outside the original specification, but is consistent with the ohmic losses in the stripline. Also shown in Figure 3.23 is the theoretical return loss response of the filter. For clarity, the measured response is not shown but is generally in good agreement with theory. The minima of the return loss

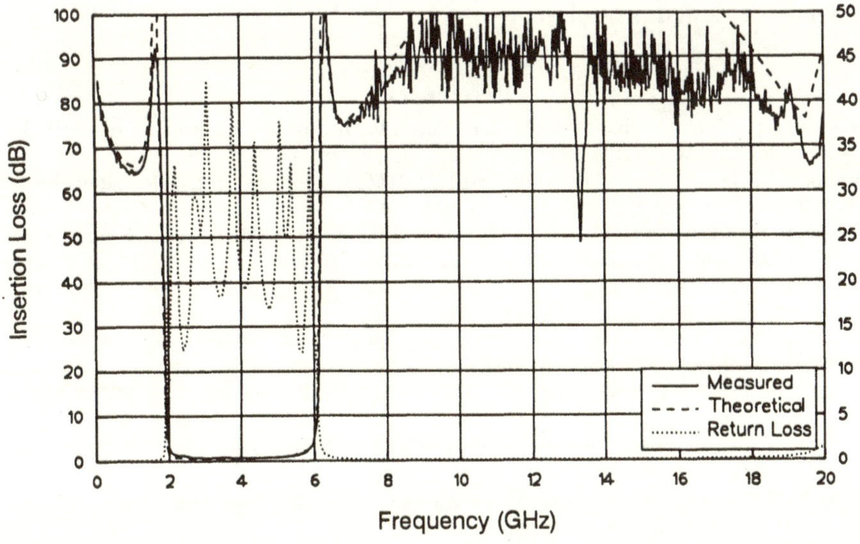

Figure 3.23 Measured and theoretical frequency responses of the noncommensurate line, 2- to 6-GHz filter.

ripples are higher than the value of 17 dB which corresponds to the 0.1 dB insertion loss ripple specified for the synthesis. This is one of the minor consequences of synthesizing a degree 16 network and then replicating the center sections.

Above 8 GHz in the stopband, it is not easy to follow the true stopband floor of the filter owing to the limitations of the network analyzer. However, except for a very narrow peak in the stopband at 13.2 GHz, the floor is generally in excess of 80 dB. The narrow peak in the stopband at 13.2 GHz is precisely at the specified commensurate frequency and is caused by slight discrepancies in the electrical lengths of some of the transmission line elements of the filter. Even if the physical lengths of the elements were exactly correct, some errors in their electrical lengths are to be expected due to spatial variations in the dielectric constant of the substrate material and the influence of frequency-dependent losses on phase velocity. This effect is common to most types of microwave transmission line filters and is nearly impossible to eliminate completely without some kind of built-in tuning. The peak is so narrow as to be unimportant for most applications. As Figure 3.23 shows, the stopband rejection remains in excess of 80 dB until, at a frequency of 18 GHz, it begins to fall toward a minimum of 68 dB at 19.5 GHz. The minimum corresponds to the lower stopband ripple associated with the next higher passband of the filter situated at just above 20 GHz.

The 2- to 6-GHz filter is an attractive component and has an impressive performance for an untuned printed circuit filter constructed in triplate stripline. Its small size of 46.0 × 36.0 mm, its ease of manufacture, its performance, and its unique circuit topology are yet more evidence of the versatility and general capability of the synthesis design procedure.

References

[1] Minnis, B. J., "Printed Circuit Coupled-Line Filters for Bandwidths up to and Greater Than an Octave," *IEEE Trans. MTT*, Vol. MTT-29, No. 3, March 1981, pp. 215–222.

[2] Minnis, B. J., "Printed Circuit Filters Covering 10%–100% Bandwidths," IEE Colloquium, *Microwave Filters*, Digest No. 1982/4, January 1982, pp. 3/1–3/4.

[3] Minnis, B. J., "Classes of Sub-Miniature Microwave Printed Circuit Filters With Arbitrary Passband and Stopband Widths," *IEEE Trans. MTT*, Vol. MTT-30, No. 11, November 1982, pp. 1893–1900.

[4] Matthaei, G. L., L. Young, and E. M. T. Jones, *Microwave Filters, Impedance Matching Networks and Coupling Structures*, New York: McGraw-Hill, 1964.

[5] HP-EEsof, *E-Syn (Series IV) User's Guide—Microwave and RF Design Synthesis*, Hewlett-Packard Ltd., Bracknell, Berkshire, U.K., February 1994.

[6] Cohn, S. B., "Shielded Coupled-Strip Transmission Line," *IRE Trans. MTT*, Vol. MTT-3, October 1955, pp. 29–38.

[7] Cohn, S. B., "Characteristic Impedance of Shielded Strip Transmission Line," *IRE Trans. MTT*, Vol. MTT-2, July 1954, pp. 52–55.

[8] Cohn, S. B., "Thickness Corrections for Capacitive Obstacles and Strip Conductors," *IRE Trans. MTT*, Vol. MTT-8, November 1960, pp. 638–644.

[9] Wenzel, R. J., "Synthesis of Combline and Capacitively Loaded Interdigital Bandpass Filters of Arbitrary Bandwidth," *IEEE Trans. MTT*, Vol. MTT-19, August 1971, pp. 678–686.

CHAPTER 4

Other Specialized Passive Components

In this chapter, exact network synthesis techniques will be applied to some useful passive components that would not normally be classified as filters. They are less complicated components than the high-selectivity filters that were the subject of the previous chapter, but nevertheless would be difficult to design by any single procedure other than exact synthesis. Section 4.1 will deal with the design of wideband bias Ts realized in triplate stripline and in microstrip transmission line media. Section 4.2 will deal with the design of novel wideband balun structures that were realized in GaAs MMIC technology. The work on both of these two types of components has been previously reported by the author in [1,2]. Finally, Section 4.3 will consider the design of wideband impedance transformers, examining the merits of the various topologies for impedance-transforming circuits, but without presenting any practical results.

With reference to the universal design procedure of Figure 2.2 in Chapter 2, design of all these three types of component involved the creation of completely new circuit topologies using mostly the elementary circuit elements contained in the lowest level of the design database. They were not knowingly based on any existing circuit configurations.

4.1 WIDEBAND BIAS Ts

Bias Ts are used for feeding direct current (dc) into active RF components in such a way that the RF behavior of the component is not adversely affected by the dc connection. They behave as diplexers, which, in the ideal case, comprise a pair of low-pass and high-pass filters with a crossover frequency somewhere just above zero (dc). Large current capacities and broad bandwidth capabilities have, in the past, been difficult to achieve without using elaborate, expensive methods of construction. However, the design described herein is capable of handling dc in excess of 3A and RF bandwidths of over a decade. Operating frequencies can reach as high as 50 GHz. The bias T is suitable for realization in microstrip or triplate stripline and can be constructed as an inexpensive individual component or, alternatively, can be embedded to form part of a more complex MIC. Two practical bias T designs will be described that cover the respective frequency bands of 1.8 to 18.2 GHz and 4.5 to 45.5 GHz.

4.1.1 General Bias T Circuit Concepts

The basic circuit concept of any bias T is that of a diplexer comprising low-pass and high-pass filters wired in parallel at one end, as illustrated in Figure 4.1(a). To achieve wide-bandwidth performance, it has been common practice for designers to concentrate on only the low-pass filter of this combination. Often the high-pass filter would comprise only a series capacitor, while the low-pass filter would be a rather more complicated mixture of resistors, capacitors, inductors, and transmission lines. For modest bandwidths, an example of a simple bias T that has seen widespread use is the stripline circuit illustrated in Figure 4.1(b). In this circuit, the low-pass filter is not significantly more complicated than the high-pass filter and is just a high-impedance transmission line decoupled to ground at its remote end by a shunt capacitor. It is the maximum realizable impedance of the line that is the chief limitation on bandwidth. Bandwidths of an octave or less are typical for this type of bias T.

In creating a new bias T design, the aim was to produce a bias T with decade bandwidth performance at microwave frequencies, which would be realizable in the form of a printed circuit. This has been achieved by giving attention to the design of the diplexer as a whole, instead of concentrating on just the low-pass filter. The conceptual changes to the filter combination required are indicated in the functional diagram of Figure 4.2(a). Most significantly, a band-pass filter has been added, which, although shown as a pair of filters in the diagram, is a single filter tapped at the center for the connection of the low-pass filter. The tapping point is nearly always in the center, but can, if necessary, be elsewhere along the length of the band-pass filter, including at either of the two ends.

One attractive stripline realization of the basic concept of Figure 4.2(a) is illustrated in Figure 4.2(b), which bears a passing resemblance to the simple circuit of Figure 4.1(b), but instead of connecting the first line of the low-pass filter directly in shunt with a uniform transmission line, the line is connected at the center of a cascade of striplines of varying width, which together with the line form a band-pass filter. The line now behaves like a short-circuit shunt stub within the band-pass filter itself. The improved bandwidth capability results from a lowering of the impedance level in the vicinity of the shunt stub by the cascaded transmission lines, thereby reducing the shunting effect of the stub. As the number of stripline sections

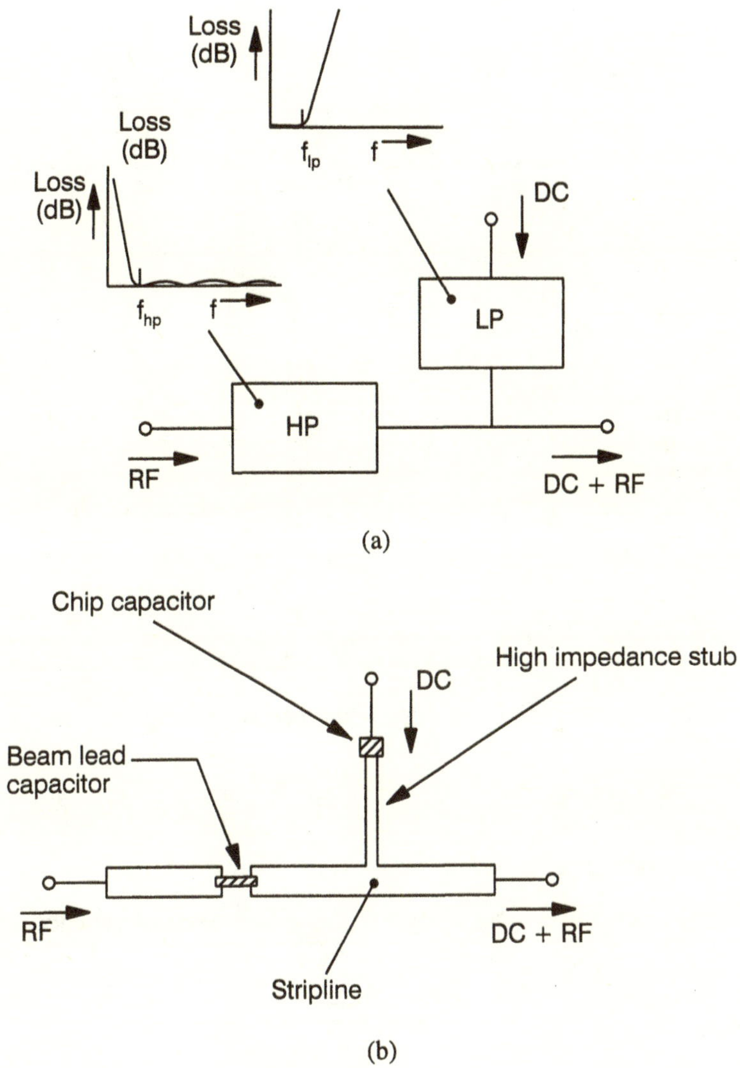

Figure 4.1 A simple bias T configuration: (a) basic concept; (b) stripline concept.

is increased, so the impedance level is lowered, and the bandwidth is therefore increased.

4.1.2 Band-Pass Filter Synthesis

For the purposes of synthesizing the band-pass filter of the modified bias T, it is assumed that within the passband of the filter, the series capacitor providing the

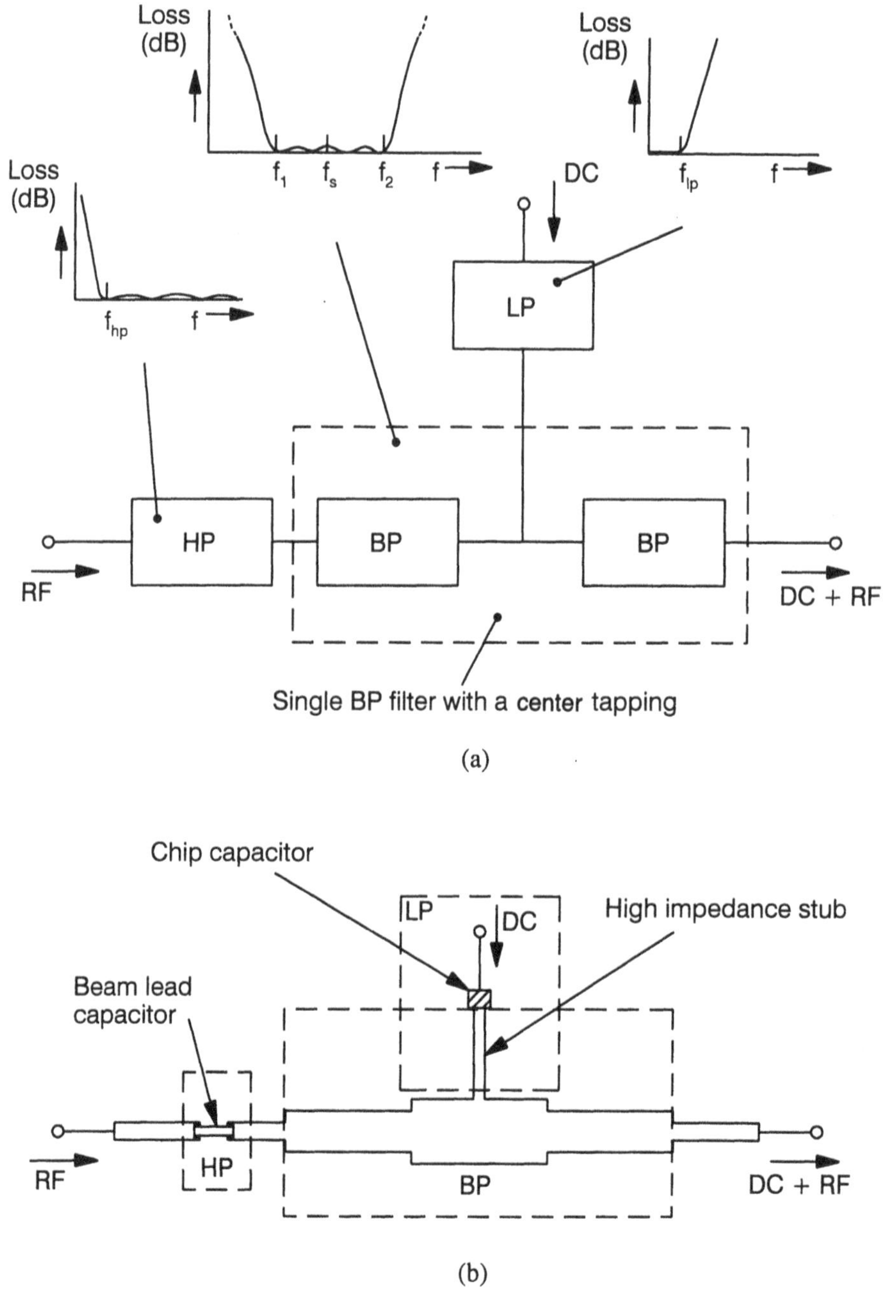

Figure 4.2 A wideband bias T concept employing band-pass filtering: (a) functional block diagram; (b) stripline realization.

dc break and the shunt capacitor at the end of the shunt stub behave like perfect short circuits. On this assumption, the *f*-plane equivalent circuit of the filter then comprises a cascade of ideal transmission lines and a single short-circuit shunt stub, as shown in Figure 4.3(a). All the lines and the stub have an electrical length equal to a quarter of a wavelength at the center frequency of the passband. Line impedances decrease toward the position at which the shunt stub is attached, although the shunt stub itself has an impedance considerably greater than that of the adjacent line. As previously mentioned, it will be usual to position the stub in the middle, since this gives equal source and load impedances. If, however, an impedance level transformation is required, then the stub can be moved away from the center toward one or the other of the terminations.

The *f*-plane frequency response of the band-pass filter is shown in Figure 4.3(b), where the frequencies f_1 and f_2 represent the edges of the passband. The commensurate frequency f_s is in the center of the passband, and the frequency $2f_s$ marks the center of the upper stopband. For any given value of f_s, the widths of the upper and lower stopbands are determined by the width of the passband, and in the limit, the stopband widths tend to zero as the passband width tends to the maximum fractional value of 200%. Clearly, this bandwidth can never be achieved in practice, and a more realistic limit is nearer 185%, with decade bandwidths (164%) being readily achieved.

Specifying a large bandwidth has the consequence of increasing the impedance of the shunt stub. However, a compensating decrease in the stub impedance can always be obtained by increasing the number of transmission line elements. An optimum number of elements is the smallest number that makes the stub readily realizable for the chosen frequency bandwidth. In practice, there are two main factors that eventually impose a limit on bandwidth. The first concerns the aspect ratio of the lowest impedance lines, which eventually becomes so small that the lines no longer behave like transmission lines. For correct behavior of the lines, it is a good rule of thumb that their lengths should always be greater than their widths (i.e., the aspect ratios should always be greater than 1). The second factor concerns the lumped capacitors. Large capacitors are needed to allow the band-pass filter to function properly at low frequencies, while at very high frequencies the parasitics of the capacitors must not become too dominant. Unfortunately, these two requirements are often mutually exclusive, and ultimately the bias T performance will depend heavily on the quality of the capacitors available.

In the *S*-plane, the equivalent circuit of the band-pass filter is a cascade of unit elements with a shunt inductor in the center, as shown in Figure 4.3(c). As indicated in Figure 4.3(d), the *S*-plane frequency response is high-pass with a passband edge at ω_1, where $\omega_1 = \tan((\pi/2) \times (f_1/f_s))$. It may not have escaped the reader's notice that these *f*-plane and *S*-plane prototypes are the same as those of the line-and-stub filter described in the previous chapter, before the splitting of the stub and the application of an admittance matrix transformation.

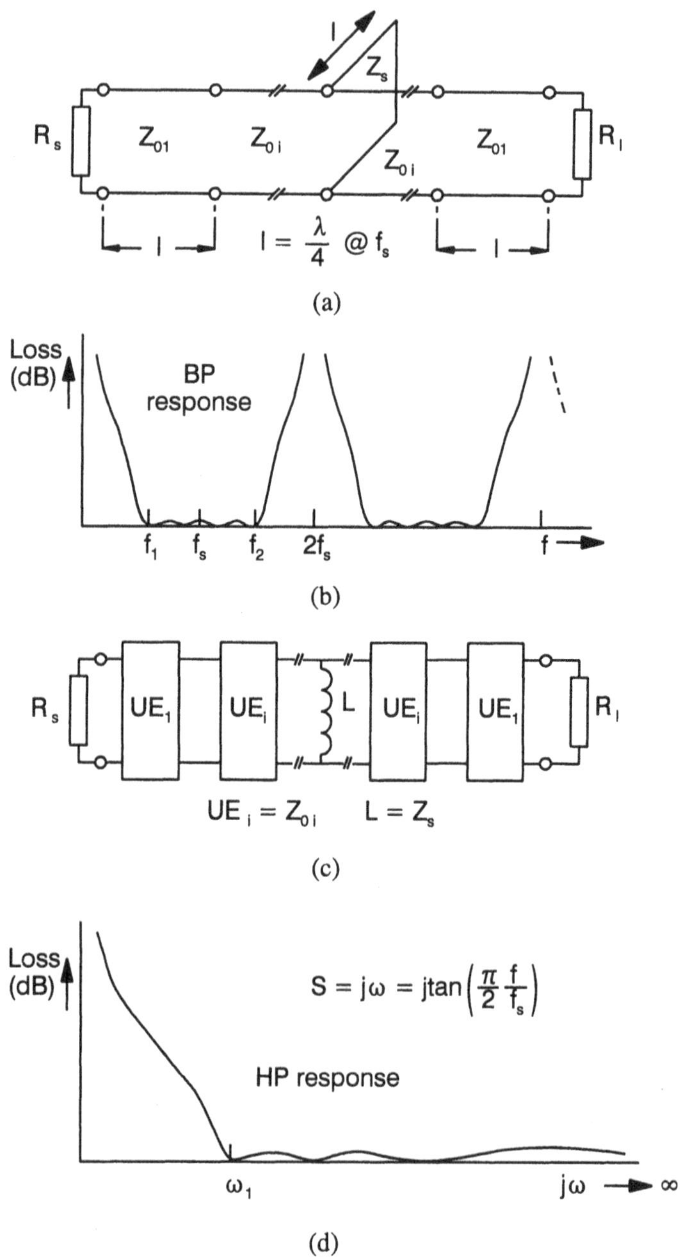

Figure 4.3 *f*-plane and *S*-plane equivalent circuits: (a) *f*-plane equivalent circuit; (b) filter response in *f*-plane; (c) *S*-plane equivalent circuit; (d) filter response in *S*-plane.

Synthesis of the *S*-plane high-pass prototype is straightforward using NETSYN or E-Syn. The essential information for the synthesis comprises the cutoff frequency ω_1, which is derived from the *f*-plane bandwidth requirement, the passband ripple value (T_p), and the number of unit elements. The degree of the filter is equal to the number of unit elements plus 1. For equal terminating impedances, the inductor is extracted in the center and the degree of the filter must be odd. The optimum number of elements (i.e., degree) is that which gives both the required bandwidth and a realizable set of element values and is easily established with a little iteration.

4.1.3 A 2- to 18-GHz Bias T in Stripline

As the first demonstration of the capability of the wideband design concept, a bias T operating over the 2- to 18-GHz band has been constructed. To allow for manufacturing tolerances, the band-pass filter was synthesized for an *f*-plane passband width of 1.8 to 18.2 GHz, which as a width ratio is more than 10:1 and as a fractional bandwidth (i.e., $2(f_2 - f_1)/(f_2 + f_1)$) is 164%. For the network synthesis, a high-pass equal-ripple response was selected with T_p set to 0.1 dB, corresponding to a worst-case input return loss of 16.43 dB. No passband offset or slope was required, so the parameters T_o and T_{sp} were set to zero. The resulting *S*-plane prototype synthesized using NETSYN is shown in Figure 4.4(a). After scaling element values for 50Ω terminations, the *f*-plane equivalent circuit becomes that of Figure 4.4(b), in which the six transmission line elements, three on each side of the central stub, reduce the impedance of the central stub to a realizable value of around 100Ω.

The relevant parameters for synthesis by E-Syn are given in Table 4.1. It should be noted that E-Syn will extract the required network working in either manual or automatic modes.

The bias T was constructed using RT Duroid 5880 material of dielectric thickness 0.25 mm and copper cladding thickness of 0.017 mm. The dielectric constant for this material is 2.2. Given the commensurate frequency of 10 GHz, the commensurate length of all the transmission line elements must be 5.056 mm (i.e., a quarter of a wavelength) before the effects of any physical discontinuities can be taken into account. Line widths were calculated using Cohn [3]. A length increase of 0.475 mm was applied to the shunt stub to take account of the reference plane shift caused by the T junction connection of the stub with the center section of the filter. All the relevant physical dimensions of the transmission line elements are given in Figure 4.4(c).

A picture showing a pair of the 2- to 18-GHz bias Ts is given in Figure 4.5. One of the bias Ts is shown fully assembled, while the other has been temporarily split down the center to reveal the printed circuit. Apart from minor differences between the types of coaxial connectors used, the two bias Ts are otherwise identical. Their printed circuits were fabricated using conventional photolithographic and

f_1 = 1.8 GHz f_2 = 18.2 GHz f_s = 10 GHz

ω_1 = 0.2401 T_p = 0.1 dB T_o = 0.0 dB T_{sp} = 0.0 dB

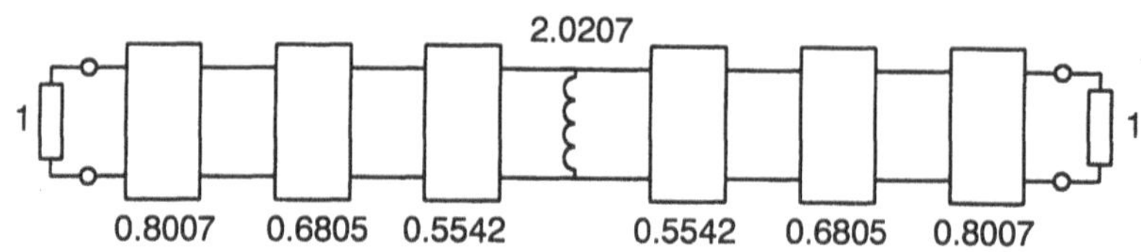

(a)

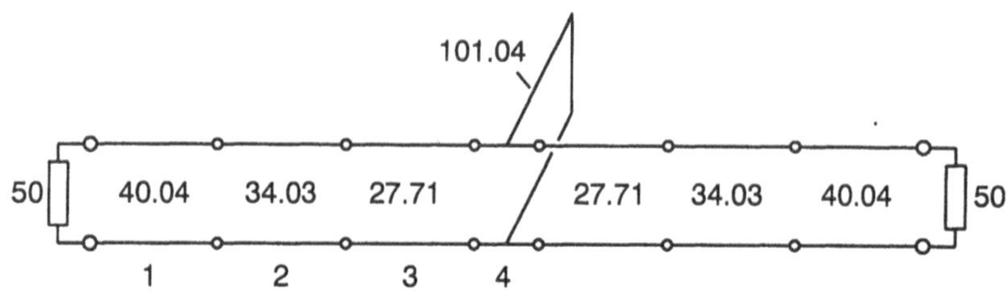

(b)

	Physical elements			
Section	1	2	3	4
f-plane element	40.04	34.03	27.71	101.04
Stripline element	l, w	l, w	l, w	w, l
Physical dimensions	w =0.555 l =5.056	w =0.697 l =5.056	w =0.913 l =5.056	w =0.082 l =5.531

Material: RT Duroid 5880, 0.25mm thickness, ½ oz copper, ε_r=2.2, t=0.017 mm, b=0.525 mm

(c)

Figure 4.4 Equivalent circuits and physical dimensions for a 2- to 18-GHz bias T: (a) *S*-plane equivalent circuit; (b) *f*-plane equivalent circuit; (c) circuit dimensions.

Table 4.1
Synthesis of the 2- to 18-GHz Bias T Prototype by E-Syn

Network type	Distributed	Tchebycheff	QW-band-pass	
Passband	f_1 = 1.8 GHz	f_2 = 18.2 GHz	T_p = 0.1 dB	MIL = 0 dB
Terminations	R_s = 1	R_l = 1	(only relevant for auto mode)	
Elements	Six transmission lines	One stub		
Mode (F0)	Manual or automatic extraction			
NxtEL (F6)	Use successively to extract next element from choice given			
Analyze (F0)	Analysis after extraction			

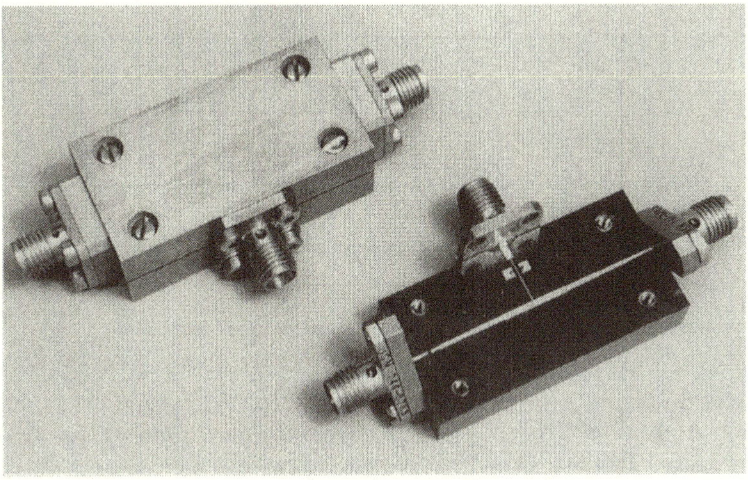

Figure 4.5 A 2- to 18-GHz bias T in triplate stripline.

etching techniques. On examination of the circuit pattern of the bias T with the top half removed, it should be apparent that RF would normally be applied at the left coaxial connector, which is isolated from dc by the series capacitor. DC would be fed into the bias T via the side connector and then both RF and dc should emerge from the coaxial connector on the right.

Three beam-lead capacitors have been mounted on the circuit. One of the capacitors is mounted across a gap in the track close to one end of the bias T, acting as the dc break (i.e., the high-pass filter). The other two capacitors are wired in parallel and act as the RF short circuit on the end of the shunt stub (i.e., part of the low-pass filter). Each of these two shunt capacitors is connected to the stub and to an adjacent copper pad earthed with a gold foil passing through the circuit board. Clearance holes have been provided at appropriate positions in the upper dielectric board for all three capacitors.

Measurements of the RF performance of the 2- to 18-GHz bias T have been made with an HP8510 network analyzer. Results are plotted together with the theoretically predicted responses in the graph of Figure 4.6. Over the band 2 to

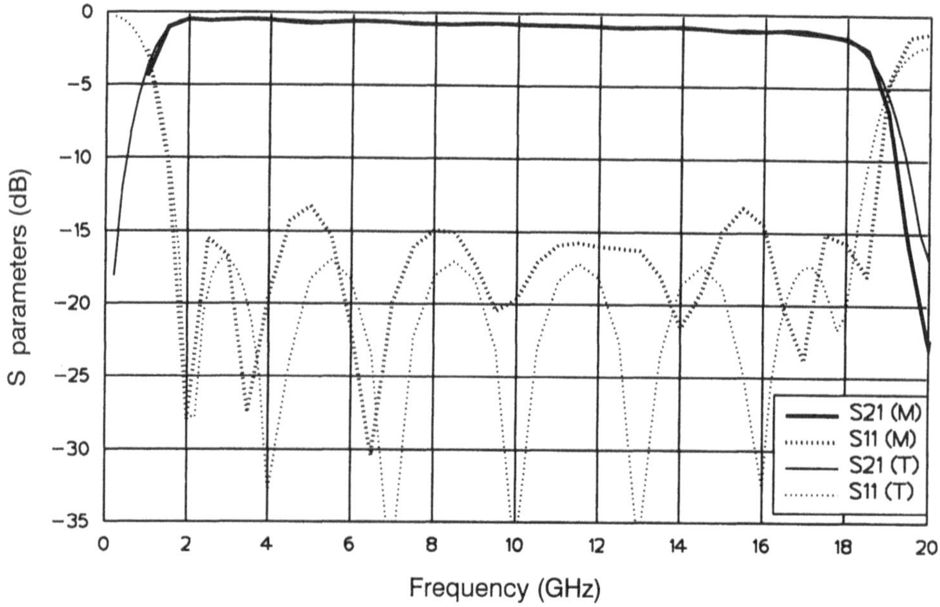

Figure 4.6 Measured and predicted frequency responses of the 2- to 18-GHz bias T.

18 GHz, the measured transmission coefficient (S_{21}) falls gently from −0.5 dB at 2 GHz to −1.6 dB at 18 GHz, in excellent agreement with theory. The passband of the bias T seems to have been realized with precisely the correct position and bandwidth. Measured peaks in the reflection coefficient (S_{11}) are typically −15 dB with worst-case peaks of −13.5 dB at 5 and 15.5 GHz. Compared with theory, the peaks are a little high, and there has also been some shift in the position of the peaks. This slight distortion in the S_{11} response is caused almost entirely by the interactions between the discontinuities of the coaxial-to-stripline transitions at the two ends of the bias T. It is an effect that would be difficult to correct without refinement of the transitions, but is, in any case, relatively insignificant. Currents in excess of 3A have been passed through the bias T, and the isolation between the dc and RF ports was found to be better than 30 dB over the 2- to 18-GHz frequency range.

4.1.4 A 4.5- to 45.5-GHz Bias T in Microstrip

In a second application of the design concept, a bias T was constructed covering the 4.5- to 45.5-GHz band. This was a scaled version of the 1.8- to 18.2-GHz bias T, the center frequency of the passband having been moved to the higher value of 25 GHz. No second synthesis was therefore needed, since the *f*-plane equivalent circuit is identical to that of the 1.8- to 18.2-GHz design except for the shorter commensurate length of the lines.

A photograph of a microstrip version of the 4.5- to 45-GHz bias T is given in Figure 4.7. The figure illustrates the microstrip circuit, the use of spark plug–type coaxial-to-microstrip transitions for the RF connections, and the provision of a large copper pad for the dc connection. Because of the higher operating frequency range, it was necessary to use a thinner (0.125 mm) dielectric substrate to maintain satisfactory aspect ratios for the transmission line elements. An MIM chip capacitor has been let into a hole in the substrate to make the RF ground connection for the shunt stub. Another MIM capacitor was let into a hole in the dc connection pad to provide some additional RF isolation. The dc break was a beam lead capacitor of the same type used in the previous 2- to 18-GHz bias T.

RF performance of the bias T has been measured using an HP8510 network analyzer operating with a 45-MHz to 40-GHz coaxial test set. Measured and theoretical frequency responses are plotted in Figure 4.8. As with the 2- to 18-GHz bias T, there appears to be excellent agreement between theory and practice for the S_{21} response. The transmission coefficient (S_{21}) at 4.5 GHz is approximately −0.6 dB, falling to −1.8 dB at the 40-GHz limit of the measurement range. The bottom edge of the passband is correctly positioned, and in the absence of measurement data, there no reason to suggest that the top edge is not also positioned at the correct frequency. Generally, the measured peaks in the reflection coefficient (S_{11}) are of

Figure 4.7 A 4.5- to 45.5-GHz bias T in microstrip.

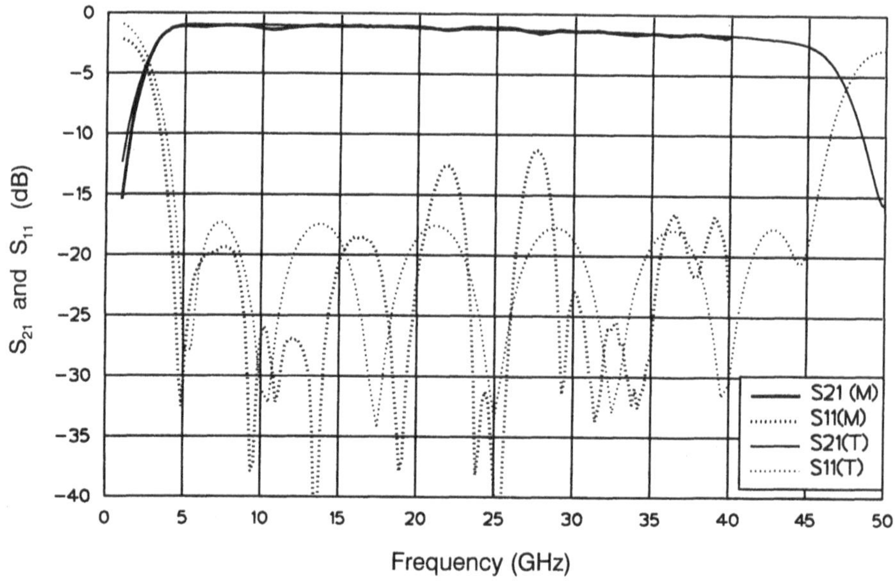

Figure 4.8 Measured and predicted frequency responses of the 4.5- to 45.5-GHz bias T.

the same level as predicted by theory except for the two peaks in the middle of the frequency range at about −13 dB. Once again, the disturbance seen in the measured S_{11} response is largely due to interactions between the coaxial connectors at the two ends of the bias T.

The current carrying capacity of this bias T was measured to be in excess of 2A.

4.2 WIDEBAND BALUN STRUCTURES

Baluns are an important group of components which are used in circuits where a transition between unbalanced and balanced modes of excitation is required. In the context of MICs, where a ground plane is invariably present throughout, they are best thought of as 3-dB power splitters whose two outputs are 180 degrees out of phase.

Baluns can be either active or passive structures, although the two types to be described in this section are both passive. They are intended primarily for use in MMICs on GaAs and offer important performance advantages over alternative active designs. Apart from requiring no dc supply, one of the most important advantages is that of power-handling capability. They are able to handle several watts of RF power and could therefore be considered for use as power combiners in push-pull power amplifiers. In fact, it was research into push-pull power amplifiers that provided the original stimulus for the investigation into these baluns. Both types of baluns can be designed for frequency bandwidths in excess of an octave anywhere in the frequency range of 1 to 20 GHz. As well as their power and phase-splitting capabilities, they are able to incorporate large impedance level transformations between input and output terminals if required.

As an introduction to the their basic circuit concepts, ideal equivalent circuits of the two baluns are presented in Figures 4.9 and 4.10, respectively. The cylindrical

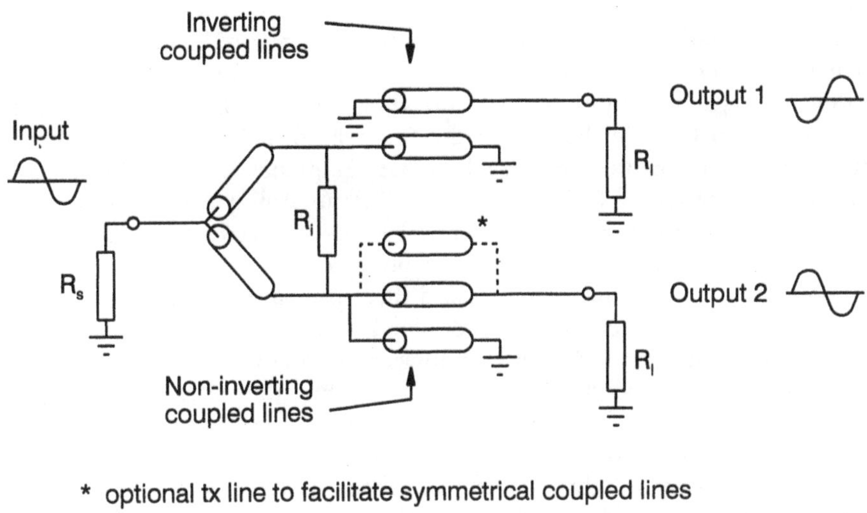

Figure 4.9 Transmission line configuration of the high-pass balun.

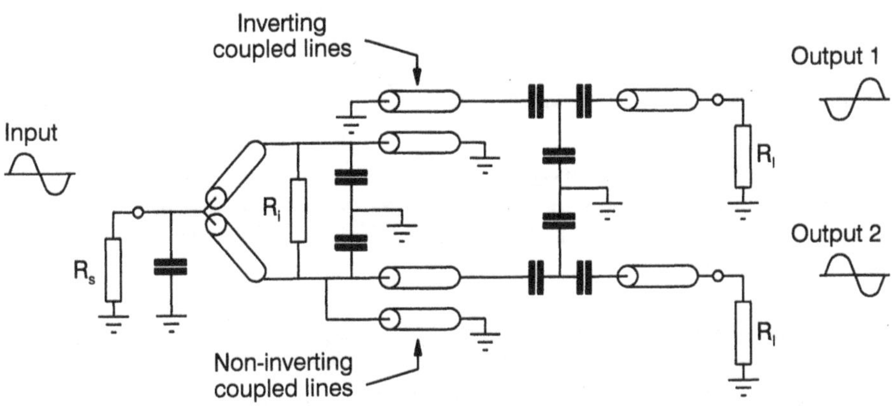

Figure 4.10 Basic lumped/distributed equivalent circuit of the compact band-pass balun.

symbols are intended to represent ideal transmission lines. With the exception of a lumped resistor, the first of the two circuits (Figure 4.9) is a purely distributed structure and is most suited to operating over fractional bandwidths of more than 100%. The second, in Figure 4.10, is a mixed lumped/distributed structure, more suited to operating bandwidths of an octave or less but also more suited to providing large impedance transformation ratios. Because of the use of lumped elements, the second is generally the smaller of the two types of baluns, but as a result of extensive folding of transmission line elements, both structures can be highly compact.

Fundamentally, each type of balun comprises a pair of band-pass filters connected in parallel at a common junction close to the input port. The two filters are identical, except that one of them introduces an extra 180-degree phase shift. As will be shown later, in the case of the first type of balun, the band-pass filter relates to an *S*-plane prototype with a high-pass frequency response. In the case of the second, the band-pass filter relates to an *S*-plane prototype with another band-pass response. For this reason they are referred to as *high-pass* and *band-pass* types, respectively. In both types, 180 degrees of phase shift is achieved by splitting the input signal into two identical samples and then inverting one of the samples. For this purpose, the baluns contain two coupled-line sections of the types illustrated in Figure 4.11. One is present in the noninverting arm of the balun and the other in the inverting arm. The two sections have the same transmission line equivalent circuits except that in the case of the first section there is an additional ideal transformer with a turns ratio of −1:1. The transformer gives the 180-degree phase shift, which, in theory at least, is entirely frequency-independent. In both types of balun, the resistor R_i (see Figures 4.9 and 4.10) is a balancing resistor that dissipates any unwanted antiphase voltages near the input port, thereby ensuring a good match at the two output ports. Some antiphase voltages near the input must always be anticipated in practice due to the effects of circuit parasitics and fabrication tolerances.

4.2.1 A High-Pass Balun for 6 to 18 GHz

The high-pass balun can be designed for various instantaneous frequency bandwidths, but it is best suited to passband widths of between 2:1 and 4:1. A 6- to 18-GHz version of the balun has been designed and constructed and this will be used as a design example.

Except for the −1:1 transformer of the inverting arm, both arms of the high-pass balun of Figure 4.9 have the same *f*-plane equivalent circuit comprising a simple length of line followed by a line-and-stub section of the type already seen in Figure 4.11. Since the two arms are wired in parallel at the input, the *f*-plane equivalent circuit must have a 2:1 ratio of source-to-load impedance if the balun is to have equal terminations on all ports. All the transmission line elements of the

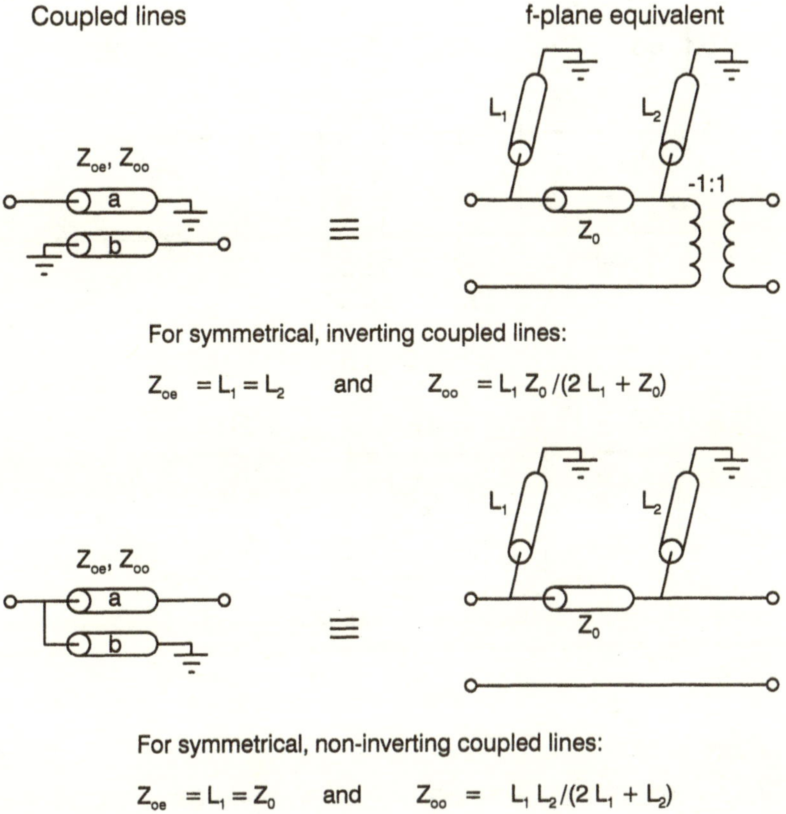

Figure 4.11 Basic coupled-line structures providing 180-deg phase shift.

equivalent circuit must be a quarter of a wavelength long at the center of the passband, which for a 6- to 18-GHz bandwidth is at a frequency of 12 GHz.

For synthesis of the *S*-plane prototype, the response type is specified high-pass, with a cutoff frequency of $\omega_1 = 1$ rad/sec. To ensure an input match of better than −15 dB, the insertion loss ripple (T_p) is set to 0.1 dB. There must be two transmission zeros at $S = (1, j0)$ corresponding to two unit elements and a single transmission zero at $S = (0, j0)$ corresponding to a pair of shunt inductors, one of which is redundant. The degree of the network is 3 and the presence of the shunt inductors forces it to be of the second canonical form.

After synthesis by NETSYN, the nonredundant network takes the form shown in Figure 4.12(a). It should be noted that for the purposes of element extraction, the resistor on the right with the value of 1Ω is the source resistor and the resistor on the left is the load. As indicated, the load resistor on the left has a value considerably greater than the required value of 2. To lower the value, the inductor

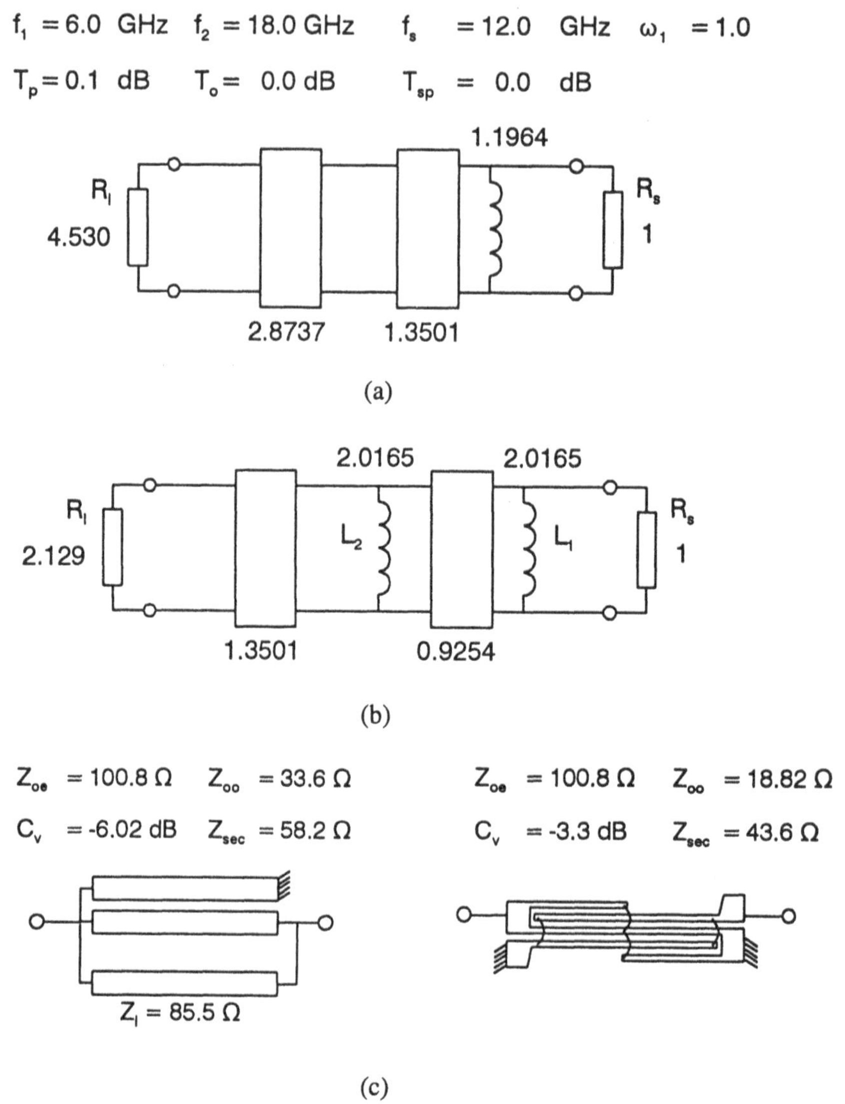

Figure 4.12 *S*-plane prototypes and coupled-line sections for the 6- to 18-GHz balun: (a) nonredundant *S*-plane prototype; (b) redundant *S*-plane prototype; (c) coupled-line sections.

must be split and a suitable part moved to the left through the adjacent unit element with a Kuroda Identity. After splitting to give inductors of the same value, the transformed equivalent circuit is that shown in Figure 4.12(b). By some good fortune, the passband width and ripple values are such that this splitting of the inductor has also produced almost exactly the required load resistor. It is important to obtain inductors of equal value, since this allows at least one of the coupled-line sections of the balun to be symmetrical. Had the load resistor not been of the required value, some modest adjustment would have been attempted by changing the width of the passband and/or the passband ripple value rather than alter the inductor values.

The parameter set for synthesis of Figure 4.12(a) by E-Syn is given in Table 4.2. A further point to note, however, is that the semiredundant network of Figure 4.12(b) can also be synthesized directly by the software if a partial instead of a complete element extraction is performed for the first inductor (i.e., the shunt *S/C* stub with E-Syn). This would avoid any need to use Kuroda Identities. With E-Syn, partial element extraction involves using the NxtEV function (shift F6) and specifying the required stub value. The only slight difficulty is the need to know the correct element value beforehand. However, even if the correct value is not already known, it can be quickly found with a little trial and error.

The rightmost unit element and the pair of surrounding inductors in Figure 4.12(b) correspond to the two coupled-line sections in the real balun. For the inverting section with short circuits on alternate ends of the two lines, the elements L_1, L_2, and the unit element relate to the odd and even mode impedances of the coupled lines by the formulas given in Figure 4.11. Making $L_1 = L_2$ gives a symmetrical coupled-line section. For the noninverting coupled-line section with a short circuit on only one end of one of the lines, the second set of formulas in Figure 4.11 gives the relevant odd and even mode impedances. In this case, equal values of L_1 and L_2 do not naturally produce a symmetrical coupled-line section. However, as previously indicated in Figure 4.9, it may be possible to make the section symmetrical by connecting it in parallel with another simple length of transmission line. The extra line whose length is the same as that of the coupled lines allows the value of Z_0 to be increased to the values of L_1 and L_2, which is the condition necessary to make the coupled lines symmetrical.

After application of the formulas given in Figure 4.11, the odd and even mode impedances for the inverting and noninverting coupled-line sections are those given in Figure 4.12(c). These impedances have also been expressed in terms of

Table 4.2
Synthesis of the 6- to 18-GHz High-Pass Balun Prototype by E-Syn

Network type	Distributed	Tchebycheff	QW-band-pass	
Passband	f_1 = 6 GHz	f_2 = 18 GHz	T_p = 0.1 dB	MIL = 0 dB
Terminations	R_s = 1	R_l = 1 (only relevant for auto mode)		
Elements	Two transmission lines	One stub		
Mode (F0)	Manual			
NxtEL (F6)	Use successively to extract next element from choice given			
Analyze (F0)	Analysis after extraction			

voltage coupling ratio C_v and section impedance Z_{sec}, which, for a pair of symmetrical coupled lines, are given by

$$C_v = 20 \log_{10}\left(\frac{Z_{oe} - Z_{oo}}{Z_{oe} + Z_{oo}}\right) \quad \text{dB} \tag{4.1}$$

and

$$Z_{sec} = \sqrt{Z_{oe}Z_{oo}} \quad \Omega \tag{4.2}$$

The drawings in Figure 4.12(c) indicate the physical shape of the two sections. In the case of the inverting section, this has to be a four-finger interdigitated structure [4] owing to the relatively high coupling ratio of −3.3 dB required. For the non-inverting section, the coupled lines have a more modest coupling ratio of only −6.02 dB, which is realizable as a pair of edge-coupled lines [5]. The impedance of the simple length of line in parallel with the coupled lines is 85.5Ω.

A direct mathematical derivation of the value of the balancing resistor R_i is a problem of considerable complexity and will not be discussed here. In practice, the value of the resistor is determined by computer optimization using the output reflection coefficient as the target electrical parameter. In the 6- to 18-GHz balun, the required value for R_i is 88Ω.

The 6- to 18-GHz baluns have been fabricated on 2-in-diameter wafers of GaAs at the Philips Microwave Limeil (PML) MMIC foundry in France. A photograph of one of the balun chips after fabrication is shown in Figure 4.13. To minimize the area occupied by the balun and also to make use of a single via hole for all the ground connections, the line elements including the couplers have been folded. Clearly, the aspect ratio of the balun can be adjusted to suit any specific application. Here the balun has been made short and wide to enable it to be positioned as the final power-combining element at the end of a power amplifier chip. Its overall size is approximately 0.7 × 2 mm. The substrate thickness is 0.1 mm.

Predicted frequency responses of the 6- to 18-GHz balun are plotted in Figure 4.14. These were calculated directly from an equivalent circuit of the form given in Figure 4.9, containing ideal transmission line elements. Phase responses of S_{21} for each of the two output arms have been plotted in the upper half of the diagram after normalization to a common reference line. As expected for the ideal circuit, the phase difference between the responses is 180 degrees at all frequencies inside and outside the passband. Similarly, in the lower half of the diagram, the two S_{21} magnitude responses are coincident, indicating a power subdivision to the output ports of precisely −3 dB over the 6- to 18-GHz band. The S_{11} response has maxima at −17 dB, which is consistent with the specified insertion loss ripple (T_p), which was 0.1 dB. The S_{22} responses for both output ports are identical and owing to the presence of resistor R_i are substantially better than −10 dB from 6 to 18 GHz.

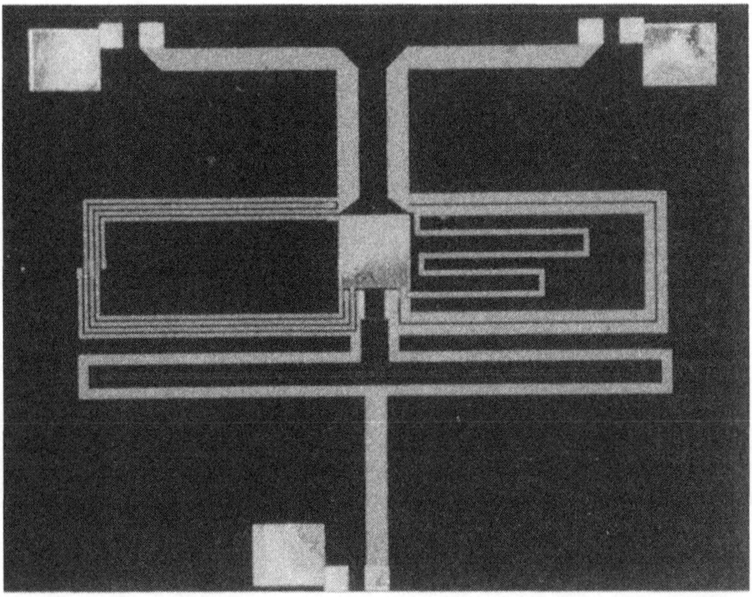

Figure 4.13 The 6- to 18-GHz MMIC balun (2 × 0.7 mm).

Several of the processed 6- to 18-GHz MMIC baluns have been measured using RF probes connected to an HP8510 network analyzer. The results for one of the baluns are presented in Figure 4.15. Amplitude responses of the two arms are smooth and track each other to within ±1 dB over the whole 6- to 18-GHz frequency band. Excess loss is on the order of 1 dB. Values of S_{11} and S_{22} for each of the output ports are better than −15 dB over most of the frequency range. The phase responses in the figure indicate that a 180-degree phase difference has been established at low frequencies, but there is a slow linear rise in phase difference with increasing frequency. This suggests that in addition to the 180-degree phase shift, there is an electrical path length difference between the two output arms of the balun. The path length difference can be attributed to the effects of circuit parasitics, particularly those associated with the folding of the couplers and the use of a common central via hole. These parasitics were not fully accounted for in the preceding simulations. When the path length difference is corrected by adding a short length of line to the noninverting arm, the absolute phase difference holds to within ±4 degrees of 180 degrees over 6 to 18 GHz.

After the fabrication of the 6- to 18-GHz balun, a 3-D field simulation was performed using the FACET program [6]. FACET and its capabilities were discussed in Chapter 2, Section 2.5.2.8. The simulation was carried out as an exercise in postfabrication diagnostics. Plots of the results of the FACET simulation for S_{21} magnitude have been plotted in Figure 4.16 for each of the two output arms.

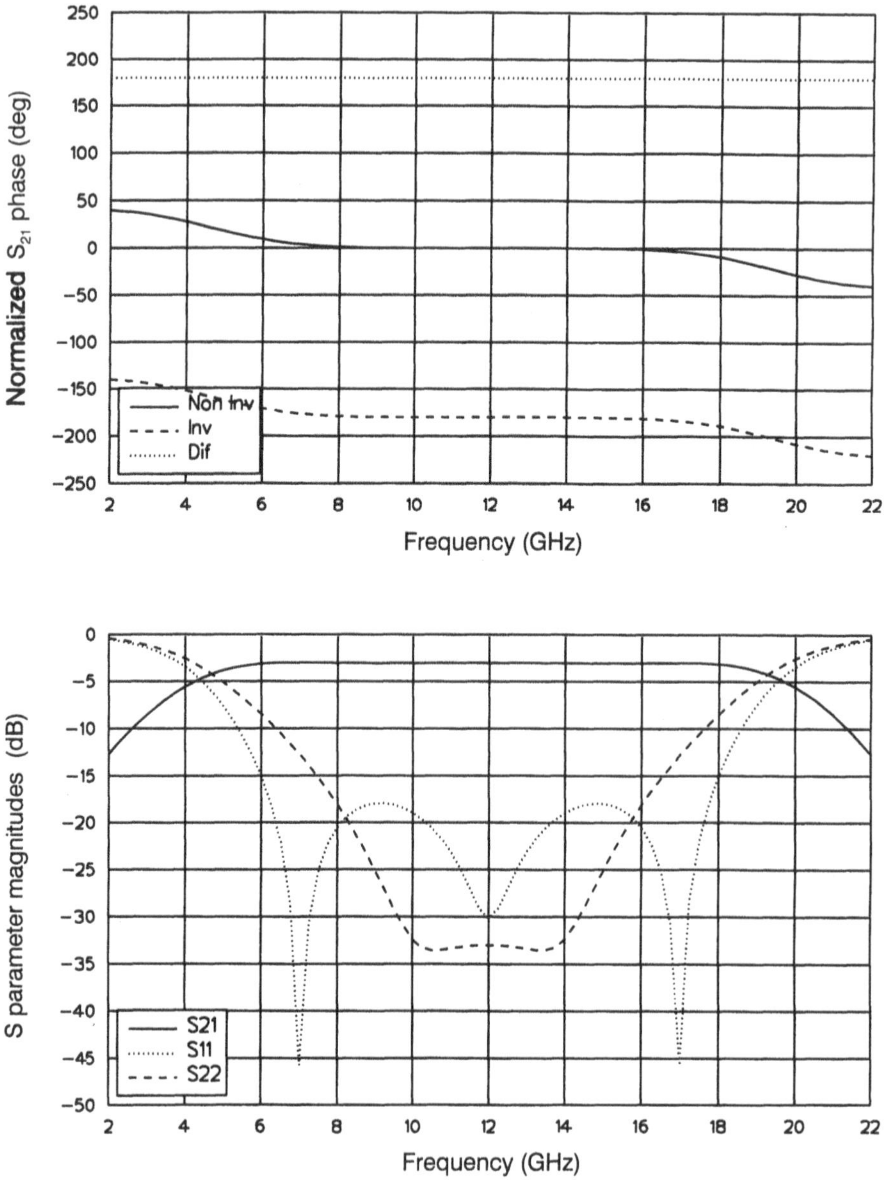

Figure 4.14 Predicted frequency responses of the 6- to 18-GHz balun.

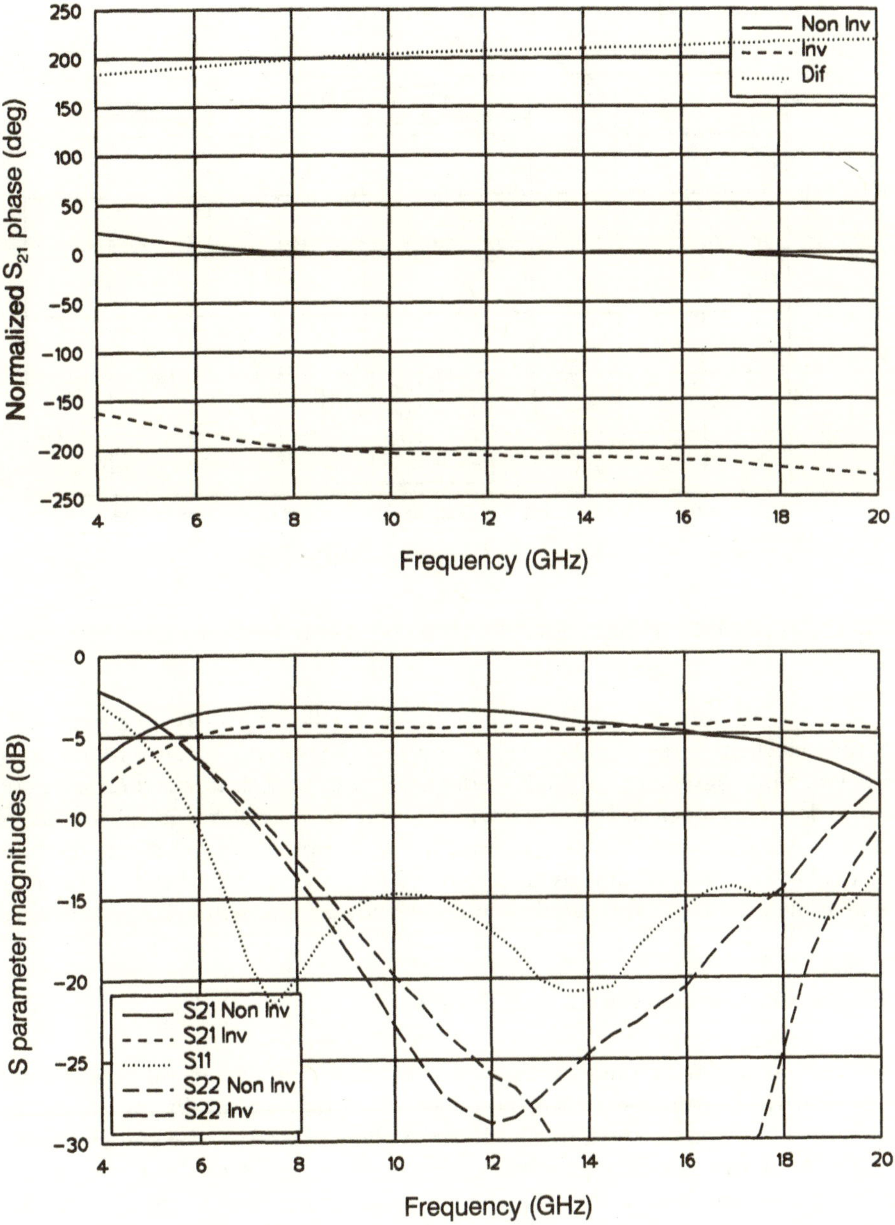

Figure 4.15 Measured frequency responses of the 6- to 18-GHz balun.

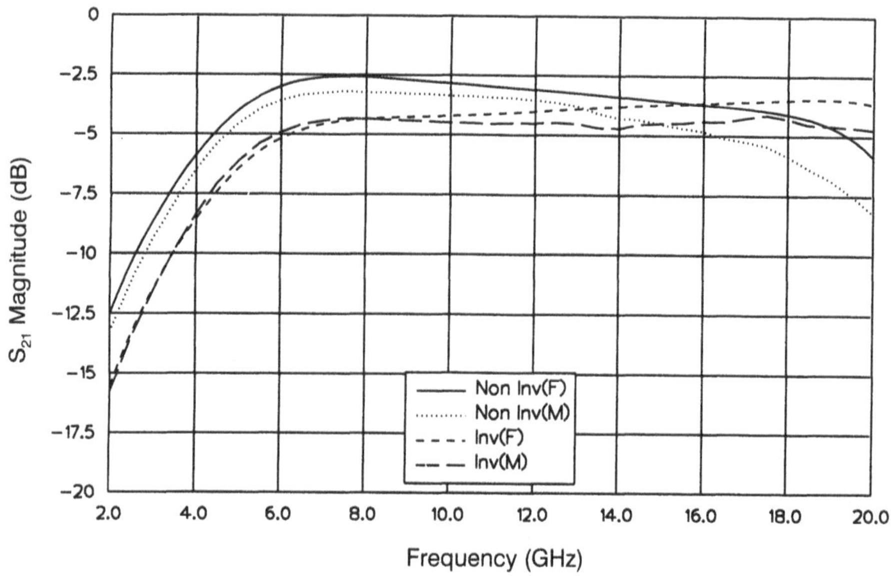

Figure 4.16 Measurement vs. FACET simulation for the 6- to 18-GHz balun.

Measured S_{21} responses have also been plotted in the figure for ease of comparison. There is excellent agreement between theory and measurement at low frequencies. However, while agreement at high frequencies is still good, it is rather less good than at low frequencies. The discrepancy is due to an optimistic calculation of losses by FACET, which in turn was caused by an optimistic specification of the electrical conductivity of the circuit metallization. Although not shown, agreement between measured and simulated S_{21} phase responses was better than ±4 degrees.

4.2.2 Band-Pass Baluns for 6.5 to 13.5 GHz

Despite the capability of the high-pass balun to deliver ultrawideband performance in an MMIC, a reduction in size could be an advantage. By basing the balun circuit on a band-pass instead of a high-pass S-plane prototype, the commensurate frequency for the f-plane equivalent circuit can be moved to the center of the stopband, resulting in a corresponding decrease in the length of all the transmission lines. Furthermore, length/impedance trade-offs can be applied to the lines and, where appropriate, stubs can be replaced by lumped capacitors. The band-pass prototype is also responsible for allowing the creation of the T arrangements of capacitors shown in the output arms of the balun of Figure 4.10. By adjusting the symmetry of these T sections, a wide range of impedance levels can be accommodated at the output ports without changing the basic balun design.

As a practical demonstration, a set of MMIC baluns based on a band-pass *S*-plane prototype have been designed for the 6.5- to 13.5-GHz frequency range, providing impedance ratios of 50Ω:50Ω, 50Ω:25Ω, and 50Ω:12.5Ω, respectively. Two of the baluns, the 50Ω:50Ω and 50Ω:25Ω versions, have been fabricated at the Philips Microwave MMIC foundry in France, and a photograph of a group of these baluns situated on a single chip of GaAs is presented in Figure 4.17. The three baluns in the top row are the 50Ω:25Ω versions, while the three in the bottom row are the 50Ω:50Ω versions.

For both types of balun circuit in Figure 4.17, a signal entering the input port first meets a single shunt capacitor connected to ground by a via hole close to the input port. It then splits into two parts and travels down a pair of 90Ω transmission lines to points where another pair of shunt capacitors are connected to ground. From here, the two signals enter the coupled-line sections of the balun, the noninverting coupled-line section being uppermost in the picture and the inverting coupled-line sections being below. Four vias provide the necessary ground connections for the coupled lines.

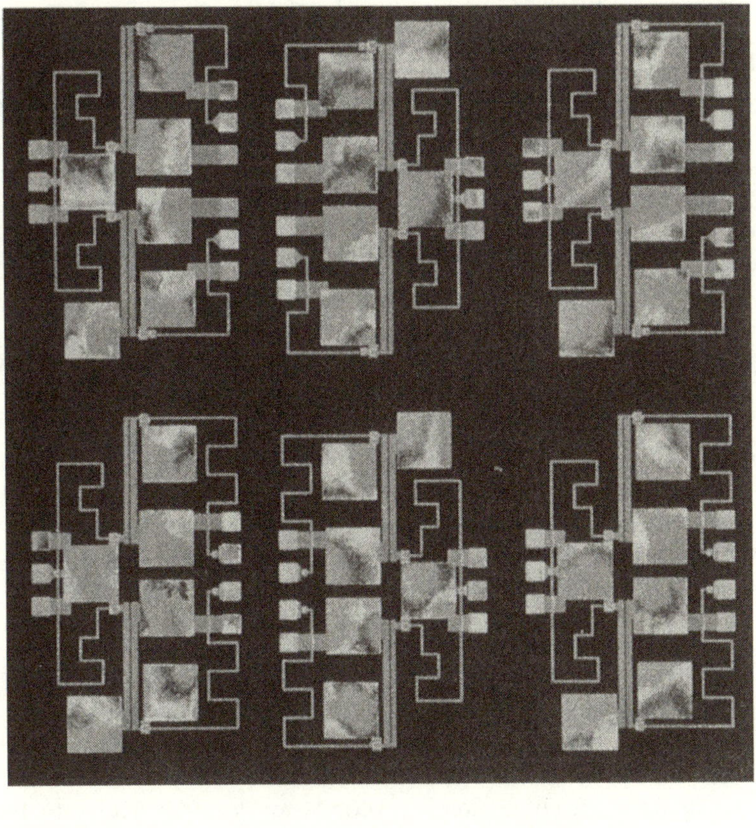

Figure 4.17 Band-pass baluns for the 6.5- to 13.5-GHz frequency range.

For the 50Ω:25Ω baluns in the top row, the output ends of the coupled-line sections are connected to another pair of shunt capacitors, followed by a pair of series capacitors, and then finally a pair of 90Ω transmission lines carry the now antiphase signals to the two output ports. Hence, with reference to Figure 4.10, the capacitor T sections are actually L sections in this version of the circuit, to achieve the desired 25Ω output impedance. In 50Ω:50Ω baluns, however, the transformation into L sections has been taken one step further such that the series capacitors have been transformed through the adjacent 90Ω transmission lines to take up their final positions next to the output ports. This has resulted in an increase in the length of the two 90Ω output lines.

The synthesis of the basic *S*-plane prototype for this set of band-pass baluns first involves specifying an *f*-plane bandwidth of 6.66 to 13.33 GHz, which corresponds to an octave bandwidth centered on a frequency of 10 GHz. This bandwidth was originally chosen to meet a particular performance requirement, but in general an octave bandwidth tends to result in prototypes with the most desirable dynamic range of element values. As a convenient starting value, which can be changed later if necessary, the commensurate frequency is set to 30 GHz. Then, using the Richards Transformation, the *f*-plane frequency specifications are translated into passband edge frequencies in the *S*-plane of $\omega_1 = 0.3640$ and $\omega_2 = 0.8391$. For a good input match, the passband insertion loss ripple value (T_p) is set to 0.1 dB, and the parameters T_o and T_{sp} are set to zero. Three transmission zeros at $S = (1, j0)$ must be specified for the three unit elements, three must be specified at $S = (0, j\infty)$ for the three shunt capacitors, and finally one specified at $S = (0, j0)$ for the single, nonredundant series capacitor. The network is doubly terminated and is of the first canonical form.

Given the above specifications, the nonredundant *S*-plane prototype can be synthesized using either NETSYN or E-Syn to take the form shown in Figure 4.18(a). It is a degree 8 network and as in the case of the high-pass balun, the input for the purposes of element extraction is on the right of the picture. To lower the value of the load resistor at the left-hand end of the network, the shunt inductor (value 1.002) must be subdivided and a part moved to the left through the adjacent unit element. As indicated in Figure 4.18(b), moving enough inductance to produce a pair of inductors of the same value lowers the load resistor value to 2.5, which although not precisely the value required (i.e., 2.0), is close enough to complete this stage of the design procedure. It is on this form of the *S*-plane prototype that the 50Ω:50Ω version of the balun is based.

As in the case of the high-pass balun, the section comprising the pair of inductors with the unit element in between corresponds to the two coupled-line sections of the balun. These elements and all those to the left of the dotted lines in Figure 4.18 are common to the whole family of band-pass baluns with the different transformation ratios. Those elements to the right, however, are specific to each balun and are the elements that must be manipulated to suit the particular output impedance requirement. To obtain the prototype for the 50Ω:25Ω balun, for exam-

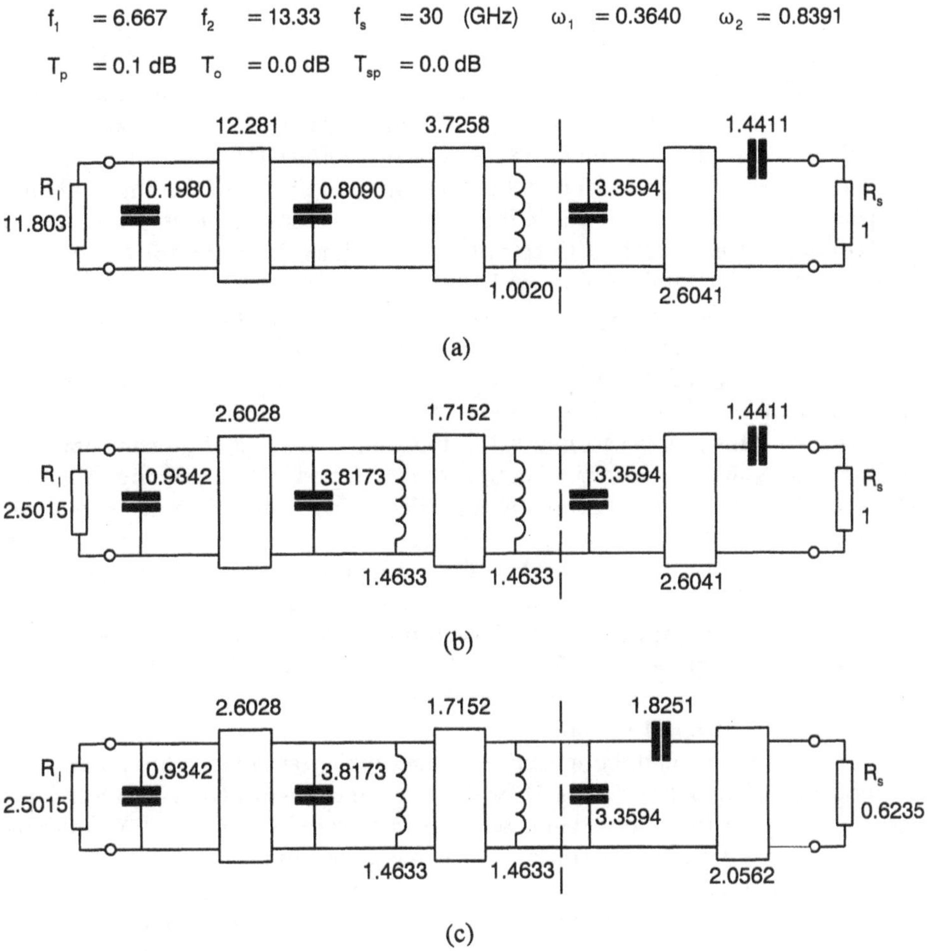

Figure 4.18 *S*-plane prototypes of the 6.5- to 13.5-GHz baluns: (a) nonredundant *S*-plane prototype; (b) redundant *S*-plane prototype for 50Ω:50Ω balun; (c) redundant *S*-plane prototype for 50Ω:25Ω balun.

ple, the series capacitor at the right (value 1.4411) is moved to the left through the adjacent unit element using a Kuroda transform, as shown in Figure 4.18(c). The ideal transformer created by the transform has the effect of lowering the value of the 1Ω termination to 0.6235Ω, which is exactly the value required to give the necessary 4:1 ratio of load/source resistance. The process can be continued still further if necessary by moving part or all of the series capacitor through the next shunt capacitor (see Chapter 2, Section 2.4.1) to produce a T section of the type that was illustrated in Figure 4.10.

The necessary parameter set for synthesis of the *S*-plane prototype by E-Syn is given in Table 4.3. It should be noted that in specifying a distributed, Tchebycheff, band-pass network, E-Syn requires that the electrical length of all the transmission lines be chosen at a frequency equal to the upper edge of the passband. This is equivalent to choosing a commensurate frequency of 30 GHz, at which all the lines have an electrical length of 90 deg. In choosing the number of transmission lines, low-pass stubs, and high-pass stubs, it is important to remember that they refer to nonredundant elements, and since this is a band-pass filter, the total number of nonredundant elements must always be even.

After synthesis of the *S*-plane prototypes, the corresponding *f*-plane prototypes are obtained in two stages. The first involves transforming the *S*-plane prototypes into purely distributed, commensurate line networks, and the second involves the application of length/impedance trade-offs of the sort described in Chapter 2, Section 2.4.3. This produces a network comprising a mixture of transmission lines and lumped capacitors whose final values are established by computer optimization. The optimization required is usually superficial, and with reference to the flow diagram of Figure 2.2 in Chapter 2, it is the last step before conversion into physical dimensions. The result for the 50Ω:50Ω balun is shown at the top of Figure 4.19. As indicated by the figure, the optimization has not only corrected the frequency responses of the network, but also adjusted the ratio of the load/source resistors to be exactly the required 2:1. The result for the 50Ω:25Ω balun is also shown, but only the modified end section has been drawn, since the rest of the circuit is the same as for the 50Ω:50Ω balun.

To minimize bend discontinuities, all the simple transmission lines were given a characteristic impedance of 90Ω, while the impedances of the line and stubs corresponding to the coupled-line sections were carefully chosen for realizable values of coupling ratio. In accordance with the above discussion on *f*-plane prototypes, the coupled-line sections of both 50Ω:50Ω and 50Ω:25Ω baluns are identical. The inverting coupled-line sections are symmetrical and have a coupling coefficient

Table 4.3
Synthesis of the 6.5- to 13.5-GHz Band-Pass Balun Prototype by E-Syn

Network type	Distributed	Tchebycheff	Band-pass	
Electrical length	40 deg at 13.33 GHz			
Passband	f_1 = 6.667 GHz	f_2 = 13.33 GHz	T_p = 0.1 dB	MIL = 0 dB
Terminations	R_s = 1	R_l = 2 (only relevant for auto mode)		
Elements	Three transmission lines	Three low-pass stubs	Two high-pass stubs	
Mode (F0)	Manual			
NxtEL (F6)	Use successively to extract next element from choice given			
NxtEV (shift F6)	Use for partial extraction of the shunt inductor if required			
Analyze (F0)	Analysis after extraction			

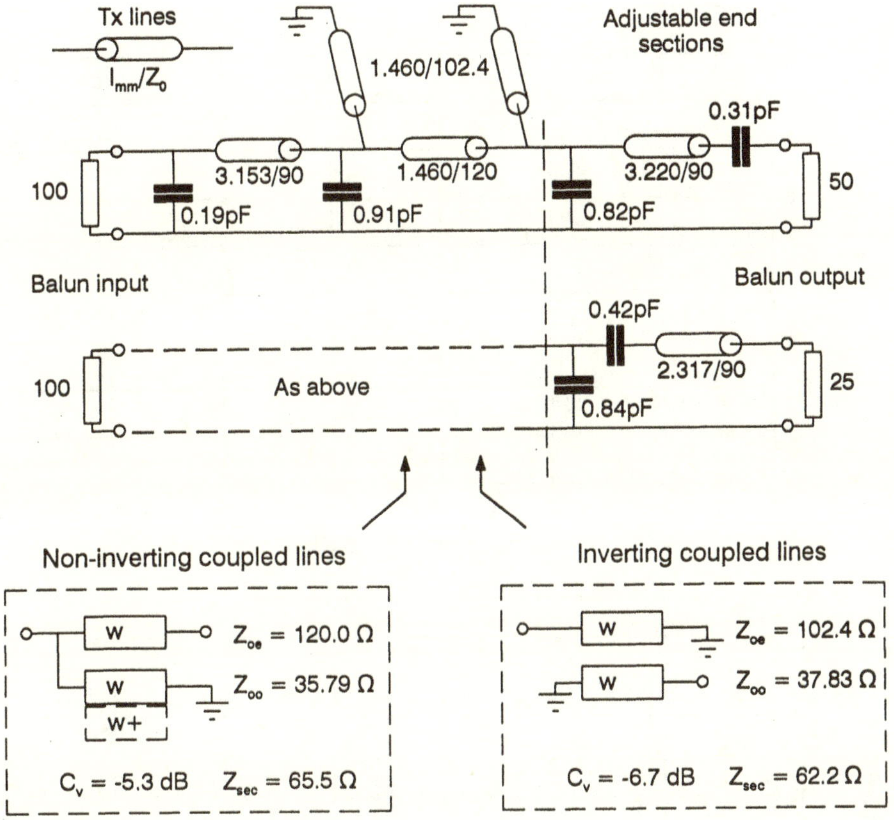

Figure 4.19 Half-balun, *f*-plane equivalent circuits (6.5 to 13.5 GHz).

of –6.7 dB. This is just realizable with a pair of edge-coupled microstrip lines and does not require an interdigitated structure. The reduced coupling is a consequence of specifying a 67% fractional bandwidth instead of the 100% bandwidth of the high-pass, 6- to 18-GHz balun. Layout constraints unfortunately force the noninverting coupled-line sections to be asymmetrical, but for the purposes of their design, they are treated as symmetrical couplers with some additional transmission line loading on one side (see Figure 4.19). As a symmetrical coupler, the relevant coupling ratio is –5.3 dB, which is also just realizable with a pair of edge-coupled lines. Line width and separation calculations for both couplers were based on [5].

When the *f*-plane equivalent circuit of Figure 4.19 (50Ω:50Ω version) is configured as a complete balun, computer simulation produces the results shown in Figure 4.20. Some transmission line losses are assumed to be present in the circuit, but no other circuit parasitics have been added. Insertion loss for each arm is shown

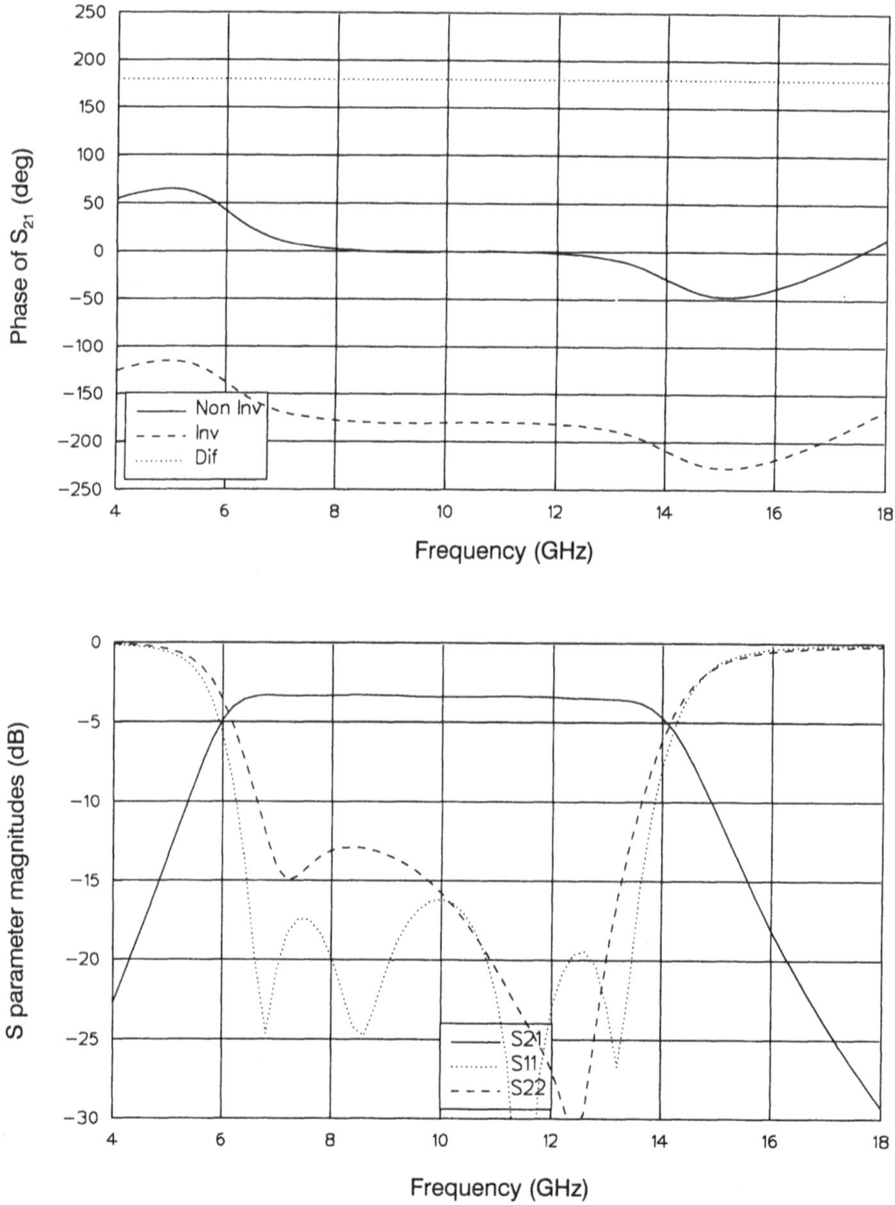

Figure 4.20 Analysis of the *f*-plane equivalent circuit of the 6.5- to 13.5-GHz band-pass balun.

to be 3.5 dB over the 6.5- to 13.5-GHz frequency band, which is a slightly extended frequency band found to be achievable during the previously mentioned optimization. After accounting for the 3-dB loss due to power splitting, the predicted excess loss is therefore 0.5 dB. Values of S_{11} across the frequency range are substantially below −15 dB and S_{22} (for both arms) is better than −13 dB as a consequence of using a 70Ω balancing resistor R_i in the circuit. Phase responses of S_{21} for each of the output ports are also plotted in the figure. In the absence of parasitics, the phase difference is exactly the required 180 deg at all frequencies, both inside and outside the passband.

In converting the f-plane equivalent circuits into the geometrical layouts of the real baluns, a great deal of care was exercised in minimizing the effect of parasitics in what had to be a highly compact structure. The most influential parasitics were associated with the relatively large vias (240 × 240 μm) and any bends in the coupled-line sections. Consequently, the coupled lines were constructed without bends and sufficient vias were used so that there was no sharing of vias by shunt capacitors and the coupled-line sections. Having completed draft layouts of the baluns, the 3-D field simulation program FACET [6] was used to analyze performance and identify any physical corrections required. The final baluns, as shown in Figure 4.17, measured only 0.7 × 1.5 mm, which represents a significant reduction in size with respect to the previous 6- to 18-GHz high-pass balun.

Results of the FACET analysis of the 50Ω:25Ω balun are presented in Figure 4.21. Compared with the responses of Figure 4.20, the effects of parasitics are clearly visible. However, the S_{21} amplitude responses of the output ports track each other to within ±0.5 dB, while excess loss is about 1 dB over most of the 6.5- to 13.5-GHz passband. Port reflection coefficients are shown to rise to just over −10 dB in the center of the passband. This is due to the detrimental effects of the grounding of the various circuit elements by large vias. Port reflection coefficients and circuit performance in general could be improved by reducing the size of the vias. As for the phase responses, the difference between the normalized phase responses is approximately 184 ± 4 deg, and there is no evidence of the path length difference in the 6- to 18-GHz balun.

With reference to Figure 4.17, the leftmost two baluns of each of the 50Ω:50Ω and 50Ω:25Ω types have been positioned as back-to-back pairs. Within each pair, the balun on the right has also been physically inverted about the horizontal. Orienting the baluns in this way allows each pair to be joined at their output ports by bond wires and then measured as a single two-port network using RF probes connected to a network analyzer. Although this is not a measurement of the precise S_{21} characteristics of the separate output ports, it does give a good indication as to whether the filters of the baluns are operating correctly. It also overcomes the problem of measuring the 50Ω:25Ω balun with a system calibrated for equal 50Ω port impedances.

Back-to-back measurements have been performed on both baluns, and the results for the 50Ω:25Ω version are plotted in Figure 4.22. For the purposes of

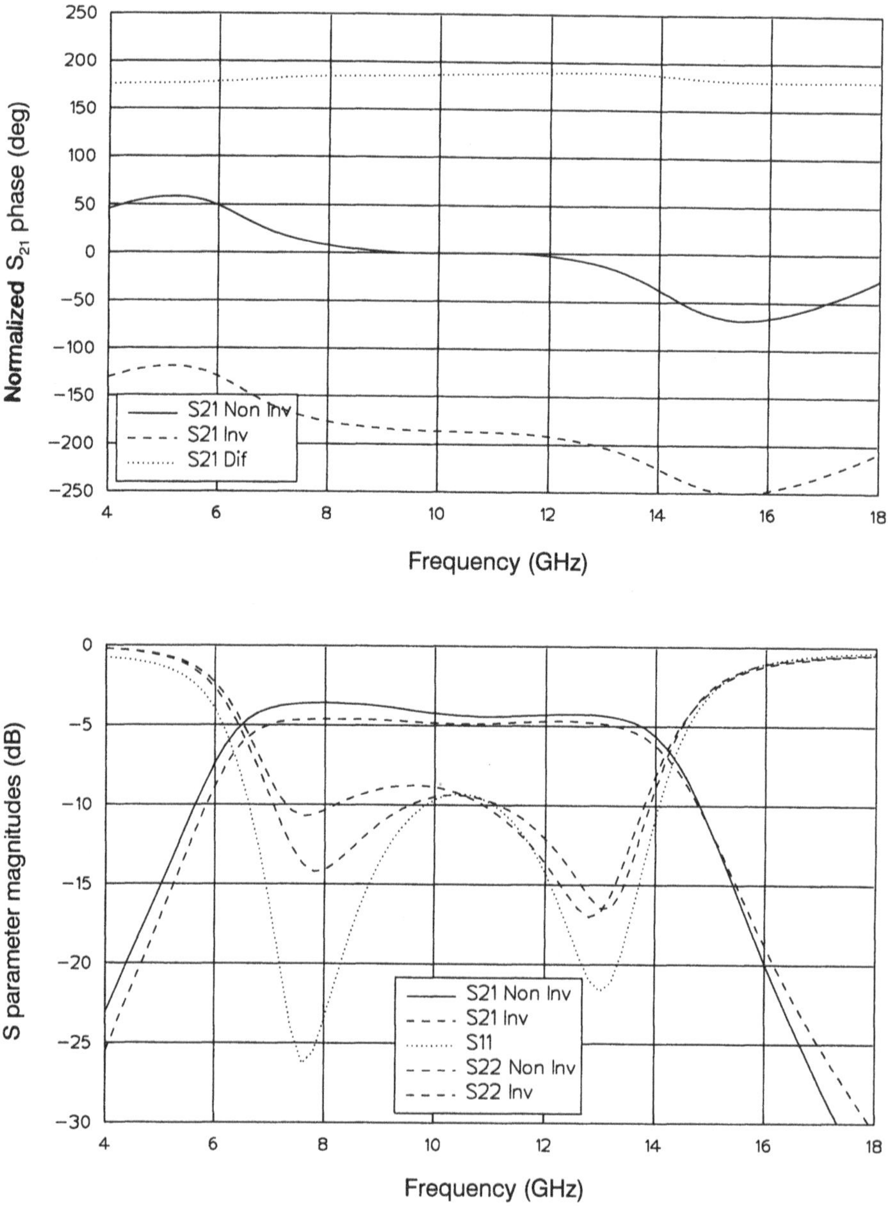

Figure 4.21 FACET 3-D field simulation of the 6.5- to 13.5-GHz band-pass balun.

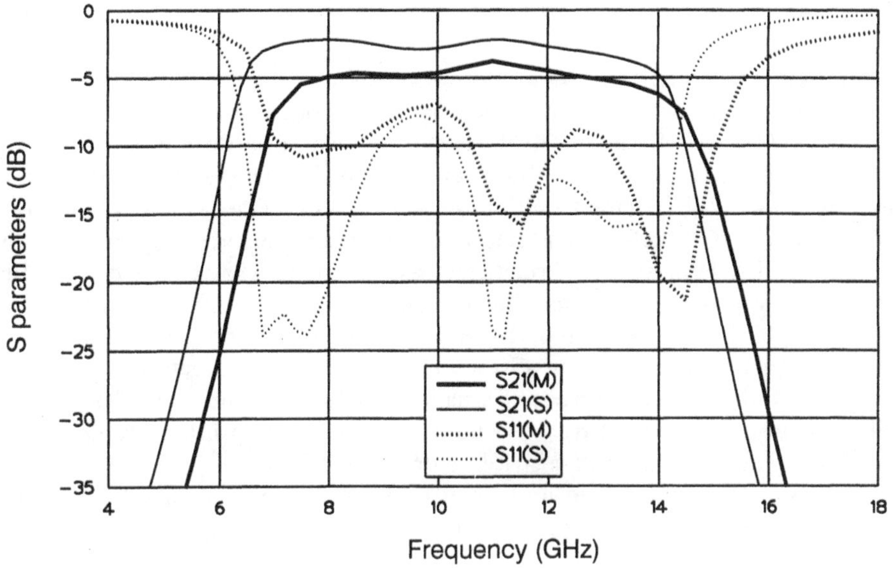

Figure 4.22 Frequency responses by measurement and by FACET simulation of two 50Ω:25Ω baluns back-to-back.

comparison, FACET simulations of the two baluns wired back-to-back are also plotted in the figure. There is generally good shape correlation between the measurements and the 3-D simulations, but there is substantially more loss in the measured S_{21} responses and a slight upward shift in the position of the passband. The shift in the passband is most likely a result of the relatively simple way in which the simulator deals with vias. The increased losses are most likely to be associated with the high concentration of currents flowing in the nonrectangular edges of the microstrip lines of the couplers.

Measurements have yet to be performed on individual baluns to establish the transmission characteristics of the two separate output ports. However, the combination of 3-D field simulations and the back-to-back measurements suggests that the phase difference between the two output ports will be acceptably close to 180 degrees over 6.5 to 13.5 GHz, and the excess losses will be close to 2 dB. The value of excess loss is higher than originally anticipated, and will make the application in push-pull power amplifiers rather unlikely. Despite this, they still have considerable potential for use in balanced mixers and phase shifters, particularly where a microstrip transmission line medium must be employed. There are very few alternative types of microwave circuit that can produce 180 degrees of phase shift over an octave band in a microstrip environment.

4.3 SOME SIMPLE IMPEDANCE-TRANSFORMING NETWORKS

Circumstances frequently arise in microwave circuit design when it is necessary to achieve a good impedance match between two resistive loads of unequal value. This is the sort of task that an ideal transformer would perform most effectively if it were possible to realize such a thing at microwave frequencies. Unfortunately, it is not possible to realize elements with pure self- and mutual inductance at microwave frequencies, and therefore it is impossible to realize an ideal transformer as a separate entity. However, it is relatively easy to create what are essentially filter networks that produce the same effect as an ideal transformer. Some of the most interesting of these have been gathered together for discussion in this section.

No practical results will be given, but the following subsections will describe the practical designs of five types of circuit that are particularly effective as impedance level transformers. All are well suited to realization in printed circuit transmission line media such as triplate stripline or microstrip.

Before dealing with the individual transformer types, it may be helpful to restate some of the basic rules governing the synthesis of impedance-transforming networks.

1. It is impossible to achieve a match between two unequal resistances at zero frequency (i.e., at dc) using any realizable, loss-free, passive network. Only an ideal transformer could achieve this and ideal transformers are not considered to be physically realizable as separate entities.
2. From statement 1 it follows that no true low-pass filter can be used as an impedance transformer. In this context, a low-pass filter is one that has perfect transmission at zero frequency. This therefore includes such structures as Tchebycheff equal-ripple ladder networks comprising an odd number of low-pass elements. However, it does not include a Tchebycheff network of even degree, since according to this definition, it is not a true low-pass filter and can perform some limited impedance level transformation.
3. An impedance-transforming network must be either high-pass or band-pass. If it is high-pass, it may be of either odd or even degree, but if it is band-pass, it must be of an even degree. Although a transformer must be a high-pass or band-pass network, there is no necessity for transmission zeros to be placed at $S = (0, j0)$. In other words, the network does not need to contain high-pass elements.

4.3.1 The Quarter-Wavelength Stepped-Impedance Transformer

The quarter-wavelength stepped-impedance transformer is the most well known type of microwave circuit used for impedance level transformation. This is probably because often only a narrow frequency bandwidth is involved, for which a single quarter-wavelength transmission line section is sufficient for the task. In these circumstances, the characteristic impedance of the transmission line is given by the very simple relation

$$Z_{01} = \sqrt{R_s \times R_l} \quad \Omega \tag{4.3}$$

where R_s and R_l are the source and load resistances, respectively. For wide bandwidths (i.e., >octave), however, a multisection stepped-impedance transformer will be required of the general form shown in Figure 4.23. The *f*-plane equivalent circuit is a cascade of transmission lines whose impedances progressively increase or decrease in the direction of the resistive load, depending on whether the load is less than or greater than the resistance of the source. The lengths of the lines are all equal to a quarter of a wavelength at the center frequency of the passband. Physical realization can be in almost any transmission line medium, such as waveguide, coax, triplate stripline, or microstrip. In microstrip, the transformer is a cascade of strips of approximately the same physical length and of either ascending or descending widths. Slight variations in physical length may be needed to compensate for the dependence of the effective dielectric constant on the width of the different strips and for the effects of the capacitive discontinuities at each step. The structure has the advantages that it is simple to model, there are no critical physical dimensions, and in the absence of capacitively coupled sections, transmis-

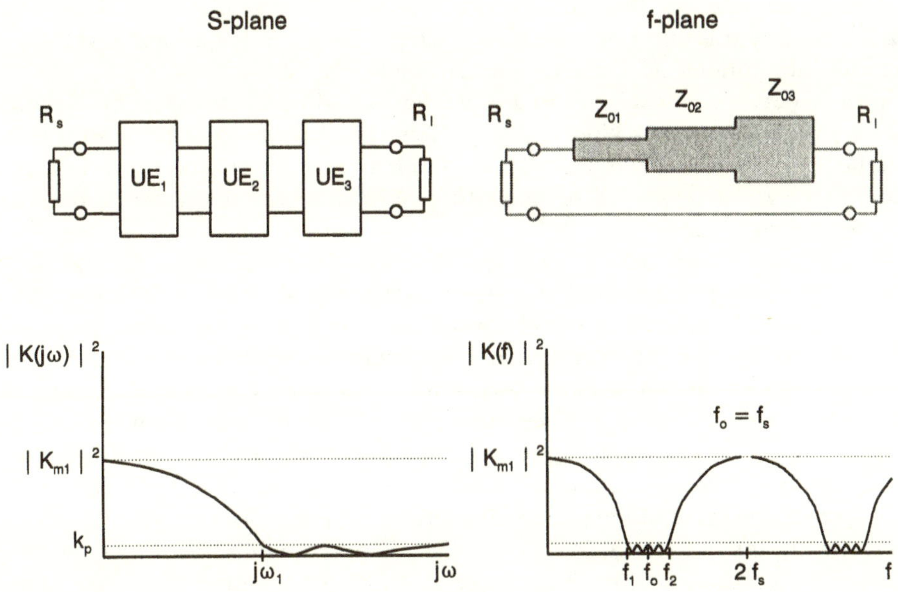

Figure 4.23 Prototypes and properties of quarter-wave stepped-impedance transformers.

sion losses are usually low. It can also operate over extremely wide bandwidths. Its major disadvantage is that for wide bandwidths and large transformer ratios, it is a long structure that cannot easily be folded.

Equations for calculating the impedances of the sections of quarter-wavelength transformers are presented in [7]. Exact formulas are given for up to four transmission line sections and approximate formulas for up to eight sections. When more than eight sections are required, a complicated procedure is offered which relies on the correlation that exists between the transformer design problem and that of the design of antenna arrays [8]. Although transformers can be designed with the help of this published information, their design by exact synthesis is simple and very much to be preferred.

With reference to Figure 4.23, the f-plane characteristic function $|K(j\omega)|^2$ for the quarter-wave transformer has a periodic filterlike shape with a passband extending from f_1 to f_2. An equal-ripple passband response is the most useful, for which k_p is the magnitude of the ripples of $|K(j\omega)|^2$. Between zero and $2f_s$, the function is symmetrical about the center of the passband f_0. Therefore, outside the passband, $|K(j\omega)|^2$ rises to the same maximum value of $|K_{m1})|^2$ on both sides. As will be seen later, $|K_{m1}|^2$ corresponds directly to a particular impedance transformation ratio. Transformed into the S-plane, the cascade of transmission lines becomes a cascade of unit elements, each one of which corresponds to a transmission zero at $S = (1, j0)$. The characteristic function is now an aperiodic high-pass response with a passband edge of ω_1 and values of k_p and $|K_{m1}|^2$, the same as in the f-plane. It is interesting to note that this S-plane response is classified as high-pass, despite the absence of any transmission zeros at $S = (0, j0)$. Because it is a high-pass S-plane network, the number of unit elements can be either odd or even.

In theory, the maximum f-plane fractional bandwidth that can be achieved with transformers of this kind is 200%. However, this would require an infinite number of circuit elements. In practice, the maximum fractional bandwidth that can be achieved depends on the magnitude of the impedance transformation, but can be as high as 195%.

Synthesis of the S-plane prototype network of the quarter-wave transformer is especially simple because all the elements are of the same type. Hence, for a given canonical form, there is only one possible circuit topology. However, to achieve a specific transformer impedance ratio, some additional mathematical formulas can be of assistance. The synthesis procedure itself will provide the values of the unit elements corresponding to specified values of the passband edge frequency ω_1, the passband insertion loss ripple T_p, and the number of transmission zeros located at $S = (1, j0)$. It will also give the value of the terminating resistor R_l and hence, since R_s is set to 1, the value of the impedance ratio for the transformer. What it will not do is give any guidance as to the values of ω_1, T_p, and the number of unit elements required to produce a given impedance ratio. In the absence of any additional information regarding the relationship between the various synthesis parameters, it would be necessary to iterate the synthesis for different values of ω_1, T_p, and the number of elements until a required value of impedance ratio is obtained. This would be perfectly acceptable. However, a more direct approach is to use a set of mathematical formulas that can be derived for transformers of this kind to calculate the precise value of ω_1 necessary to give the impedance ratio required for a given value of T_p and a given number of unit elements. The synthesis then needs to be performed only once. The relevant mathematical formulas are derived follows.

Using the work of Horton and Wenzel [9], it can be shown that the characteristic function of an equal-ripple high-pass network with n transmission zeros at $S = (1, \mathrm{j}0)$ is of the form

$$|K(\mathrm{j}\omega)|^2 = k_p \cos^2\left[n \cos^{-1}\left(\sqrt{\frac{1+\omega_1^2}{1+\omega^2}}\right)\right] \tag{4.4}$$

for all values of $\omega \geq \omega_1$, where $\omega = S/\mathrm{j}$ and where ω_1 is the lower edge of the passband. For values of ω outside the passband, where $\omega < \omega_1$, the expression must be modified such that

$$|K(\mathrm{j}\omega)|^2 = k_p \cosh^2\left[n \cosh^{-1}\left(\sqrt{\frac{1+\omega_1^2}{1+\omega^2}}\right)\right] \tag{4.5}$$

and if this is transposed for ω_1 after setting $\omega = 0$, we can write

$$\omega_1 = \sqrt{\cosh^2\left[\frac{\cosh^{-1} A}{n}\right] - 1} \tag{4.6}$$

where the constant A is given by

$$A = \sqrt{\frac{|K(\mathrm{j}\omega)|^2_{\max}}{k_p}} = \frac{1 - \frac{R_s}{R_l}}{2\sqrt{\frac{R_s}{R_l}\left(10^{\left(\frac{T_p}{10}\right)} - 1\right)}} \tag{4.7}$$

Hence the combination of equations (4.6) and (4.7) gives the value of ω_1 which must be specified during the synthesis of a transformer comprising n unit elements, having a transmission loss ripple of T_p dB, and providing an impedance ratio between source and load of R_s/R_l, where $R_s/R_l \leq 1$. Values of impedance ratio less than unity correspond to the first of the dual canonical forms of the network. For values of impedance ratio greater than unity, it is necessary to take the reciprocal value for substitution into (4.7) and then extract the second dual form of the network.

In practice, the recommended procedure for designing the transformer is to choose a value of T_p that is less than or equal to the performance specification, select a small value of n, and then calculate the value of ω_1 that corresponds to the required impedance ratio. If ω_1 is higher than what is acceptable, giving insufficient f-plane frequency bandwidth, then n should be increased and the calculation repeated until ω_1 meets the specification. Network synthesis can then follow directly after a satisfactory combination of parameters has been established. If the transformer is to be synthesized using E-Syn, the network must be specified as distributed, Tchebycheff, and QW-band-pass. The passband edges must be chosen to be consistent with the S-plane cutoff frequency ω_1, the number of transmission lines must be set equal to the value of n above, and the number of high-pass stubs must be set to zero. E-Syn is then capable of extracting the elements of the network automatically.

4.3.2 Stepped-Impedance Transformers from Band-Pass Prototypes

The second type of impedance transformer to be described is not particularly well known or in common usage. This is perhaps because no known explicit formulas or tables of element values are available to help with its design. Essentially, the only way to design these transformers is to use exact synthesis.

The general structure of the transformer is indicated in Figure 4.24. It is a cascade of transmission lines whose lengths are all the same but substantially less than a quarter of a wavelength in the region of the passband. In fact, they are all a quarter-wavelength long at the center frequency of the stopband, f_s, which becomes an extra parameter to be specified as part of the design procedure. Another physical distinction from the previous quarter-wave transformer is that the characteristic impedances of the transmission lines, while having a general decline or ascent in value toward the load, alternate between high and low values. An advantage of transformers of this type over the quarter-wave transformer is that despite requiring typically twice as many elements for a given electrical requirement, the elements are much shorter, giving an overall reduction in length. Another advantage is that there is some independent control available over the width of the stopband. A disadvantage is that the structure depends heavily on the precise step in impedance between adjacent transmission lines and is therefore more susceptible to any slight errors in physical modeling than the quarter-wave transformer.

A sketch of the general shape of the f-plane characteristic function $|K(f)|^2$ for the modified stepped-impedance transformer is shown in Figure 4.24. This shows an equal-ripple passband response of magnitude k_p extending between f_1 and f_2, with stopbands on either side that are no longer mirror images of each other. This is because the vertical axis of symmetry (i.e., the commensurate frequency) is now at the center of the stopband. Also, the stopband maxima now have different values $|K_{m1}|^2$ and $|K_{m2}|^2$ on the two sides of the passband, but as will be shown, only $|K_{m1}|^2$ is relevant to the calculation of transformer impedance ratio.

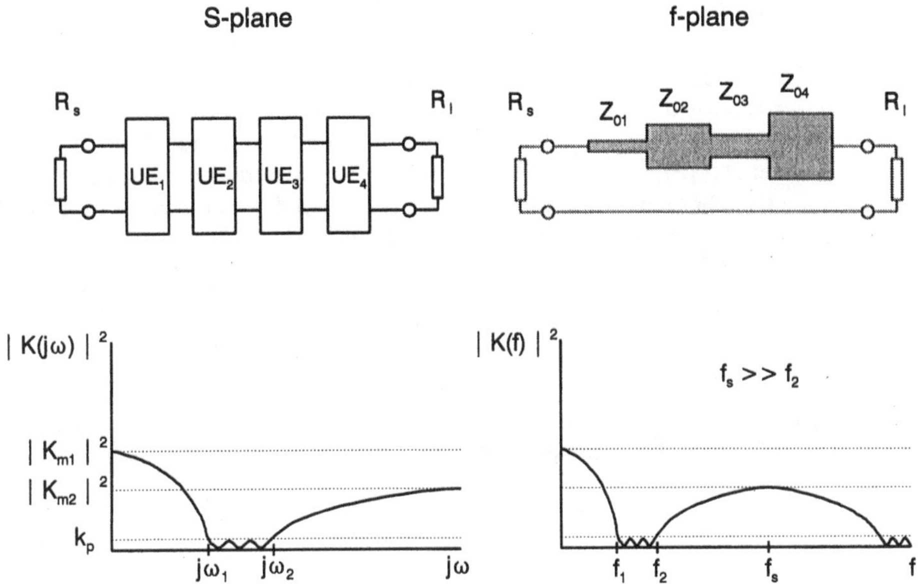

Figure 4.24 Prototypes and properties of band-pass stepped-impedance transformers.

Transformed into the *S*-plane, the characteristic function is a single band-pass response on the jω axis, with unchanged values of k_p, $|K_{m1}|^2$, and $|K_{m2}|^2$. The function is asymptotic to the finite values of $|K_{m1}|^2$ and $|K_{m2}|^2$, which is consequence of there being no transmission zeros at either $S = (0, \text{j}0)$ or $S = (0, \text{j}\infty)$. The only transmissions zeros are at $S = (1, \text{j}0)$, one for every unit element in the network. In accordance with the general rule of synthesis for all band-pass *S*-plane networks, the degree of this network must be even, and since there are no redundant elements, the number of transmission zeros must also be even.

Realization of the band-pass stepped-impedance transformer is most convenient in coaxial line, stripline, or in microstrip transmission line media. It is a structure capable of operating over ultrawide frequency bandwidths of a decade or more and providing large impedance ratios as high as 10:1.

As in the case of the quarter-wave transformer, synthesis of the *S*-plane prototype will give element values and an impedance ratio (i.e., load resistance) for a specified set of design parameters. The parameters include n, the number of unit elements T_p, the passband insertion loss ripple ω_1, the lower passband edge, and an extra parameter ω_2, the upper edge of the passband. Commensurate frequency is not a synthesis parameter, but it must be taken into consideration because it affects how the *S*-plane bandwidth maps into the *f*-plane. It also determines the length of the transmission lines and the range of their characteristic impedances.

Commensurate frequency needs to be chosen at the beginning of the design sequence.

To alleviate the need for iteration during synthesis, another set of equations has been derived for this transformer that allows the essential synthesis parameters to be determined for any given bandwidth and impedance ratio. From the basic theory of Wenzel [9], it is possible to show that within the passband where $\omega_1 \leq \omega \leq \omega_2$, the characteristic function of an equal-ripple band-pass network with n transmission zeros at $S = (1, j0)$, where n is even, is of the form

$$|K(j\omega)|^2 = k_p \cos^2\left[n \cos^{-1}\left(\sqrt{\frac{(1+\omega_2^2)(\omega^2-\omega_1^2)}{(1+\omega^2)(\omega_2^2-\omega_1^2)}}\right)\right] \tag{4.8}$$

Outside the passband, the trigonometric functions become hyperbolic, and for the special case of $\omega = 0$, the expression for $|K(j\omega)|^2$ becomes

$$|K(j0)|^2 = |K_{m1}|^2 = k_p \cosh^2\left[n \sinh^{-1}\left(\sqrt{\frac{(1+\omega_2^2)\omega_1^2}{(\omega_2^2-\omega_1^2)}}\right)\right] \tag{4.9}$$

At the other extreme of the frequency range, where $\omega = \infty$, the expression is

$$|K(j\infty)|^2 = |K_{m2}|^2 = k_p \cosh^2\left[n \cosh^{-1}\left(\sqrt{\frac{(1+\omega_2^2)}{(\omega_2^2-\omega_1^2)}}\right)\right] \tag{4.10}$$

For the purposes of designing a transformer, only the $\omega = 0$ case is relevant, and in the following two steps, (4.10) is transposed for ω_1:

$$\frac{\omega_1^2(1+\omega_2^2)}{(\omega_2^2-\omega_1^2)} = \sinh^2\left[\frac{1}{n}\cosh^{-1}A\right] = x \tag{4.11}$$

and therefore

$$\omega_1 = \omega_2\sqrt{\frac{x}{1+x+\omega_2^2}} \tag{4.12}$$

where A is given by

$$A = \sqrt{\frac{|K_{m1}|^2}{k_p}} = \frac{1-\dfrac{R_s}{R_l}}{2\sqrt{\dfrac{R_s}{R_l}\left(10^{\left(\frac{T_p}{10}\right)}-1\right)}} \tag{4.13}$$

To design the modified stepped-impedance transformer, starting values of f_s, T_p (dB), f_2, ω_2, and n should be chosen for a required transformer ratio R_s/R_l, and (4.11), (4.12), and (4.13) used to calculate a corresponding value of ω_1. If ω_1 is greater than the value required, corresponding to insufficient frequency bandwidth, then n should be increased accordingly, making sure that it remains even. Alternatively, some fine adjustment of the value of ω_1 can be made by changing T_p, as long as this does not take T_p outside the performance specification. When an acceptable set of parameters has been obtained giving the necessary impedance ratio and bandwidth, the parameters can be passed to the synthesis software to obtain the prototype element values.

There are at least two approaches to synthesizing stepped-impedance transformers of this type with E-Syn. The first is to select a distributed, Tchebycheff, band-pass network and to set the electrical length, the passband edges, and the passband ripple to be consistent with the above *S*-plane parameter set. After choosing an even value of n transmission lines, the network can be synthesized in a fully automatic mode for the first canonical form. If the second canonical form is required, which would give the reciprocal transformer ratio, the network has to be synthesized in a manual mode. The second approach is to select a distributed, Tchebycheff, transformer network. In this case, the electrical length of the elements is fixed by the program to be 60 degrees at the upper edge of the passband. When the passband edge frequencies and the values of the source and load resistances have been specified, the program then requires an *order* parameter to be set that is equal to half the number of transmission lines. The order parameter can therefore be odd or even. After the number of optional low-pass stubs has been set to zero, the program then computes the insertion loss ripple value T_p, which gives exactly the required transformer ratio. The network can then be synthesized automatically.

As far as the physical realizability of the transformer is concerned, it is important to ensure that the aspect ratio of individual transmission line sections does not fall significantly below unity. If this is a problem, then it may be possible to find a more satisfactory solution by changing the value of f_s and repeating the design process.

4.3.3 Transformers Using Line and Open-Circuit Shunt Stubs

This third type of transformer can be a useful alternative to the stepped-impedance transformer when overall physical length needs to be reduced. However, the main reason for its presentation here is that it is a good example of what appears to be at first sight a low-pass network performing an impedance-transforming function. It is, in fact, not a low-pass but a band-pass network, despite being composed of only low-pass elements. The bandwidth and impedance-transforming capabilities of the circuit are similar to those of the stepped-impedance transformers, but there is a tendency to need more elements for a given performance specification. Physical realization is most suited to stripline or microstrip transmission line media.

The f-plane equivalent circuit of the transformer is shown in Figure 4.25 and comprises an alternating cascade of shunt open-circuit stubs and interconnecting transmission lines. In this particular type of transformer, the interconnecting transmission lines are redundant elements that have been introduced from the source end of the network to act as element separators. All the lines and stubs are a quarter of a wavelength long at the center frequency of the stopband. In the f-plane, the characteristic function is a periodic band-pass frequency response, symmetrical about the commensurate frequency f_s. At zero frequency, $|K(f)|^2$ has a turning point as in the case of the stepped-impedance transformers, but above the passband the function tends to infinity at f_s.

In the S-plane, the equivalent circuit is an LC (inductor/capacitor) ladder network with an even number of elements and therefore an even degree. Its frequency response is band-pass and it is of the first canonical form. Below the passband, the value of $|K(j\omega)|^2$ tends to the finite value of $|K_{m1}|^2$ due to the absence of any high-pass elements, but above the passband, $|K(j\omega)|^2$ tends to infinity due to the presence of transmission zeros at $S = (0, j\infty)$, one for every low-pass element. As previously explained, there are no nonredundant unit elements. However, as the first step toward transformation into the f-plane, redundant unit elements are moved across the network from the source using Kuroda Identities to separate the elements and to change all the reactive branches into shunt capacitors.

The relevant equations for determining the synthesis parameters for a particular transformer ratio will now be given. Within the passband, the expression for $|K(j\omega)|^2$ is

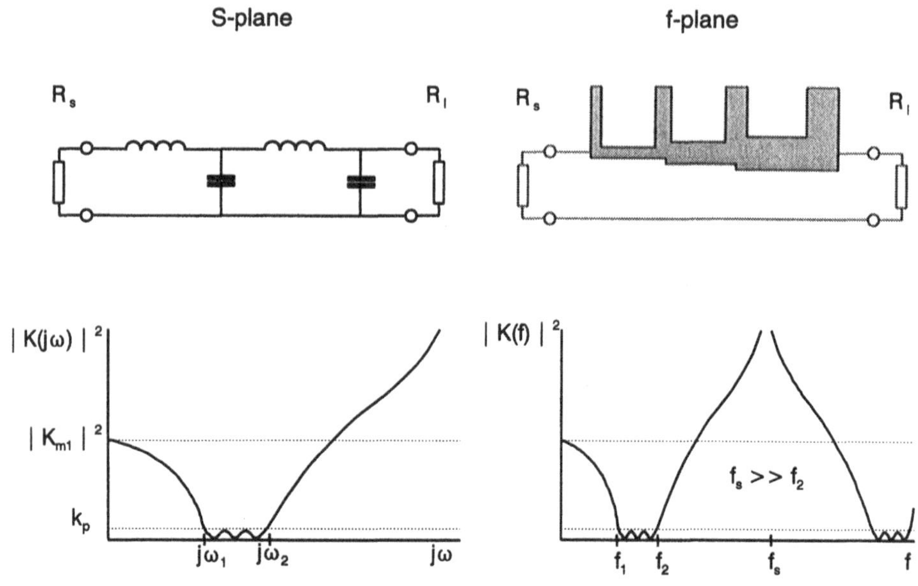

Figure 4.25 Prototypes and properties of band-pass, line, and open-circuit stub transformers.

$$|K(j\omega)|^2 = k_p \cos^2\left[n \cos^{-1}\left(\sqrt{\frac{(\omega^2 - \omega_1^2)}{(\omega_2^2 - \omega_1^2)}}\right)\right] \tag{4.14}$$

where n must be even. At $\omega = 0$, this simplifies to

$$|K(j0)|^2 = |K_{m1}|^2 = k_p \cosh^2\left[n \sinh^{-1}\left(\sqrt{\frac{\omega_1^2}{(\omega_2^2 - \omega_1^2)}}\right)\right] \tag{4.15}$$

and transposing for ω_2, we can write

$$\omega_2 = \omega_1\sqrt{\frac{1}{\sinh^2\left[\frac{1}{n}\cosh^{-1} A\right]} + 1} \tag{4.16}$$

The value of A is given by (4.13).

The use of the above equations and the synthesis of the *S*-plane prototype are virtually the same as for the modified stepped-impedance transformer except for working with the different types of elements. Once again, E-Syn can synthesize the network in two ways, the first involving the selection of a distributed, Tchebycheff, band-pass network and the second the selection of a distributed, Tchebycheff transformer. The main difference for both is that low-pass stub elements are specified instead of transmission lines. However, another difference is that in a manual mode, the program will offer the option of extracting the necessary redundant unit elements, which saves having to apply Kuroda Identities later. In an automatic mode, the program assumes that the redundant unit elements will always be required and extracts them without asking.

4.3.4 Transformers Incorporating High-Pass Elements

It has already been stated that any passive microwave circuit with a high-pass *S*-plane prototype can act as an impedance transformer, and in general the higher the proportion of high-pass elements in the network, the larger the transformer ratio that can be achieved for a given number of elements. Networks containing *only* high-pass elements would be highly effective transformers, but unfortunately would not often be physically realizable in microwave transmission line media. The best compromise is usually to have a mix of high-pass, low-pass, and all-pass elements (i.e., unit elements), which is not only more likely to be realizable, but will also give maximum flexibility in controlling the impedance ratio.

The basis for one of the most effective impedance-transforming circuits is the line-and-stub band-pass filter of the type already described in Chapter 3 and used in the bias Ts of Section 4.1. When used as a transformer, only one stub is usually used, which is moved to either one or the other end of the circuit, as illustrated in Figure 4.26. All the transmission lines and the stub are a quarter of a wavelength at the center frequency of the passband, corresponding to a band-pass frequency response in the *f*-plane and a high-pass response in the *S*-plane. Each transmission line corresponds to a unit element with a transmission zero at $S = (1, j0)$ in the *S*-plane, and the stub corresponds to a shunt inductor producing a single transmission zero at $S = (0, j0)$. As a high-pass *S*-plane network, its degree may be either odd or even.

There are two aspects of this and any other network containing high-pass elements that make the preparation for synthesis different from that of the preceding transformers. First, the transmission zero at $S = (0, j0)$ results in an infinite value of $|K(j\omega)|^2$ instead of a finite value, and therefore the value of the characteristic function at zero frequency cannot be related to the transformer ratio as before. Second, the topology of this network is not completely fixed by the specification of the transmission zeros. It is possible to move the inductor through the unit elements and so modify the transformer ratio for any arbitrary value between upper and lower limits. Therefore, the derivation of a set of equations to help determine the synthesis parameters for a given transformer ratio is very much more difficult and not really worthwhile. The most effective way to design these transformers is to estimate the number of elements that might be required for a given bandwidth and transformer ratio and then to iterate the synthesis until an optimum is found.

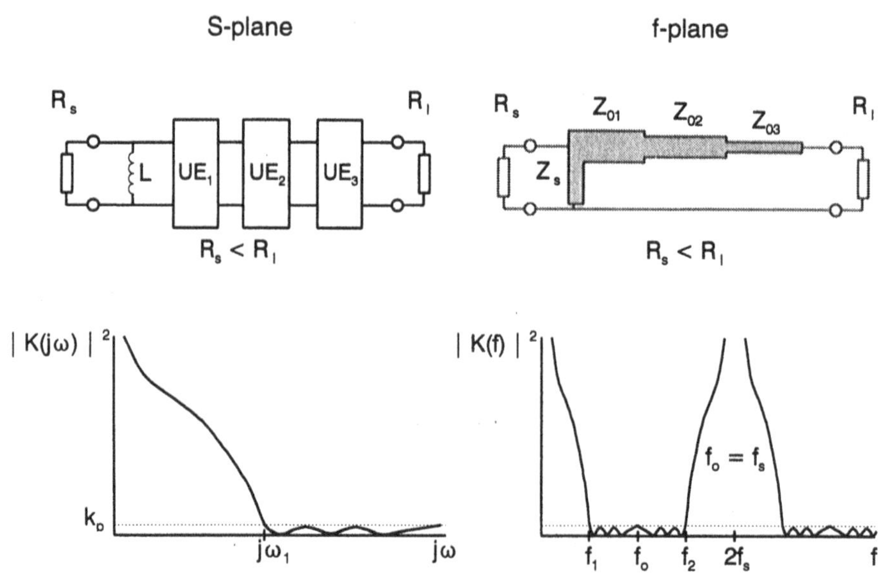

Figure 4.26 Prototypes and properties of line and short-circuit stub transformers.

During this procedure, the stub should be extracted at one extreme end of the circuit depending on whether a transformation up or down in impedance is required.

Realization of transformers of this kind is convenient in triplate stripline or in microstrip transmission line media, wherein impedance ratios of more than 10:1 and bandwidths of more than 20:1 are relatively easily achieved.

The final type of transformer deserving a brief description is shown in Figure 4.27. It is a derivative of the coupled-line band-pass filter and is the dual of the line-and-stub transformer. It has the advantage that no short circuits are required, allowing the circuit to be completely planar and potentially less expensive to fabricate. Its disadvantage is that bandwidths of greater than 100% are difficult to realize without an unacceptably narrow gap between the coupled lines.

Synthesis of the *S*-plane prototype follows the same procedure as for the line-and-stub transformer except for the selection of a first instead of a second canonical form. The nonredundant network would be synthesized with a single series capacitor at one end and then the capacitor would be split and transformed into two equal parts, one on either side of the first unit element, to allow for the creation of the coupled-line section.

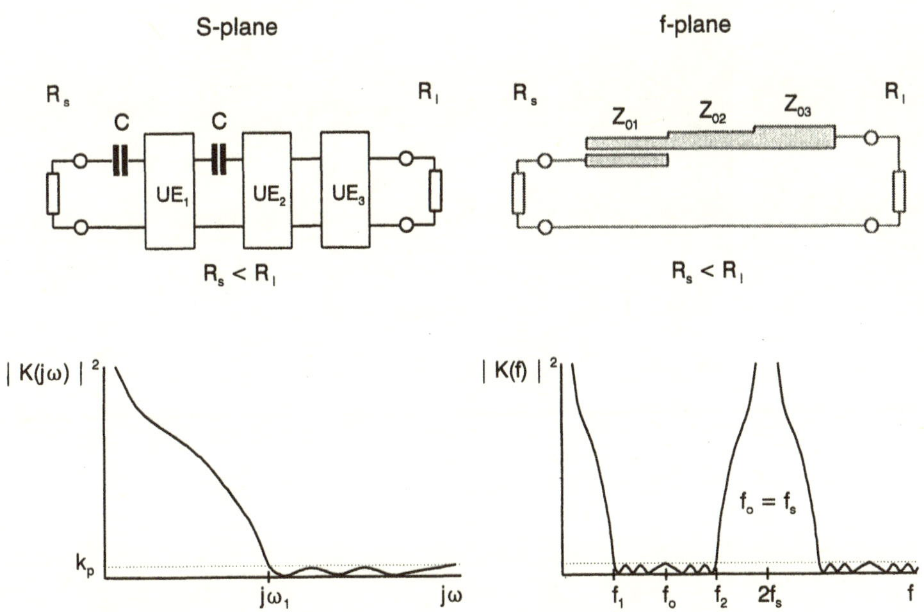

Figure 4.27 An impedance transformer using capacitively coupled lines.

References

[1] Minnis, B. J., "Decade Bandwidth Bias Ts for MIC Applications up to 50 GHz," *IEEE Trans. MTT*, Vol. MTT-35, No. 6, June 1987, pp. 597–600.

[2] Minnis, B. J., and M. Healy, "New Broadband Balun Structures for Monolithic Microwave Integrated Circuits," *IEEE MTT-S International Microwave Symposium Digest*, Vol. 2, June 1991, pp. 425–428.

[3] Cohn, S. B., "Characteristic Impedance of Shielded Strip Transmission Line," *IRE Trans. MTT*, Vol. MTT-2, July 1954, pp. 52–55.

[4] Lange, J., "Interdigitated Strip-Line Quadrature Hybrid," *IEEE Trans. MTT*, Vol. MTT-17, December 1969, pp. 1150–1151.

[5] Kirshning, M., and R. H. Jansen, "Accurate Wide-Range Design Equations for the Frequency-Dependent Characteristics of Parallel Coupled Microstrip Lines," *IEEE Trans. MTT*, Vol. MTT-32, No. 1, January 1984, pp. 83–90.

[6] Milsom, R. F., "FACET—A CAE System for RF Analogue Simulation Including Layout," *26th ACM/IEEE Design Automation Conf.*, 1989, pp. 622–625.

[7] Matthaei, G. L., L. Young, and E. M. T. Jones, Chap. 6 in *Microwave Filters, Impedance Matching Networks and Coupling Structures*, New York: McGraw-Hill, 1964, pp. 255–354.

[8] Dolph, C. L., "A Current Distribution for Broadside Arrays Which Optimizes the Relationship Between Beam Width and Side-Lobe Level," *Proc. IRE 34*, June 1946, pp. 335–348.

[9] Horton, M. C., and R. J. Wenzel, "General Theory and Design of Optimum Quarter-Wave TEM Filters," *IEEE Trans. MTT*, Vol. MTT-13, No. 3, May 1965, pp. 316–327.

CHAPTER 5

Active Circuit Design

In this final chapter concerning the practical application of exact synthesis procedures, particular attention is given to active microwave circuits. Over the last 10 years, advances in the fabrication of high-performance, three-terminal solid-state devices, such as MESFETs and high-electron mobility transistors (HEMT), and the development of MMIC technology on GaAs have made active circuit design the main occupation of microwave circuit designers. Most make use of modern, integrated CAD software to perform simulation, optimization, and layout of new circuits. However, few make use of any synthesis software, relying heavily on the modification and optimization of existing circuit designs as the basis for their design procedure. This is a pity, because synthesis techniques enable new circuits with novel topologies to be created that achieve performances as close as possible to the theoretical maximum. They give the designer much more control over the process of circuit creation and are the basis of the Universal Design Procedure that was formulated and proposed in Chapter 2.

The first of the main sections of the chapter will deal with the fundamental limits governing the synthesis of two-port matching networks. It will present the formulas for determining the bandwidth over which a particular quality of match can be achieved for several basic types of complex impedance terminations. This will then be followed by five practical sections describing various types of microwave amplifiers designed by exact synthesis. Section 5.2 will deal with the practical design of the important class of amplifiers known as *distributed*, or *traveling-wave*, amplifiers. Section 5.3 will deal with some reactively matched MMIC power amplifiers that achieve output power as high as 2.5W. Section 5.4 will deal with the design of a new type of cascadable amplifier circuit that uses a combination of traveling-wave and reactive matching networks. Finally, Sections 5.5 and 5.6 will together describe the design of a multistage amplifier that produces a very worthwhile combination of 25-dB gain and 1.5W output power.

5.1 PRINCIPLES OF MATCHING INTO COMPLEX TERMINATIONS

In designing most types of active microwave circuits by exact synthesis, the major part of the process will be the synthesis of two-port matching networks. Most of these matching networks will have to operate between a resistive termination at one end and a complex termination at the other, but occasionally, as in the case of interstage matching networks, a complex termination will be present at both ends. Before any synthesis begins, it is essential that the theoretical limits on frequency bandwidth be identified. These can be calculated using some of the simple theory presented below.

Once the overall concept of a circuit has been established at the start of a design exercise, the circuit is partitioned into separate sets of active and passive two-ports. For each active device, an accurate equivalent circuit will be needed for the purposes of general circuit simulation, but it will also be needed to help derive simpler, one-port equivalent circuits representing the complex impedances seen looking into the two ports of the device. Each of these equivalent circuits must comprise a single resistor in combination with one or more reactance elements, the aim being to absorb the reactance elements into the adjacent matching network so that they can be synthesized as part of the prototype matching network. The resistor is then treated as if it were one of the two resistive terminations at the ends of a conventional two-port reactance network. Because the reactance elements are to be absorbed into the matching network, for the purposes of the synthesis, the topology and combination of elements in the one-port equivalent circuit must obey the same rules that apply to any reactance two-port (see Chapter 2, Section 2.3.6.3).

The basic frequency responses of the matching network will be largely determined by the types of reactance elements present in the one-port equivalent circuits. Often, only a single low-pass or high-pass element will be involved in each one-port circuit. Occasionally there may be two elements, but it would be rarely necessary to model a complex termination with more than two elements. A single low-pass element will force the matching network to have either a low-pass or a band-pass response, and a high-pass element will force either a high-pass or band-pass response. In both cases, however, the preferred choice of frequency response will almost always be band-pass. Band-pass matching networks have much greater versatility than either of the other two types. Low-pass networks have especially limited scope because of their inability to achieve any impedance level transformations.

While the types of elements in the one-port equivalent circuits determine the type of frequency response (i.e., band-pass or high-pass) of the matching network, the values of the elements determine the quality of match that can be achieved over a given bandwidth. By being aware of this limiting bandwidth before the start of synthesis, the need for design iteration can be minimized. Fano [1], in his classic paper on broadband matching, provided the necessary theory for the calculations. It will now be applied to the four most commonly encountered types of complex loads.

According to Bode [2], all passive one-port impedance networks have what could be loosely described as a *match-bandwidth* product. That is, for any given complex, terminating impedance, the match that can be achieved between the impedance and a power source with a real internal resistance is a function of bandwidth; that is, the wider the bandwidth, the worse the match, or conversely, the narrower the bandwidth, the better the match. More precisely, Bode states that

$$\int_{\omega=0}^{\infty} \ln \frac{1}{|\rho(\omega)|} d\omega \leq \text{constant} \tag{5.1}$$

where $\rho(\omega)$ is the reflection coefficient at the input of a loss-free, passive matching network terminated by the complex impedance. In other words, the area under the curve $\ln(1/|\rho(\omega)|)$ is a constant and therefore, as shown in Figure 5.1, it is possible to achieve many different combinations of match and bandwidth for a given load impedance.

5.1.1 Shunt Capacitor and Resistor

If the impedance looking into one of the ports of an active device is approximated by a shunt capacitor (C) and resistor (R) as shown in Figure 5.2, then for this low-pass combination, Fano gives

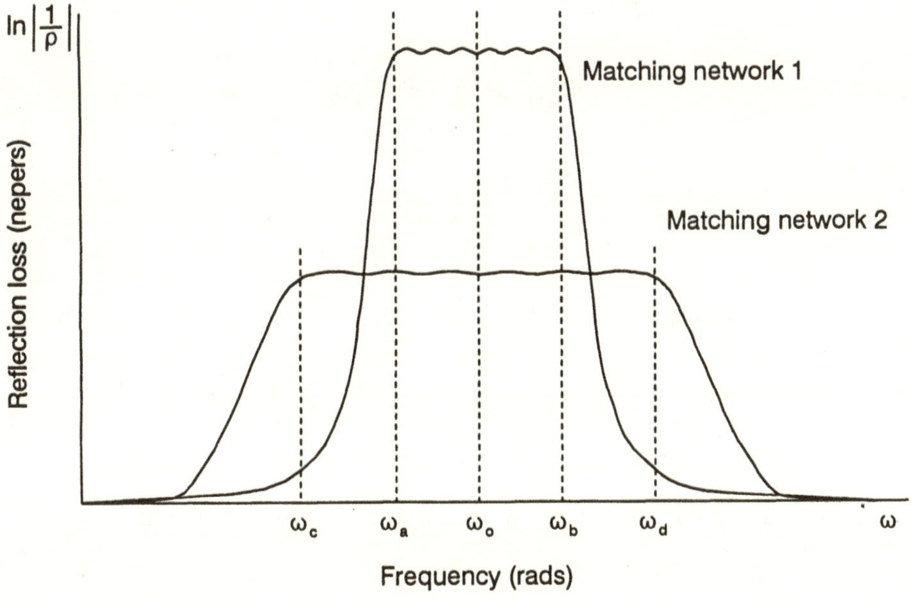

Figure 5.1 Different match vs. bandwidth functions for a given complex load.

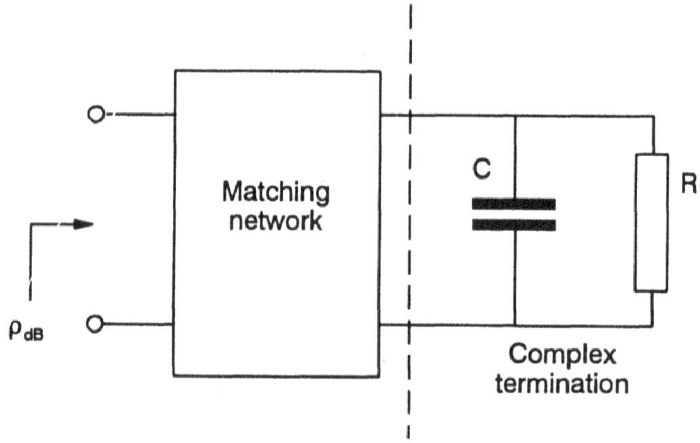

Figure 5.2 Shunt *R/C* termination.

$$\int_{\omega=0}^{\infty} \ln \frac{1}{|\rho(\omega)|} d\omega \leq \frac{\pi}{RC} \tag{5.2}$$

If $|\rho|$ is set to a constant value of $|\rho_x|$ between frequencies ω_1 and ω_2, where $\omega_1 < \omega_2$, then (5.2) simplifies to

$$\ln\left(\frac{1}{|\rho_x|}\right)(\omega_2 - \omega_1) = \frac{\pi}{RC} \tag{5.3}$$

and taking inverse logarithms we find

$$|\rho_x| = e^{\frac{-\pi}{RC(\omega_2 - \omega_1)}} \tag{5.4}$$

Alternatively, if the value of the reflection coefficient is expressed in decibels (ρ_{dB}), then using

$$\rho_{dB} = 20 \log_{10}|\rho_x| = 20\frac{\log_e|\rho_x|}{\log_e 10} \tag{5.5}$$

equation (5.2) gives an extremely convenient expression for the reflection coefficient:

$$\rho_{dB} = \frac{-8.6859\pi}{RC(\omega_2 - \omega_1)} \tag{5.6}$$

This is a surprisingly simple but valuable result. It shows that for given values of R and C, there is a maximum, absolute frequency bandwidth over which a given constant value of ρ_{dB} could be maintained by the very best of passive matching networks. In fact, the maximum could never be achieved in practice, since this would require a network with an infinite number of elements. However, a well-designed network will achieve close to the maximum, and the value is a good guide as to whether a particular performance specification can be met at all.

If the combination of match and bandwidth predicted by (5.6) does satisfy the given performance specification, then the figures can be used as good starting values for the purposes of synthesis. If they do not meet the specification, then there would be no point in starting any synthesis because the resulting network would be incapable of absorbing the shunt capacitor of the external complex load. Unless some relaxation of the specification can be tolerated in terms of either match or bandwidth, the only way the specification could be met would be to lower the Q of the complex load. This would entail adding an additional shunt resistor to lower the effective value of the resistor R, but would carry the disadvantage of wasting what may be scarce signal power.

It is important to emphasize that the bandwidth figure given by (5.6) refers to an absolute and *not* a fractional bandwidth. Hence, the maximum, absolute passband width will be the same whether the lower edge is set to zero (i.e., for a low-pass response) or set to a much higher frequency (e.g., for a band-pass response). Furthermore, the equation applies whether the equivalent circuit contains lumped, s-plane reactive elements or S-plane elements. Only the transformation into the f-plane is different. To establish the achievable f-plane bandwidth for a lumped-element matching network, the linear frequency transformation $s = j\omega = 2\pi f$ should be used, whereas the transformation $S = j\omega = j\tan((\pi/2)\cdot(f/f_s))$ should be used for a distributed matching network. For the lumped-element, s-plane case, the exact result of the transformation into the f-plane is

$$\rho_{dB} = \frac{-8.6859}{2\,RC(f_2 - f_1)} \quad \text{if } \omega = 2\pi f \tag{5.7}$$

However, this is also a good approximation to the achievable bandwidth for a distributed matching network if, as in most circumstances, the elements of a distributed matching network are electrically short in the frequency range of the passband.

5.1.2 Series Inductor and Resistor

A series inductor and resistor combination of the form illustrated in Figure 5.3 is the dual of the preceding example. Hence, by using the simple properties that apply to any pair of network duals, (5.2) and (5.7) can be transformed into equivalent equations for the series inductor and resistor.

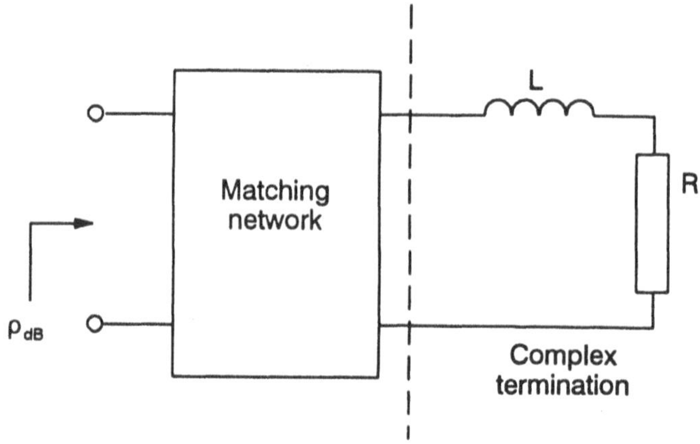

Figure 5.3 Series R/L termination.

Thus, from Fano,

$$\int_{\omega=0}^{\infty} \ln \frac{1}{|\rho(\omega)|} \mathrm{d}\omega \leq \frac{\pi R}{L} \tag{5.8}$$

and

$$\rho_{\mathrm{dB}} = \frac{-8.6859R}{2L(f_2 - f_1)} \tag{5.9}$$

where L and R are the values of the inductor and resistor, respectively.

5.1.3 Shunt Inductor and Resistor

A shunt inductor and resistor combination (Figure 5.4) is the high-pass equivalent of the previous shunt capacitor and resistor combination.

Thus, by applying a simple low-pass-to-high-pass frequency transformation, Fano's basic equation becomes

$$\int_{\omega=0}^{\infty} \ln \frac{1}{\omega^2 |\rho(\omega)|} \mathrm{d}\omega \leq \frac{\pi L}{R} \tag{5.10}$$

and using the same development as in Section 5.1.1,

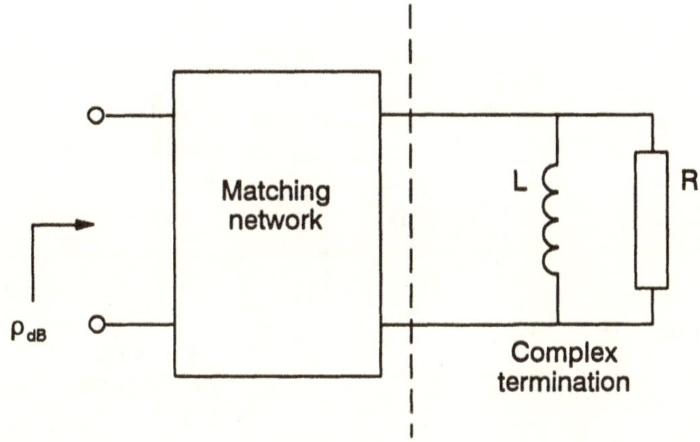

Figure 5.4 Shunt *R/L* termination.

$$\rho_{dB} = \frac{-8.6859\ \pi L}{R\left(\dfrac{1}{\omega_1} - \dfrac{1}{\omega_2}\right)} \tag{5.11}$$

where *L* and *R* are the values of the inductor and resistor, respectively.

5.1.4 Series Capacitor and Resistor

A series capacitor and resistor combination (Figure 5.5) is the dual of the shunt inductor and resistor combination.

The relevant bandwidth-determining equations are thus

$$\int_{\omega=0}^{\infty} \ln \frac{1}{\omega^2 |\rho(\omega)|} d\omega \le \pi RC \tag{5.12}$$

and

$$\rho_{dB} = \frac{-8.6859\ \pi RC}{\left(\dfrac{1}{\omega_1} - \dfrac{1}{\omega_2}\right)} \tag{5.13}$$

where *C* and *R* are the values of the capacitor and resistor, respectively.

All of the four preceding examples of complex loads have been assumed to be present at one end of a matching network and terminated at its opposite end

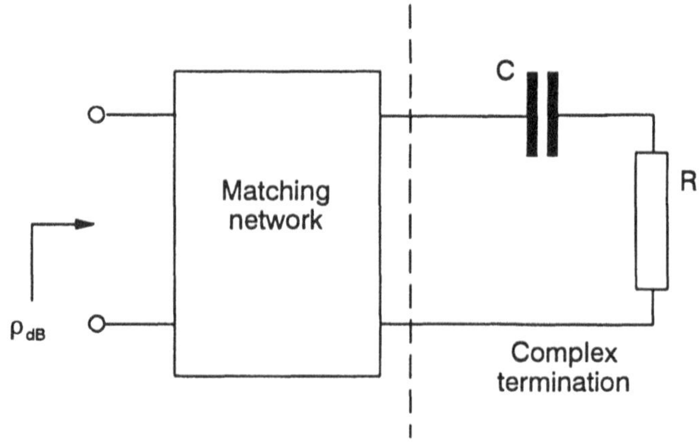

Figure 5.5 Series R/C termination.

in a pure resistance. When complex loads are present at both ends, then the network must be capable of absorbing the reactive elements of both the loads, the maximum achievable frequency bandwidth being determined by the load with the highest Q. In general, the distribution of the transmission zeros for the matching network should be chosen to ensure the creation of the correct types of reactive elements at the two ends of the network. However, while observing this general principle, it is advisable to select at least one but not more than two transmission zeros at $S = (0, j0)$ (i.e., high-pass elements). This gives the network the necessary effectiveness as an impedance level transformer without causing the problems that can be associated with realizing high-pass elements in a microwave circuit. Selecting roughly equal numbers of zeros at $S = (1, j0)$ (unit elements) and $S = (0, j\infty)$ (i.e., low-pass elements) ensures adequate separation of elements and tends to give the most desirable circuit topology. The total number of elements should be kept as small as possible and should correspond to a network with an even degree if the response is band-pass.

5.2 DISTRIBUTED AMPLIFIERS

The concept of a distributed amplifier was first reported by Percival [3] in 1937. In those early days, before the appearance of solid-state amplifying devices, the technique was applied to RF amplifiers employing thermionic devices, enabling them to achieve operating frequency bandwidths significantly greater than anything previously demonstrated. Following this success, very little interest was shown in distributed amplifiers until microwave MESFETs started to become available in the early 1980s. Since then, interest has grown considerably, and most companies in the business of designing microwave amplifiers have designed and fabricated distributed amplifiers in MIC and/or MMIC technology. Some notable references on the subject are [4–7].

The basic concept of the solid-state distributed amplifier is illustrated by the simple diagram in Figure 5.6. Essentially, the amplifier comprises a pair of uniform filter structures lying adjacent to each other, between which are connected a row of FETs in a common source configuration. The parasitic capacitances of the gates of the FETs form an integral part of the gate filter structure, and the capacitances of the drains form an integral part of the drain filter. In the normal mode of operation, an input signal enters from the left of the gate filter structure, propagates along the structure, suffering relatively little loss until it is absorbed at the far end by the terminating resistor. As the signal travels past the gates, the induced gate voltage is transformed into a current by each FET, which is then injected into the drain filter structure. Half of the injected current from each FET propagates toward the left and half toward the right. Those currents propagating to the left interfere destructively and largely cancel each other except at very low frequencies. Those propagating to the right arrive in phase at the output load, where they add constructively to produce the output signal.

To ensure the correct phasing of the currents in the drain filter, both the gate and drain filters need to be electrically identical. Their frequency characteristics are usually low-pass and determine the frequency characteristics of the whole amplifier. In this regard it is particularly important that the filters have the same passband cutoff frequencies. Since the gate capacitance is always greater than the drain capacitance, the maximum cutoff frequency is determined by the gate capacitance,

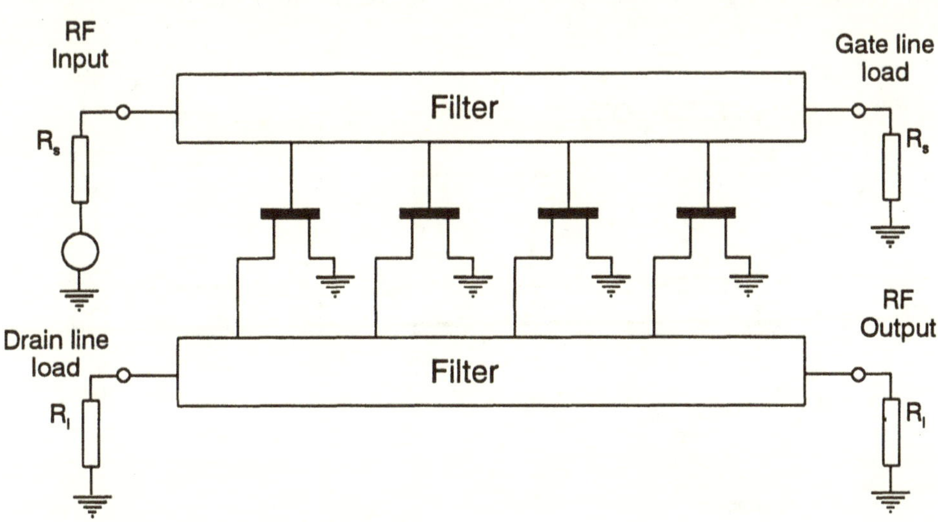

Figure 5.6 Basic distributed amplifier concept.

which in turn is determined by the gate width of the FETs. The number of FETs, as will be explained in more detail later, affects the general level and slope of the gain/frequency characteristics. The circuit designer can select any number, but four FETs is invariably the optimum for a flat gain response.

The distributed amplifier has numerous advantages over other types of microwave amplifiers. The most obvious advantage is the extremely wide bandwidth capability, which can often be in excess of a decade. Over this very wide bandwidth, the gain of the amplifier can be held to an almost constant value, and because of the presence of the internal terminating resistors, the return losses at the input and output terminals are usually excellent (e.g., typically better than 15 dB). Another major advantage is the high tolerance of the amplifier to variations in the characteristics of both its active and passive parts. Variations are an inevitable aspect of the fabrication process of any circuit and arise from the uncertainties associated with the equivalent circuit models used to represent the various circuit elements. By spreading out the active device or "distributing" the active device along the length of two matched filter structures, the distributed amplifier acquires unrivaled tolerance to these circuit variations.

5.2.1 Basic Design Concept and Prototype Synthesis

Figure 5.7 shows a more realistic circuit diagram of a distributed amplifier comprising four FETs connected between a pair of low-pass filters. The filters are uniform along most of their length and can be thought of as a pair of artificial transmission lines. Along the gate transmission line, capacitors at the gates of the FETs are

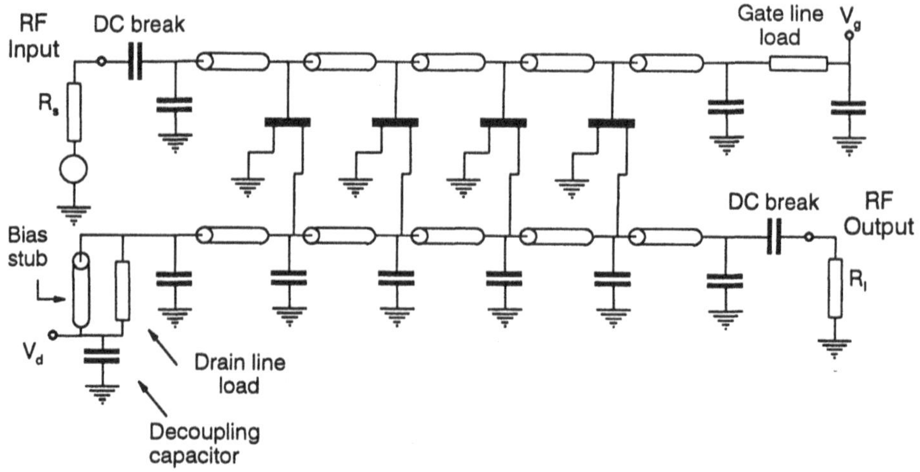

Figure 5.7 A common configuration for the distributed amplifier.

omitted because the FETs provide all the necessary shunt capacitance. Along the drain transmission line, the drains of the FETs cannot provide enough capacitance, and therefore extra capacitors need to be added to make up the shortfall. At the input and output ports of the amplifier, series coupling capacitors have been added to provide dc isolation. Also, at the remote end of the gate line, a shunt capacitor decouples the far end of the load resistor to facilitate the injection of gate dc bias. Drain dc bias is injected at the remote end of the drain line through a quarter-wavelength shunt stub. Two more capacitors decouple the end of the stub and the end of the load resistor. Except for extremely wide bandwidth performance, these additional bias filtering elements can be ignored for the purposes of designing the low-pass filters.

It will be apparent from Figure 5.7 that the distributed amplifier circuit is a mixture of lumped and distributed circuit elements, which presents the designer with two fundamental design options. He may choose to base the design of the amplifier on the synthesis of lumped-element prototypes, in which case he will have to perform a subsequent transformation of all the inductors into high-impedance transmission lines. Alternatively, he may choose to base the design on the synthesis of distributed-element prototypes, in which case he will have to transform some shunt open-circuit stubs into lumped capacitors. The better choice depends on the relative numbers of lumped and distributed elements in the finished circuit and on the relative accuracy of the lumped-to-distributed or distributed-to-lumped element conversions. Most designers have previously opted for the lumped-element approach, probably because of its apparent simplicity. This, however, is not the better choice, because transmission lines are generally poor approximations to series inductors. Furthermore, it is sometimes physically advantageous to use shunt stubs instead of lumped capacitors in the main body of the circuit, and if this is taken into account, there are substantially more distributed elements than lumped elements.

If the distributed element approach is adopted, the *f*-plane equivalent circuit of the artificial transmission lines will usually take the form shown in Figure 5.8(a). This is an almost uniform cascade of shunt open-circuit stubs and interconnecting transmission lines. Only the outermost stubs and lines at the two ends of the filter differ significantly in impedance from the rest. In this form it is a band-stop commensurate line network with all the elements a quarter of a wavelength long at the center frequency of the stopband. The corresponding *S*-plane prototype is of the form shown in Figure 5.8(b). This is a cascade of shunt capacitors and unit elements, each capacitor corresponding to a transmission zero specified at $S = (0, j\infty)$ and each unit element to a zero specified at $S = (1, j0)$. For a symmetrical network, the number of transmission zeros, and therefore the degree of the network, must be odd. As indicated by the frequency transformation shown in Figure 5.8(c), the *S*-plane response is low-pass. It is usual to synthesize the network for a Tchebycheff, equal-ripple passband response with a ripple value low enough to give good

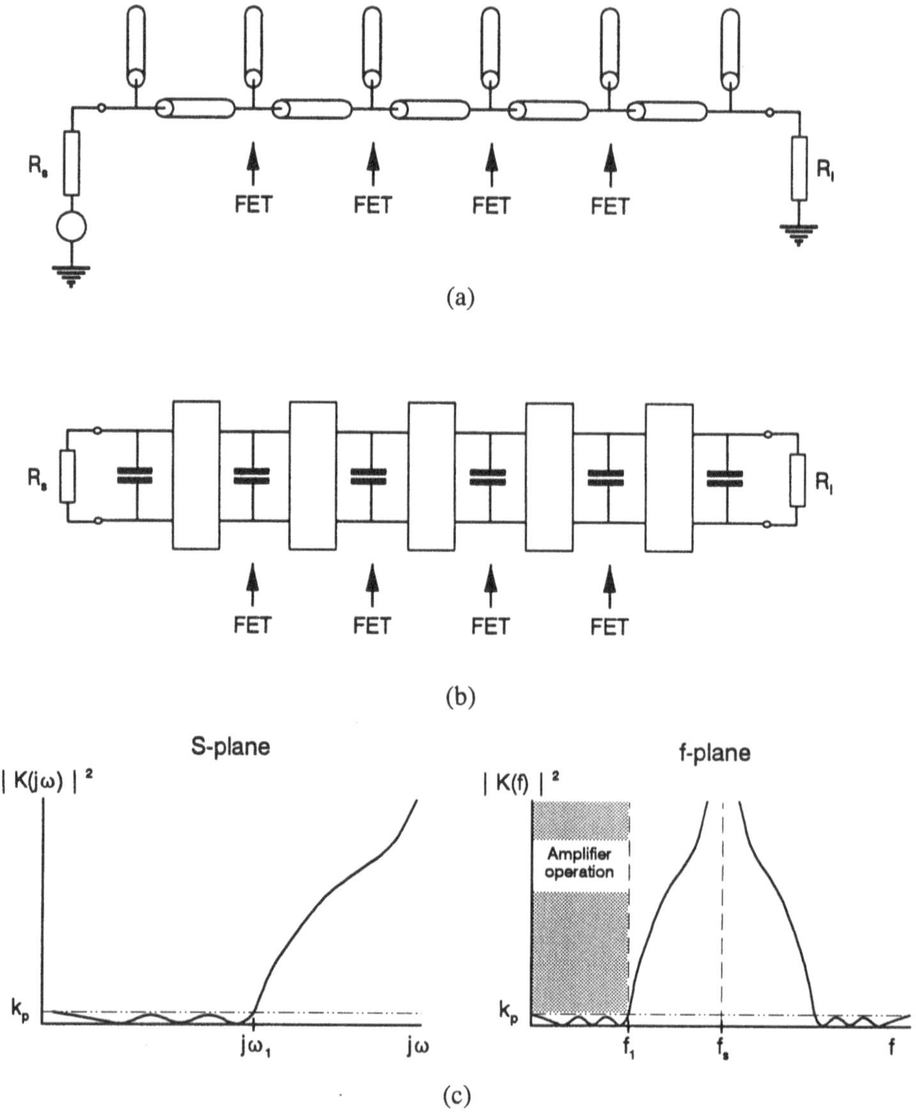

Figure 5.8 *S*-plane and *f*-plane prototype networks for the distributed amplifier: (a) *f*-plane equivalent circuit; (b) *S*-plane equivalent circuit; (c) *S*-plane and *f*-plane frequency responses.

return losses at the output terminals (e.g., a good choice is T_p = 0.1 dB for an equivalent return loss of 17 dB).

Figure 5.8(b) shows the prototype *S*-plane network in its second canonical form (i.e., the second dual). In this form, the impedance in the central, uniform region of the network is significantly lower that at the two ends, making the inner capacitor values larger than the two outer capacitor values. For this reason, FETs should normally be connected only at the positions of the inner capacitors and not at the positions of the two outermost capacitors (see Figure 5.8(b)). In turn, this allows the largest possible FET devices to be employed for a given cutoff frequency. To find the maximum size of a FET, the maximum gate capacitance (C_{gs}) must be calculated using the formula

$$2\pi f_1 C_{gs} = C_S \omega_1 \tag{5.14}$$

where

$$\omega_1 = \tan\left(\frac{\pi}{2}\frac{f_1}{f_s}\right) \tag{5.15}$$

C_S is the value of the *S*-plane capacitors in the middle of the network and f_1 is the intended maximum frequency of operation of the amplifier.

It may be interesting to note that an alternative form of prototype exists for the distributed amplifier, which is almost, but not quite, the dual of the previous prototype. The network is shown in Figure 5.9 and is of the first canonical form. However, in contrast to the previous prototype, it has fewer capacitors than unit elements, and the impedance in the middle region of the network is generally higher that at the two ends. The absence of capacitors at the two ends of the network suggests that a physically smaller amplifier could be constructed for a given number of FETs. This may be true, but the higher network impedances force the FETs to be smaller, which has an adverse effect on gain and on power output. Generally, the prototype of Figure 5.8(b) is to be preferred.

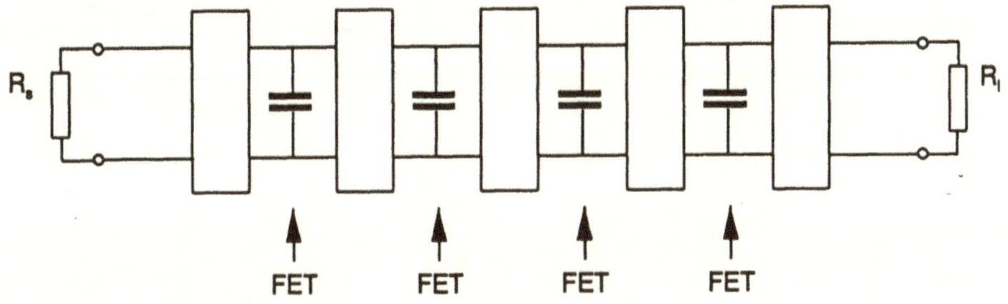

Figure 5.9 An alternative *S*-plane prototype.

5.2.2 A Simple Theory Based on Constant K Sections

Most designers of distributed amplifiers have in the past tended to use constant K section low-pass filters as the basis for their amplifier design [8]. An example of a constant K section low-pass filter in its second canonical form is shown in Figure 5.10. Several reasons are believed to be the cause of this preference. First, being a cascade of identical sections except at the two ends, the filter network is naturally constrained to be uniform—a condition mistakenly thought to be essential for the distributed amplifier. Second, the constant K section network is the network most commonly associated with the lumped-element approximation of a transmission line. Finally, the formulas given in Figure 5.10, which relate the L and C values of the network to the cutoff frequency and characteristic impedance, are attractively simple. Together, these factors have helped promote the constant K section design approach.

The main criticism of the constant K section approach is that, compared to exact synthesis, it is much less precise and the resulting network requires more computer optimization. Because the prototype is lumped-element, all the inductors must be converted into transmission lines by what can be rather poor approximations. Furthermore, a constant K section filter requires the addition of special end sections [9] (m-derived sections) that allow it to operate properly into its resistive terminations R_s and R_l. These end sections are not easily realizable in distributed form and are usually omitted, which is one of the reasons why the circuit tends to need substantial optimization. As a last comment, the constant K section approach is rather inflexible. Some of the circuits to be described later in this chapter are novel derivatives of the distributed amplifier and could not have been designed using constant K sections. Their design was dependent on the very much greater topological flexibility offered by exact synthesis techniques.

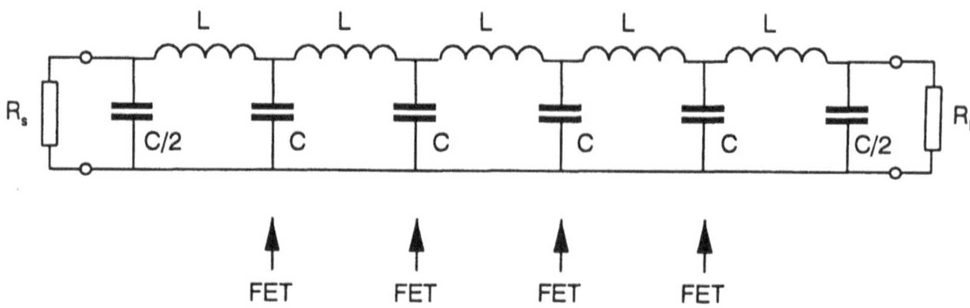

Basic equations: $f_c = \dfrac{1}{\pi\sqrt{LC}}$ and $Z_o = \sqrt{\dfrac{L}{C}}$

Figure 5.10 A constant K section low-pass filter.

It is the firm belief of the author that distributed amplifiers are most effectively designed by exact synthesis. However, while synthesis produces accurate circuit solutions that require little or no optimization, the synthesis process itself gives no indication as to the size of FETs required, the number of FETs, or the likely gain of the resulting amplifier. For this purpose and for the purpose of giving an insight into the effect of changing certain key circuit parameters, a simple theory of distributed amplifiers based on constant K section filters can be helpful.

Figure 5.11(a) shows a single constant K section in which some resistive loss has been added in series with the shunt capacitor. The resistor and capacitor together are intended to approximate the input impedance to the gate of an FET connected at the junction of the two series inductors. If the effects of the loss resistor are small, the expressions for the cutoff frequency f_c and the characteristic impedance Z_0 are given by

$$f_c = \frac{1}{\pi\sqrt{LC}} \tag{5.16}$$

and

$$Z_0 = \sqrt{\frac{L}{C}} \tag{5.17}$$

where L and C are the inductance and capacitance per unit cell, respectively. Combining the two equations, we can also write for f_c

$$f_c = \frac{1}{\pi Z_0 C} \tag{5.18}$$

This equation shows that for a given value of Z_0, which is usually set to the value of the load terminations R_s and R_l, the cutoff frequency is fixed by the value of C. In other words, if an amplifier is to operate up to a maximum frequency of f_c, the equivalent shunt capacitance of the gate of the FET must not exceed the value C, which in turn determines the maximum gate periphery of the FET.

The starting point for an expression giving the power gain of a distributed amplifier is

$$G \approx \frac{n^2 g_m^2 R_s R_l}{4} \tag{5.19}$$

where g_m is the mutual conductance of an individual FET, n is the number of FETs, R_s is the source resistance, and R_l is the load resistance. Here it is assumed that R_s and R_l are also equal to the characteristic impedances of the gate and drain artificial

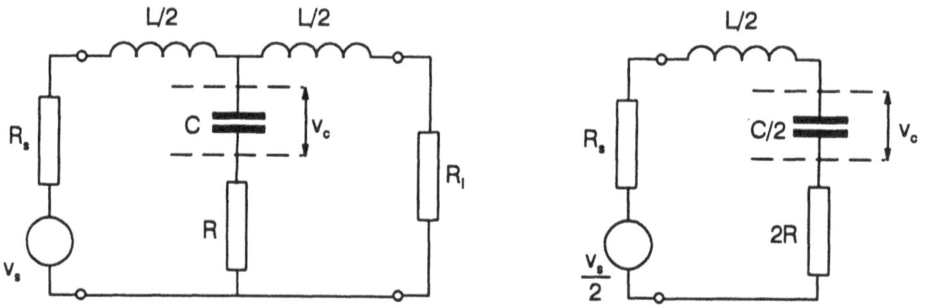

(a) (b)

$$Z_o = \sqrt{\frac{L}{C}} \quad ; \quad f_c = \frac{1}{\pi\sqrt{LC}} \quad ; \quad f_c = \frac{1}{\pi Z_o C}$$

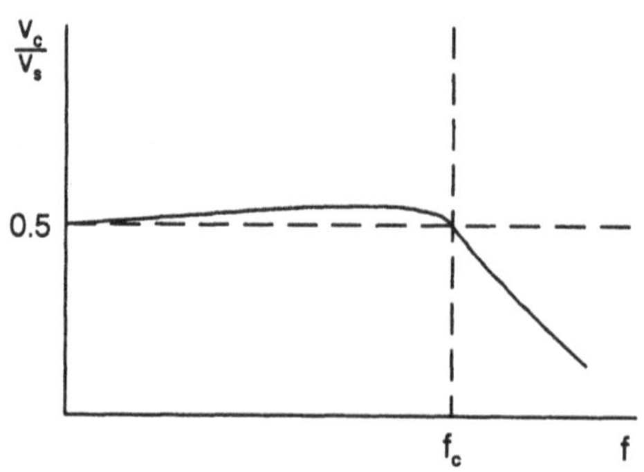

$$\frac{V_c}{V_s} = \frac{Z_o}{2\,(R_s + 2\,R)} \approx 0.5$$

$$Q = 2\,\frac{V_c}{V_s} \approx 1.0$$

at f_c
if $R << R_s$
and $Z_o = R_s$

(c)

Figure 5.11 Characteristics of a low-pass constant K section: (a) constant K section; (b) half-constant K section; (c) frequency response.

transmission lines. The expression shows that increases in g_m or the line impedances will give an increase in gain. However, the expression should be used with caution for calculating absolute values of gain, since in its derivation, the effects of ohmic losses have been ignored. Without losses, (5.19) implies that an unlimited gain can be achieved by, for example, just adding more FETs. In practice, the maximum achievable gain of a distributed amplifier is always less than the maximum available gain (MAG) of the FETs.

Equation (5.19) can be developed into another interesting expression for gain that is linked to the fundamental parameter f_t of the FETs. The parameter f_t is the frequency at which the short-circuit current gain of the FET falls to unity. From (5.19), setting $R_s = R_l = Z_0$ and substituting for Z_0 using (5.18), the expression for the available gain of the amplifier becomes

$$G \approx \left[\frac{n g_m}{2\pi f_c C}\right]^2 \tag{5.20}$$

On the assumption that the capacitance C is equal to the gate capacitance C_{gs} of the FET, this equation gives the first indication that the *maximum* achievable gain of an amplifier with a given number of FETs is largely independent of the size of the FETs, because both g_m and C_{gs} are directly proportional to FET size. Further confirmation is obtained when the expression

$$f_t = \frac{g_m}{2\pi C_{gs}} \tag{5.21}$$

is used in (5.20) such that the gain is given by

$$G \approx n^2 \left(\frac{f_t}{f_c}\right)^2 \tag{5.22}$$

Equation (5.22) indicates that the gain of an amplifier is proportional to the square of the ratio of f_t and f_c. The f_t of a transistor is, to a first approximation, independent of size, and f_c, the upper operating frequency of the amplifier, is also independent of size if the appropriate impedance transformations are allowed inside the filter networks. Gain is therefore inversely proportional to the square of the cutoff frequency. As a simple example, a distributed amplifier with n FETs and a cutoff frequency f_c will have a *maximum* gain that is 6 dB less than an amplifier with the same number and type of FETs, but with a cutoff frequency $f_c/2$. Like (5.19), (5.22) has not taken the effects of ohmic losses into account. It will give rather optimistic values of absolute gain, particularly for n greater than the order of 4.

The simple constant K section theory can be used to explain why distributed amplifiers tend to have a nominally flat gain/frequency characteristic rather than the sloping −6-dB/octave response of common reactively matched amplifiers. With reference to Figure 5.11, if the shunt R/C branch is assumed to be a representation of the FET gate connection, it follows that the gain of the amplifier will be proportional to the square of the normalized capacitor voltage v_c/v_s. To derive the frequency dependence of the function and therefore that of the gain of the amplifier, it is convenient to split the single section of Figure 5.11(a) into half sections of the form shown in Figure 5.11(b). The process of splitting is based on the superposition principle and on the application of odd and even mode excitations at both ends of the single section. Then v_c/v_s can be expressed as

$$\frac{v_c}{v_s} = \frac{-jX_c}{2[(R_s + 2R) + j(X_l - X_c)]} \tag{5.23}$$

where X_l and X_c are the reactances of the inductor and capacitor, respectively. Substituting for X_l and X_c, this becomes

$$\frac{v_c}{v_s} = \frac{1}{j\omega C(R_s + 2R) - \dfrac{\omega^2 LC}{2} + 2} \tag{5.24}$$

and if $2R \ll R_s$, (5.16) and (5.18) can be used to simplify the expression such that the modulus of v_c/v_s is approximated by

$$\left|\frac{v_c}{v_s}\right| \approx \frac{1}{2\left|1 - \dfrac{f^2}{f_c^2} + \dfrac{jf}{f_c}\right|} \tag{5.25}$$

By evaluating (5.25) over a range of frequencies, the variation of v_c/v_s with frequency is found to be of the form indicated in Figure 5.11(c). As shown, the function is predominantly flat, rising slightly from a value of 0.5 at dc before falling again to 0.5 at f_c. This slight positive slope corresponds to the positive gain slope often seen in a distributed amplifier employing only one or two FETs. When more FETs are used, increased circuit losses counterbalance the positive slope to produce a flat gain in an amplifier with an optimum number of FETs.

At the cutoff frequency f_c, the loaded Q of the section must be given by

$$\left|\frac{v_c}{v_s}\right| = \frac{Q}{2} \tag{5.26}$$

and hence another expression for v_c/v_s is

$$\left|\frac{v_c}{v_s}\right| = \frac{1}{2}\left\{\frac{2}{\omega C(R_s + 2R)}\right\} \tag{5.27}$$

If $\omega = 2\pi f_c$ and $R_s = Z_0$, then also using (5.16) and (5.17), the expression simplifies to

$$\left|\frac{v_c}{v_s}\right| = \frac{Z_0}{2(Z_0 + 2R)} \approx 0.5 \qquad \text{if } R \ll Z_0 \tag{5.28}$$

The loaded Q of the section is therefore approximately equal to 1 at $f = f_c$.

5.2.3 Practical MMIC Distributed Amplifiers With Various Gain Slopes

In this section, exact synthesis techniques will be applied to the design of a practical MMIC distributed amplifier. However, the section begins by discussing in a little more practical detail how distributed amplifiers can be designed for sloping gain/frequency characteristics.

Most amplifiers are designed for a flat gain/frequency characteristic, but there are several applications in microwave systems where it can be useful to have an amplifier with a positive or a negative gain slope or even an adjustable gain slope. The most obvious general need for a positive gain slope comes from the area of loss compensation. This is particularly true in communication systems where frequency-dependent losses in long transmission paths can be a problem. A more specific example where amplifiers with different gain slopes have been useful is in the multistage, high-gain amplifier about to be described. Here, several MMIC amplifier modules with various gain slopes were connected in cascade to create an amplifier with a high and exceptionally flat gain characteristics. In effect, selecting amplifier modules with the different gain slopes was used as a means of overcoming the effects of process variations and improving processing yield.

For a reactively matched amplifier, a positive, constant gain slope is particularly difficult to achieve over a wide frequency range and involves sacrificing terminal match. A distributed amplifier, in contrast, can produce the positive gain slope over very wide frequency ranges and still maintain an excellent input and output match. It is equally capable of flat or negative gain/frequency characteristics.

The gain slope of a distributed amplifier is largely dependent on the loaded Q of the gate and drain artificial transmission lines. As was seen in the preceding section and Figure 5.11, if there are small losses in the artificial transmission lines and the load impedances are chosen for good terminal match, a positive gain slope will be produced. However, as more ohmic losses are introduced, the loaded Q of the lines decreases, causing a decrease in the gain slope until eventually it becomes negative. In practice, the parasitic losses introduced by a typical microwave MESFET are small enough such that in an amplifier containing just one or two FETs, the gain slope is still positive. When three or four MESFETs are used, the accumulated losses become large enough to cancel the positive slope and produce a flat gain response. With more than four FETs, the losses tend to dominate and the amplifier has a negative slope. Adjustment of the number of FETs is therefore the main method by which gain slope is preset.

There is at least one other useful method of controlling the gain slope, however. Since any change to the circuit that modifies the loaded Q of the artificial transmission lines will change the gain slope, changing the gate bias supply can have a significant effect. Altering the gate bias supply alters the gate capacitance, which in turn alters the Q value. As will be seen in the practical examples to follow, the gate dc bias can be an effective means of control, having the advantage that it can be applied more than once.

Three versions of an MMIC distributed amplifier have been designed giving positive, flat, and negative gain slopes, respectively, over the 6- to 18-GHz frequency band. They were fabricated at Philips Research Laboratories using an MMIC process developed jointly by Philips Research Laboratories and Philips Microwave (Hazel Grove). This process incorporated MESFET devices with gate lengths of 0.5 μm defined by deep UV photolithography. The three amplifiers employed three, four, and five FETs, respectively, and were all based on the same filter prototype. Hence, the synthesis procedure was only performed once and addressed specifically the design of the four-FET version of the amplifier.

The first step in the design procedure is to choose the gate width of the FETs. For a given upper operating frequency, FETs with the largest possible gate width must be chosen to maximize gain. According to (5.18), the capacitor values in the middle region of the gate artificial transmission line must be no greater than 0.32 pF for a cutoff frequency of 20 GHz and a load impedance of 50Ω. With a prior knowledge of the capacitance/unit gate width of the FETs (1.25 pF/mm), this can be related directly to a FET gate width. However, without taking into account the effects of series inductance at the gate terminal, the value of gate width obtained will be too large.

After a little trial and error, a gate width of 150 μm was determined to be the most suitable. The equivalent circuit of a FET with a 150-μm gate width is shown in Figure 5.12, corresponding to the dc bias conditions of V_{ds} = 3V and V_{gs} = 0V. By analyzing the network over the 1- to 20-GHz frequency range, the S_{11} and S_{22} frequency responses are modeled by the much simpler one-port equivalent circuits also shown in the figure. These show that despite the relatively low value of C_{gs} in the FET equivalent circuit (0.188 pF), the effective capacitive loading produced by the FET on the gate artificial transmission line will actually be 0.28 pF at 20 GHz, which is close to the maximum of 0.32 pF for the 20-GHz cutoff frequency. This increase is caused largely by the presence of the 47-pH gate parasitic inductance. On the drain side of the device, the effective capacitive loading of the drain will be the much lower value of 0.054 pF.

On an initial assumption that four FETs would be most appropriate for a flat gain response, an S-plane prototype was synthesized of the form shown in Figure 5.13. Here the f-plane cutoff frequency was set to 20 GHz, which gives a good working margin against the required maximum operating frequency of the amplifier

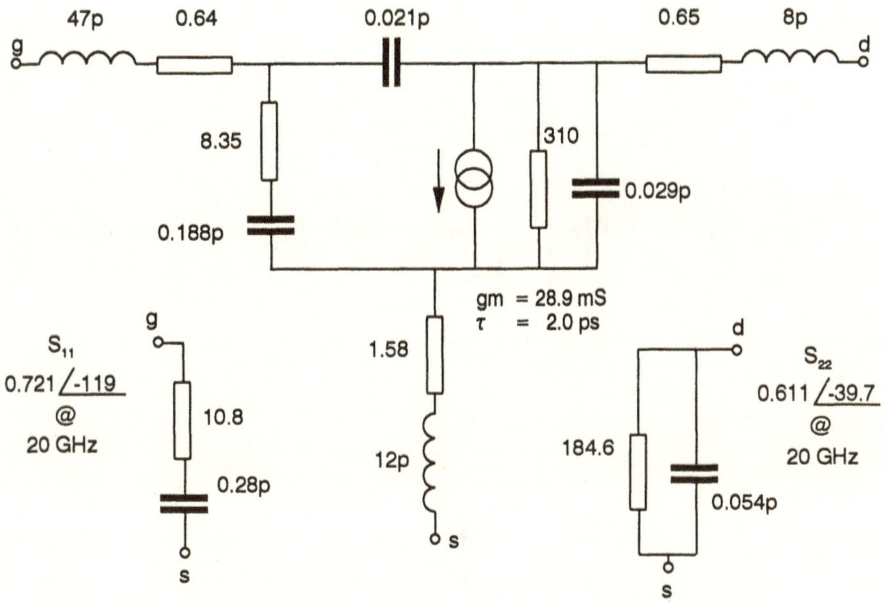

Figure 5.12 Equivalent circuit of 150 μm FET at V_{ds} = 3V and V_{gs} = 0V.

of 18 GHz. A commensurate frequency of 30 GHz was chosen to minimize the length of the subsequent transmission line elements without making their impedances unrealizable. These correspond to a low-pass cutoff frequency in the *S*-plane of 1.7321. The *S*-plane prototype contains six shunt capacitors and five unit elements corresponding to six transmission zeros at S = $(0, j\infty)$ and five at S = $(1, j0)$. It is of the second canonical form and of degree 11. Passband insertion loss ripple was set to 0.1 dB for port return losses of approximately 17 dB. Passband offset and slope parameters were both set to zero.

The *S*-plane prototype as shown in Figure 5.13 was synthesized using NETSYN, but is also easily synthesized using E-Syn. A distributed, Tchebycheff, QW-band-stop network must be selected and the stopband edges set to 20 and 40 GHz, respectively. Terminal impedances should be set to 1Ω, the passband ripple to 0.1 dB, and the offset parameter to zero. E-Syn will generate networks matching this specification automatically, but unfortunately none will be found to have a topology of the form required. Hence, the network elements should be extracted in a manual mode. A summary of the parameter settings for synthesis by E-Syn is given in Table 5.1.

As seen in Figure 5.13, except for the capacitors at the two ends, those in the middle region of the network are all of approximately the same value. Therefore, it causes an insignificant disturbance to the performance of the network to round

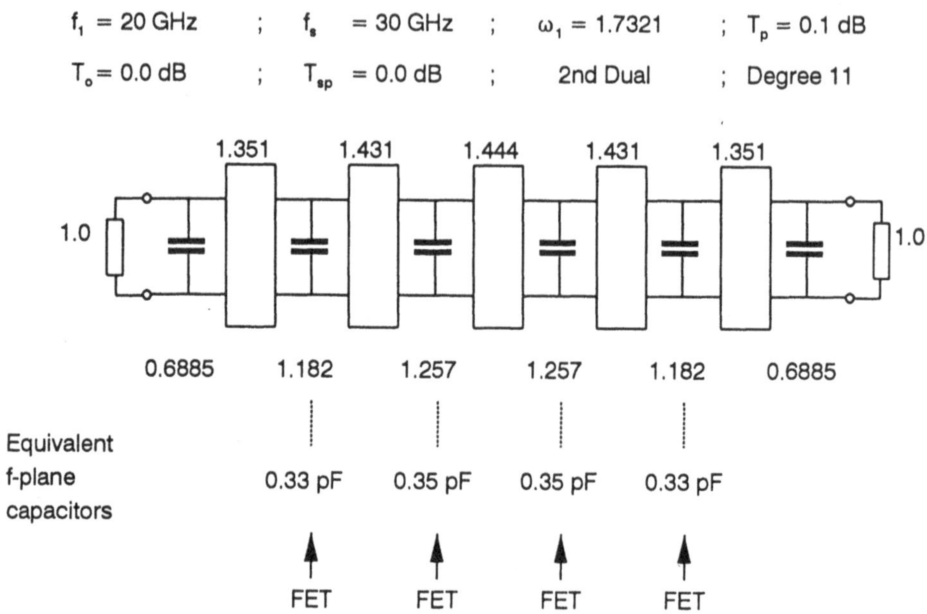

Figure 5.13 *S*-plane prototype of a four-FET, 6- to 18-GHz distributed amplifier.

Table 5.1
Prototype Synthesis by E-Syn for the 6- to 18-GHz Distributed Amplifier

Network type	Distributed	Tchebycheff	QW-band-stop	
Stopband	f_1 = 20 GHz	f_2 = 40 GHz	T_p = 0.1 dB	MIL = 0 dB
Terminations	R_s = 1	R_l = 1		
Elements	Five transmission lines	Six low-pass stubs		
Mode (F0)	Use a manual mode of element extraction			
NxtEL (F6)	Use successively to extract next element from choice given			
Analyze (F0)	Analysis after extraction			

the values of all these capacitors to a value of 1.2. Equating susceptances using the equation

$$\omega_1 C_s = 2\pi f_c C_f 50 \tag{5.29}$$

this *S*-plane capacitance (C_S) in a 1Ω system is equivalent to an *f*-plane capacitance (C_f) of 0.33 pF in a 50Ω system, which is almost exactly the figure given previously by (5.18). As indicated by the FET equivalent circuit, the gates of the FETs can provide most of this capacitance, and for the purposes of an initial assembly of the amplifier circuit, it can be assumed that no additional capacitors or stubs will be necessary at the gate connections. The drains, however, cannot provide enough capacitive loading. Using (5.29), an effective drain capacitance of 0.054 pF corresponds to an *S*-plane capacitance of 0.2. According to the prototype, a total of 1.2 is required, which means that the equivalent of an *S*-plane capacitor of value 1.0 must be added at each drain connection. This can be provided in a 50Ω system by either a shunt stub of characteristic impedance 50Ω or, alternatively, by a real capacitor of value 0.28 pF.

Taking the capacitive loading effects of the FETs into account, the S-plane equivalent circuit of Figure 5.13 can be transformed into the simple amplifier circuit shown in Figure 5.14. In making this transformation, a decision was made to use open-circuit stubs wherever possible instead of lumped capacitors to reduce the number of via hole ground connections required in the circuit. With all its stubs, the circuit appears rather different from the more conventional form of circuit shown in Figure 5.7. However, in minimizing the number of capacitors and via holes, a worthwhile improvement in processing yield can be obtained. No dc bias components have been included in the circuit of Figure 5.14, and at this stage, all the transmission lines are of the same length (i.e., $\lambda/4$ at 30 GHz ≡ 2.5 mm in air).

On first analysis, the circuit produced approximately 6 dB of gain with good input and output return losses, but the upper cutoff frequency was slightly below 18 GHz. The problem was traced to the values of the shunt stubs at the FET drain connections, the characteristic impedance of which needed to be increased to 60Ω. With this modification but with no circuit parasitics included, the circuit achieved

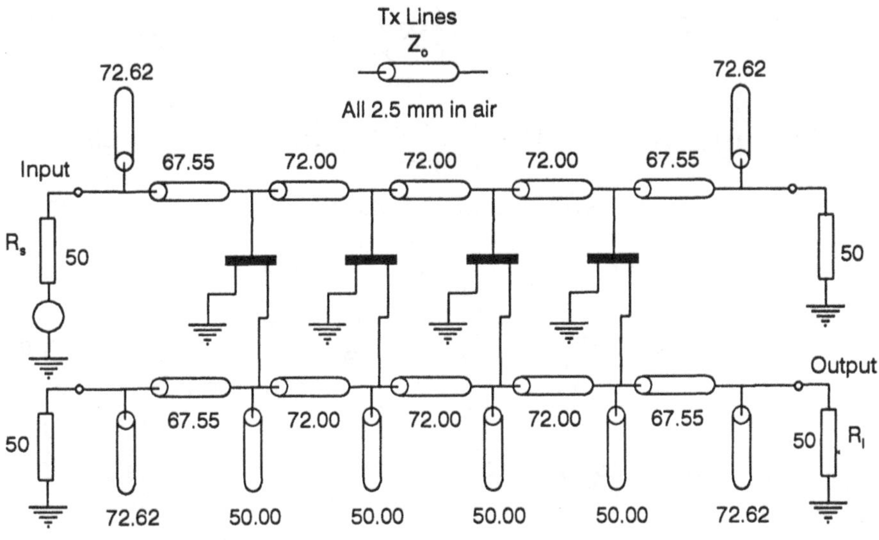

Figure 5.14 *f*-plane equivalent circuit without dc bias components.

6 ± 0.25 dB from 4 to 20 GHz. It is to the credit of the synthesis approach that this result was achieved without any computer optimization.

Despite the good performance, however, the circuit of Figure 5.14 has some topological difficulties. From the point of view of realization as an MMIC on GaAs, the stubs at the FET drains are rather long and there is a problem of gaining access to the FET drains by both the stubs and by the interconnecting lines between the drains. To overcome the problem, the whole drain artificial transmission line was effectively moved out, away from the FET drains, making the connections to the shunt stubs approximately half way along their length. The modification is illustrated in Figure 5.15. This not only solved the topological problems, but also gave some slight improvement in gain. The effect of the extra lines inserted between the FETs and the drain artificial transmission lines is to lower the effective output impedance of the FETs. Hence, at the end of the lines where they connect to the artificial transmission line, the effective drain capacitance of the FETs is increased, which allows the shunt stubs to be reduced in length. The effective g_m of the FETs is also increased, which accounts for the increase in gain.

As can be seen in Figure 5.15, all the dc bias components have now been added and there have been some minor changes to the lengths of some of the transmission line elements as a result of some superficial optimization. For the most part, line impedances were set to round numbers close to the original values and then line lengths were adjusted for optimum performance. The values of the coupling and decoupling capacitors were all set to 6 pF to maintain good performance at low frequencies, without being too large for practical realization on an

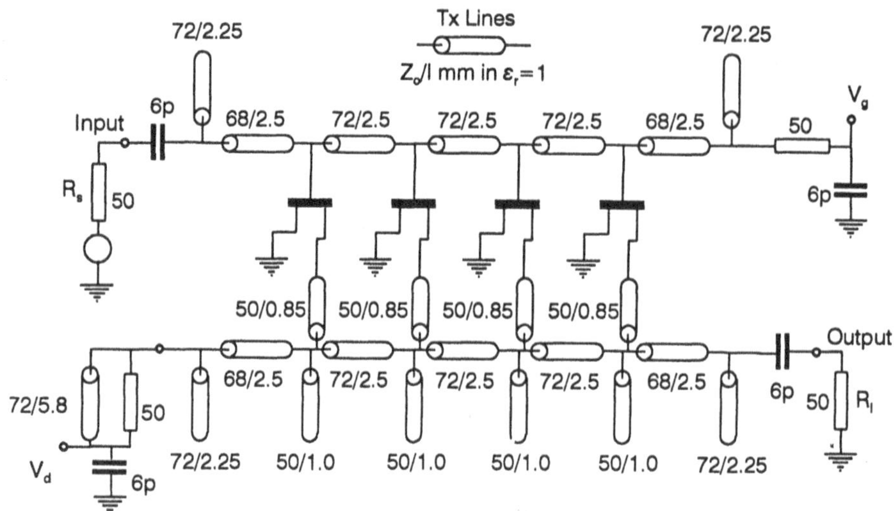

Figure 5.15 Complete equivalent circuit of a 6- to 18-GHz distributed amplifier.

MMIC. Gate bias involves no current flow, so the bias can be fed through the 50Ω resistor at the remote end of the gate line. However, drain bias is fed through an inductive stub at the remote end of the drain line. The stub length is not critical but has been set to slightly less than a quarter of a wavelength at the center of the operating frequency range (12 GHz).

From the element values given in Figure 5.15, the dimensions of all the microstrip lines and capacitors were calculated using the relevant references cited in Chapter 2, Section 2.5.2. Then the geometrical layout of the amplifier was completed ready for fabrication. A photograph of the resulting four-FET amplifier MMIC is shown in Figure 5.16. An input signal applied at the left enters the circuit on a short length of 50Ω line before meeting the 6-pF coupling capacitor and the first shunt stub. It then moves along the gate transmission line past the FETs, meeting the second shunt stub before being terminated by a 50Ω ion-implanted resistor. The four FETs are arranged in a straight line and are separated by a row of ion-etched via holes. The drain stubs below the FETs are clearly visible, as is the roughly central connection to the stubs by the lines of the drain artificial transmission line. Output signal appears on the right of the chip after passing through the second 6-pF coupling capacitor. Pads for the connection of dc bias supplies are located at the top (gate) and bottom (drain) right-hand corners. The long line at the bottom of the chip is the drain bias stub. The complete circuit measures 2.2 × 1.8 mm.

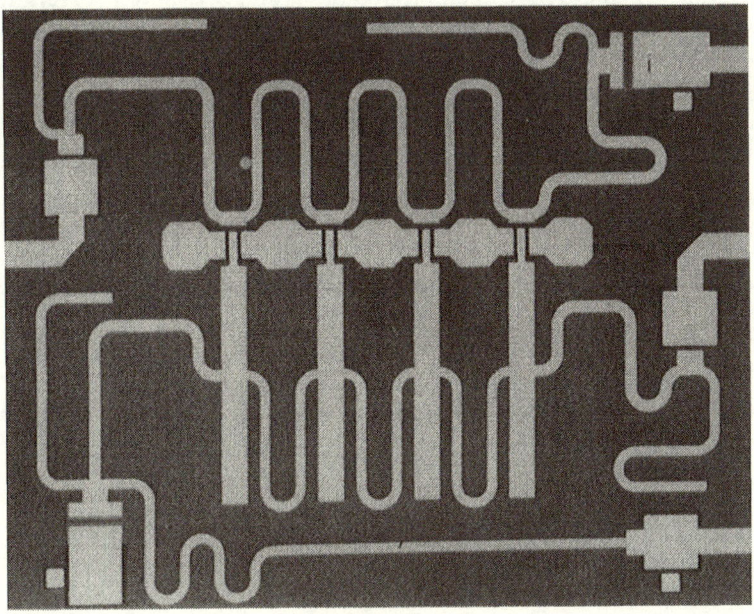

Figure 5.16 The four-FET distributed amplifier MMIC (2.2 × 1.8 mm).

A plot of the simulated performance of the four-FET amplifier complete with circuit parasitics is presented in Figure 5.17 for a gate bias of $V_{gs} = 0$ and a drain bias of $V_{ds} = 3V$. Gain is 6.15 ± 0.15 dB over the required 6- to 18-GHz frequency range, but remains at a useful level over the much wider range of 2 to 20 GHz. Terminal matches are better than 16 dB over 6 to 18 GHz. Corresponding measured performances of a real four-FET amplifier are plotted in Figure 5.18. The measurements were made with an HP8510 network analyzer for a single chip soldered to a carrier and mounted in a coaxial test fixture. Gain is 5.25 ± 0.25 dB over 6 to 18 GHz and the terminal matches are better than 16 dB over most of the same band. Comparing the gain responses of the measurement and simulation, the measured gain of the amplifier is a little lower and the bandwidth a little wider than predicted. The reason for this has been traced to a variation in FET characteristics on this particular wafer, caused by a slight variation in implant activation. Actual values of both gate capacitance and g_m were lower than the values used in the equivalent circuit model on which the amplifier design was based. Notwithstanding this variation, the amplifier design was considered to be a success.

Fabricated on the same wafer as the four-FET amplifiers were the three-FET and five-FET versions, offering positive and negative gain slopes, respectively. They were created by taking the four-FET amplifier and either removing or adding a FET together with its associated elements in the center of the circuit. For the sake of brevity, the three-FET and five-FET amplifier circuits will not be shown, but their measured performances have been plotted in Figures 5.19 and 5.20. As anticipated, the three-FET amplifier does have a positive gain slope, rising from 3.5 dB at

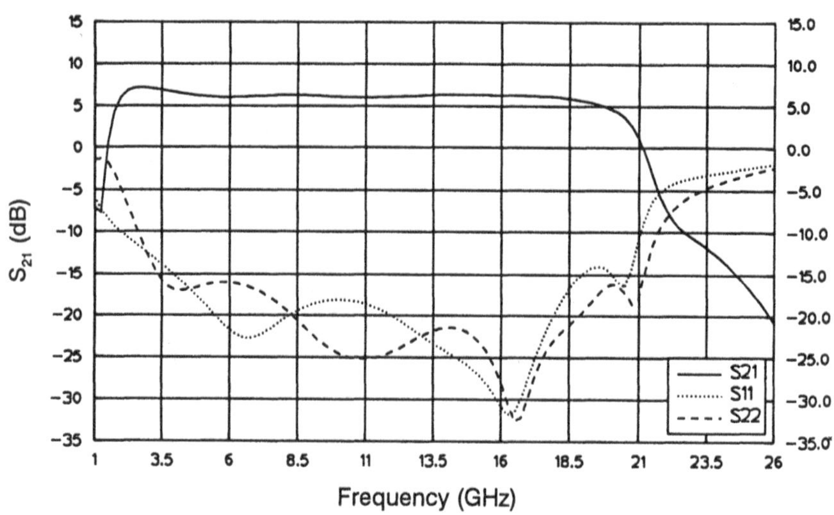

Figure 5.17 Simulated frequency responses of the four-FET distributed amplifier.

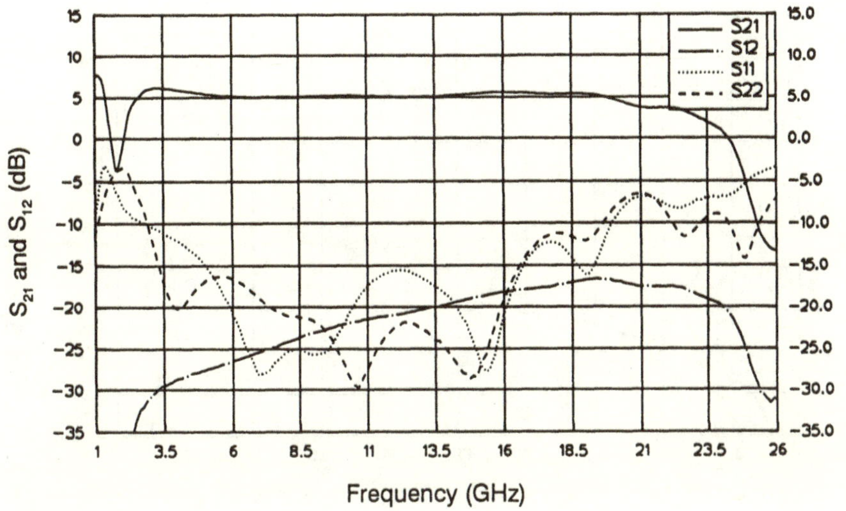

Figure 5.18 Measured frequency responses of the four-FET distributed amplifier.

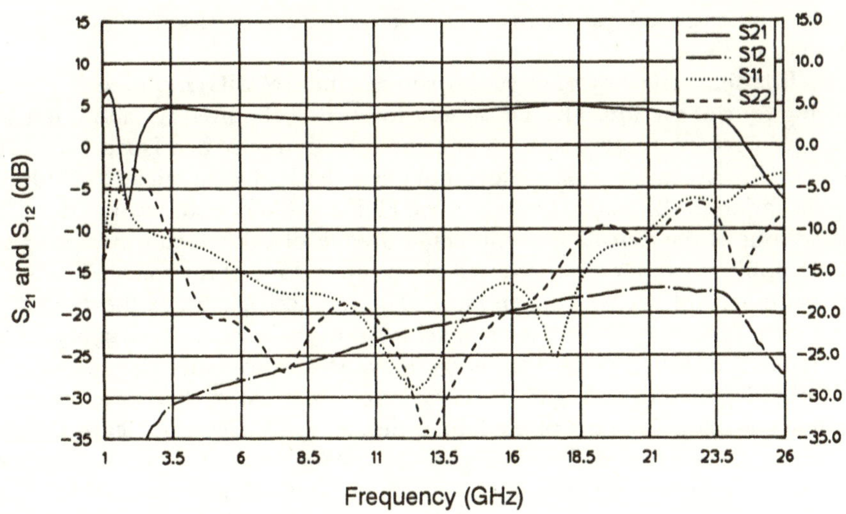

Figure 5.19 Measured frequency responses of the three-FET distributed amplifier.

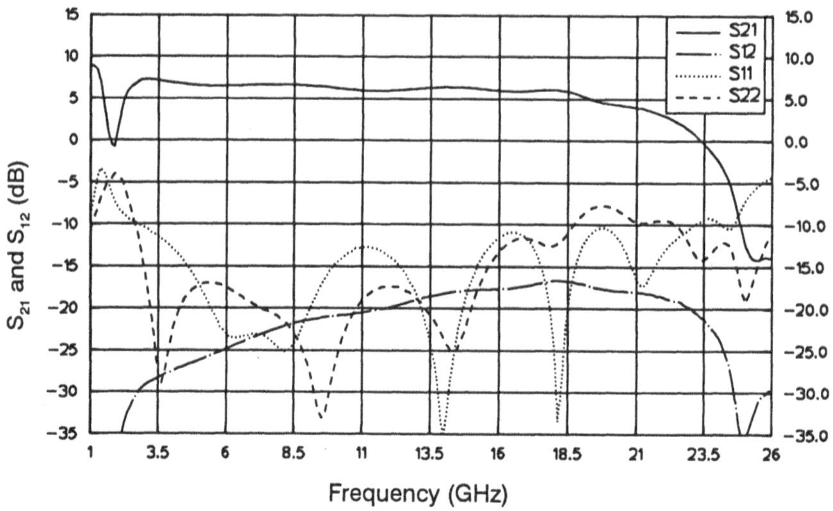

Figure 5.20 Measured frequency responses of the five-FET distributed amplifier.

6 GHz to 5 dB at 18 GHz. The five-FET amplifier has a negative slope falling from nearly 7 dB at 6 GHz to just under 6 dB at 18 GHz. This is a clear demonstration, therefore, that the gain slope of a distributed amplifier can be modified by changing the number of FETs.

As a first demonstration of cascading some of these MMIC amplifiers, a module comprising four of the four-FET amplifiers and two of the five-FET amplifiers has been constructed. A photograph of the module is shown in Figure 5.21. The objective in this case was to produce an amplifier with a high gain but an exceedingly small gain variation. By using two of the five-FET amplifiers at the two ends of the chain and four of the four-FET amplifiers in the middle, the negative slope of the five-FET amplifiers overcomes the very slight positive slope of the four-FET amplifiers. Measurement of the six-chip module (Figure 5.22) illustrates the success of the approach. A gain of 32 dB has been achieved over the 6- to 18-GHz band with a gain ripple of less than ± 0.5 dB. Return losses were also better than 15 dB.

Application of negative gate bias was also suggested earlier as a means of controlling gain slope in a distributed amplifier, and this has been clearly demonstrated using the six-chip module of Figure 5.21. With a negative gate bias applied so as to reduce drain current by 50%, the amplifier produced a pronounced positive gain slope, as indicated by the measured responses plotted in Figure 5.23. Gain is 21 dB at 6 GHz, rising to the higher value of 26 dB at 18 GHz.

5.2.4 Increasing the Maximum Operating Frequency of Distributed Amplifiers

A novel approach will now be described that can be used to extend the maximum operating frequency of a distributed amplifier. The approach relies heavily on the

Figure 5.21 Assembly of four 4-FET and two 5-FET distributed amplifier MMICs.

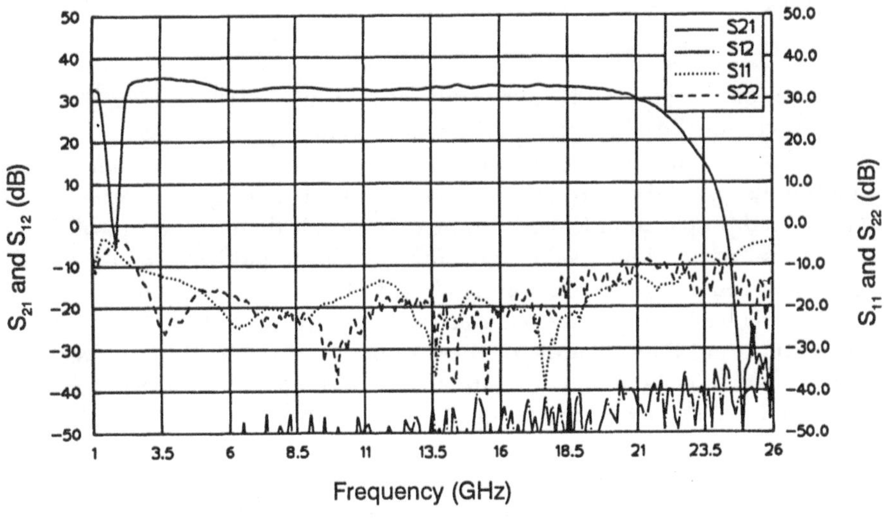

Figure 5.22 Measured frequency responses of the cascade of six MMIC amplifiers.

use of exact synthesis techniques and would be difficult to implement using any alternative, approximate procedure.

When the need arises for an increase in the maximum frequency of operation of a conventional distributed amplifier, some of the obvious options are: (1) to use better quality FETs with a shorter gate length to increase the value of f_t (see (5.22)), (2) to use smaller FETs to give a smaller value of C_{gs}, or (3) to lower the characteristic

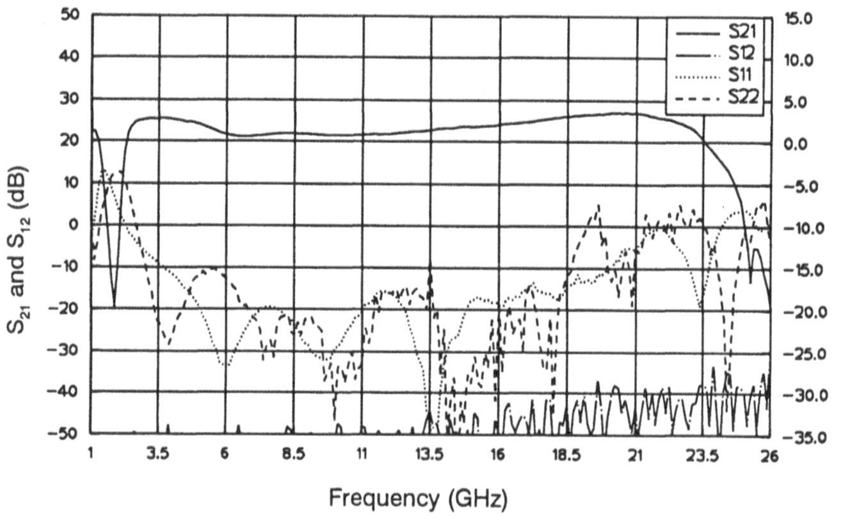

Figure 5.23 The effects of negative gate bias on gain slope.

impedances of the gate and drain artificial transmission lines (see (5.18)). In practice, better quality FETs may not be readily available, and if smaller FETs were to be used, they would have to be used in larger numbers to try to maintain gain. As the number of FETs increases, however, a limit is quickly reached when gate and drain line propagation losses become excessive. If the impedances of the gate and drain transmission lines were to be lowered, this would have an adverse effect on gain and would also cause a mismatch with the 50Ω source and load resistors unless impedance-transforming sections could be introduced as integral parts of the gate and drain lines. If none of these options are viable, then the technique to be described offers a potential solution. The technique involves using band-pass instead of low-pass filters for the artificial transmission lines of the amplifier. It will be illustrated with the help of a practical example.

In Section 5.2.3, a FET with a 0.5-μm gate length and a 150-μm gate width was used in a conventional distributed amplifier with a cutoff frequency of approximately 18 GHz. As indicated by the FET equivalent circuit in Figure 5.12, the effective input capacitance of the FET at 20 GHz was 0.28 pF, which, according to (5.18), permits a maximum operating frequency of approximately 22 GHz in a 50Ω system. This was broadly confirmed by the synthesis of the low-pass prototype in Figure 5.13, in which the maximum equivalent f-plane capacitance that could be absorbed by the network was shown to be 0.33 pF. If, therefore, the need arose to extend the frequency range of the conventional amplifier to, say, 30 GHz, the gate capacitance of the 150-μm FETs would be too great to be absorbed by the scaled version of the gate line filter. In order to be absorbed by the filter, either the gate capacitance would have to be reduced in the ratio of approximately 3:2 (i.e., to 0.22 pF) or some means must be found of increasing the values of the capacitors in the filter without reducing the cutoff frequency or changing impedance levels.

It is helpful to consider the consequences for the *S*-plane low-pass prototype of the increase in cutoff frequency to 30 GHz. Although the *f*-plane filter network will be modified by the increased cutoff frequency, the low-pass *S*-plane prototype of Figure 5.13 would stay the same as long as the commensurate frequency was also increased to 45 GHz. This is because the cutoff frequency in the *S*-plane would still have the same value of 1.7321. However, to be able to absorb the *S*-plane equivalent of the FET gate capacitance at 30 GHz (i.e., at least 0.33 pF), the *S*-plane capacitors in the middle of the network must be raised by the factor 3/2. In other words, the capacitors in the *S*-plane network must be raised to a value greater than or equal to 1.9.

The solution to the problem is to synthesize the gate and drain filters of the amplifier as band-pass networks of the form shown in Figure 5.24. Compared with the networks shown in Figure 5.8, the *f*-plane prototype now has extra short-circuit shunt stubs distributed along its entire length and correspondingly the *S*-plane prototype has a set of extra shunt inductors. As in the case of the low-pass networks, FETs would normally be located only at the junctions of the inner stubs and not at the position of the outer stubs. The corresponding frequency responses of the *S*-plane and *f*-plane networks are plotted in Figure 5.24(c). Not surprisingly, the effect of the band-pass network on the *f*-plane response has been to introduce an extra cutoff frequency at the bottom end of the passband and to confine the operation of the amplifier to within this smaller passband. This could be an unwelcome restriction for some applications. However, there will be many applications where the low-frequency performance is not essential and where the increase in the upper operating frequency that can be achieved is more important. It should also be remembered that, in practice, conventional low-pass distributed amplifiers do not usually operate down to dc because of the inevitable presence of the dc bias circuitry.

To synthesize a suitable *S*-plane prototype for an amplifier capable of operating to 30 GHz, the upper cutoff frequency in the *S*-plane (ω_2) was set to the previous value of 1.7321. This corresponds to the *f*-plane cutoff frequency of 30 GHz and a commensurate frequency of 45 GHz. Then a likely starting value of 0.5774 (equivalent to 15 GHz in the *f*-plane) was chosen for the lower cutoff frequency (ω_1), enabling the synthesis to proceed and the values of the internal capacitors of the network to be determined. As previously stated, the minimum value required is approximately 1.9 and the best approach is to iterate the synthesis, changing the value of ω_1 until the capacitors exceed the value of 1.9. In this example, the most appropriate value for ω_1 was found to be 0.6745, which by the Richards Transformation corresponds to an *f*-plane cutoff frequency of 17 GHz. The expected bandwidth of the new amplifier was therefore 17 to 30 GHz.

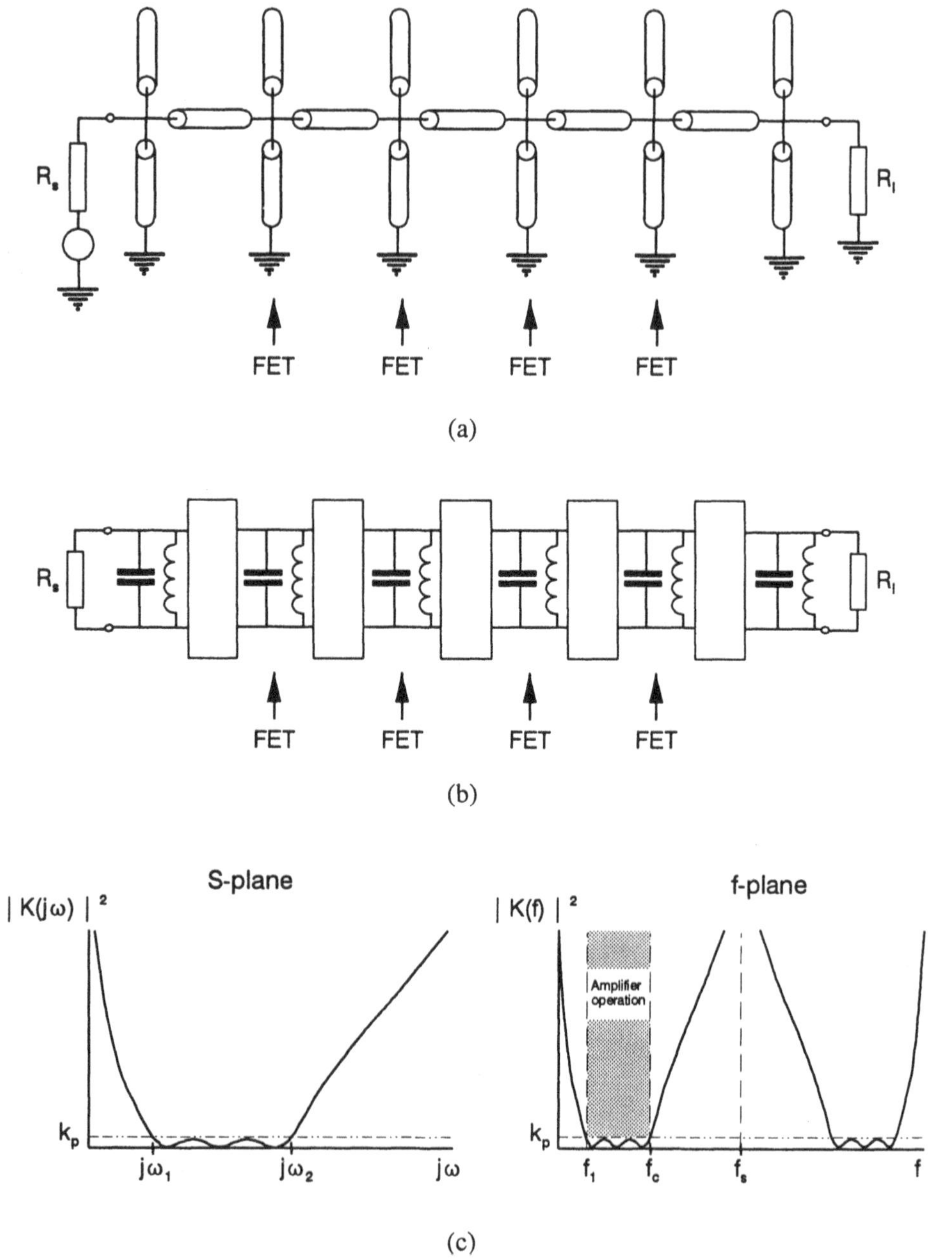

Figure 5.24 *S*-plane and *f*-plane prototype networks for a distributed amplifier using band-pass networks: (a) *f*-plane equivalent circuit; (b) *S*-plane equivalent circuit; (c) *S*-plane and *f*-plane frequency responses.

The nonredundant form of the resulting *S*-plane prototype is given in Figure 5.25(a). This was synthesized using NETSYN with a passband ripple value of 0.1 dB and sufficient capacitive elements to accommodate a total of four FETs. Hence, there is a total of six shunt capacitors. Each of the capacitors corresponds to a transmission zero at $S = (0, j\infty)$ and each of the unit elements to a zero at $S = (1, j0)$. There is a single inductor that, for convenience, has been extracted close to the center of the network and that corresponds to a single transmission

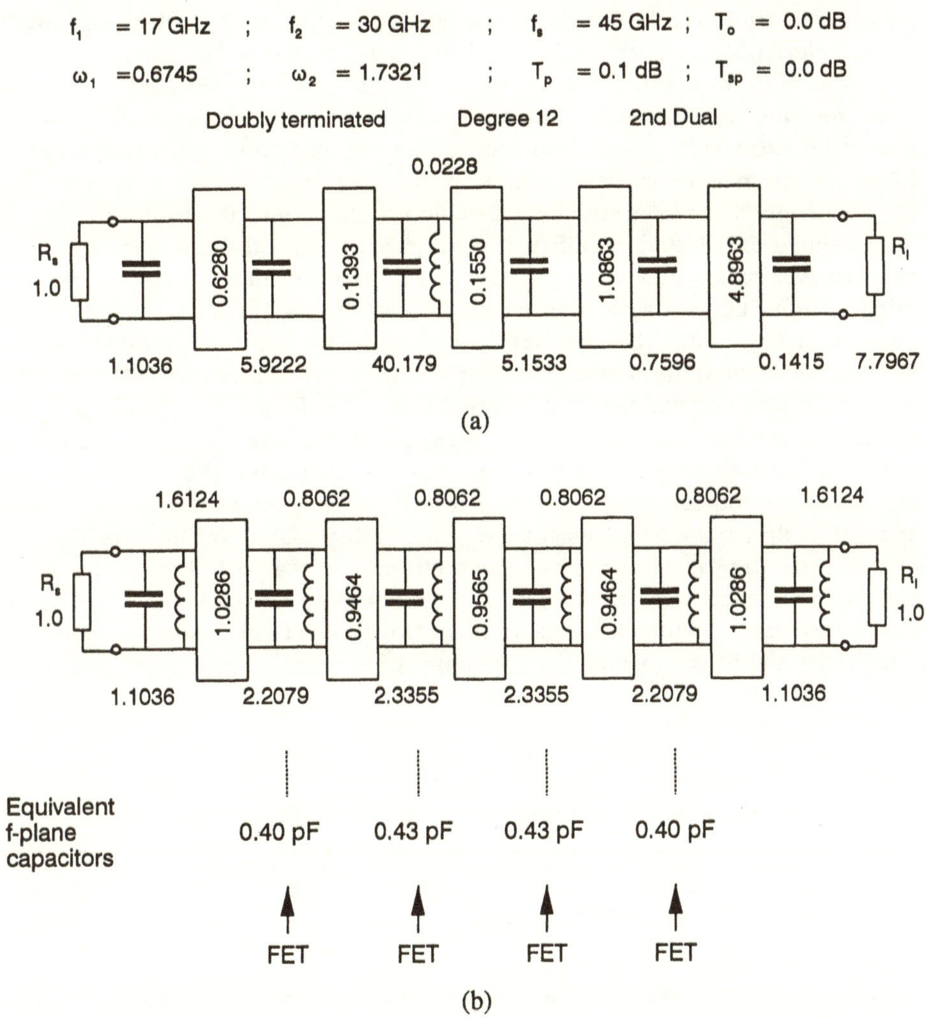

Figure 5.25 *S*-plane prototype for the 17- to 30-GHz distributed amplifier: (a) no redundant elements; (b) with redundant elements.

zero at $S = (0, j0)$. The network is of degree 12, is doubly terminated, and is of the second canonical form.

It is interesting to note that the impedance levels in the nonredundant network are lowest at the position of the shunt inductor and that the network is not symmetrical. The asymmetry is a consequence of not being able to extract the inductor precisely in the center of this particular network and is something that will be corrected automatically during the subsequent introduction of redundancy. Synthesis of the network using E-Syn is straightforward. A distributed, Tchebycheff, band-pass network must be selected, the passband edges set to 17 and 30 GHz, respectively, and the electrical length at 30 GHz set to 60 degrees. Input and output terminations should be both set to unity, five transmissions lines should be selected, six low-pass stubs and one high-pass stub. For a satisfactory network topology, the elements should be extracted in a manual mode. The set of parameters for synthesis by E-Syn are summarized in Table 5.2.

To convert the nonredundant S-plane prototype into the form required for the modified distributed amplifier, it is necessary to split the single inductor into six parts and redistribute the six parts among the other elements of the network using an admittance matrix transformation (see Chapter 2, Section 2.4.2). In the network of Figure 5.25(b), this has been done in such a way as to make all the internal inductor values equal to half the value of the outer two inductors. The basic transformation will allow the internal inductor values to be set arbitrarily, and in theory it is possible to lower impedances such that any value of FET gate capacitance could be absorbed. However, lowering impedance has the undesirable side effect of lowering amplifier gain, and it is important only to lower impedance enough to absorb the gate capacitance of the FET available. In this case, the ratio of 1:2 lowers the capacitor values to an equivalent of 0.4 pF in the f-plane, which gives a comfortable working margin above the minimum value of 0.33 pF. It also ensures consistency with the low-pass S-plane prototype of the previous conventional distributed amplifier in which the internal capacitor values were twice that of the outer capacitors.

Table 5.2
Prototype Synthesis by E-Syn for the 17- to 30-GHz Distributed Amplifier

Network type	Distributed	Tchebycheff	Band-pass	
Electrical length	60 deg at 30 GHz			
Passband	$f_1 = 17$ GHz	$f_2 = 30$ GHz	$T_p = 0.1$ dB	MIL = 0 dB
Terminations	$R_s = 1$	$R_l = 1$		
Elements	Five transmission lines	Six low-pass stubs	One high-pass stub	
Mode (F0)	Use a manual mode of element extraction			
NxtEL (F6)	Use successively to extract next element from choice given			
Analyze (F0)	Analysis after extraction			

The effect of the lower cutoff frequency on the values of the capacitors in the band-pass filters can be illustrated by considering a classical low-pass-to-band-pass frequency transformation applied to a lumped element LC ladder network. For the capacitors of the low-pass ladder (C_{lp}), these are transformed into capacitors in the band-pass ladder (C_{bp}) such that

$$C_{bp} = C_{lp} \times \frac{f_c}{f_c - f_l} \tag{5.30}$$

where f_l is the lower edge of the passband for the band-pass network and f_c is the common upper edge of both passbands. More details of the effects of this transformation will be given in Section 5.4. However, it can be seen that as f_l is increased in value, reducing the width of the passband, the value of C_{bp} is increased. For an octave bandwidth, the increase in C_{bp} is a worthwhile factor of 2, which is also broadly consistent with the increase that was achieved in the network of Figure 5.25. Hence, the simple formula can be used to give a first estimate of the value of the lower cutoff frequency that will be required to achieve a given increase in capacitance value, as long as the commensurate frequency (f_s) for the f-plane network is significantly higher than f_c.

With the exclusion of circuit parasitics, the S-plane prototype of Figure 5.25(b) was transformed into an f-plane circuit of the form shown in Figure 5.26. Along the central region of the gate line, the FETs provide all the necessary shunt capacitance, and therefore only the shunt short circuit stubs need to be present. At the ends of the gate and drain lines, where there are no FETs, the pairs of short- and open-circuit stubs have been exchanged for single stubs of double the commensurate length to help with the circuit topology. The relevant network transformation can be found in Appendix A. Along the entire drain line, all the shunt elements are connected to ground via decoupling capacitors to facilitate the application of drain dc bias. Those capacitors connected to the internal load resistor and the double-length stubs are large capacitors (50 pF) giving effective decoupling at low frequencies. However, the capacitors connected to the central stubs of the drain line are of a small value (0.2 pF). This is because, unlike the gates of the FETs, the drains do not provide sufficient shunt capacitance to eliminate the open-circuit stubs of the basic prototype. Together with the transmission line elements E5, the capacitors approximate the combination of inductive and capacitive susceptance required at each FET. They also have the advantage of being easily realizable.

The values of the elements of the f-plane equivalent circuit are given in the table at the bottom of Figure 5.26. They were derived directly from the S-plane prototype of Figure 5.25, but have been subjected to some computer optimization to correct for the modifications that were necessary in respect to elements E5 and E7. The resulting simulated frequency responses of the amplifier are plotted in Figure 5.27. Gain is a nearly flat 5 dB over 18 to 29 GHz, and S_{11} and S_{22} are better than −11 dB over the same band. Despite every effort with the circuit optimization,

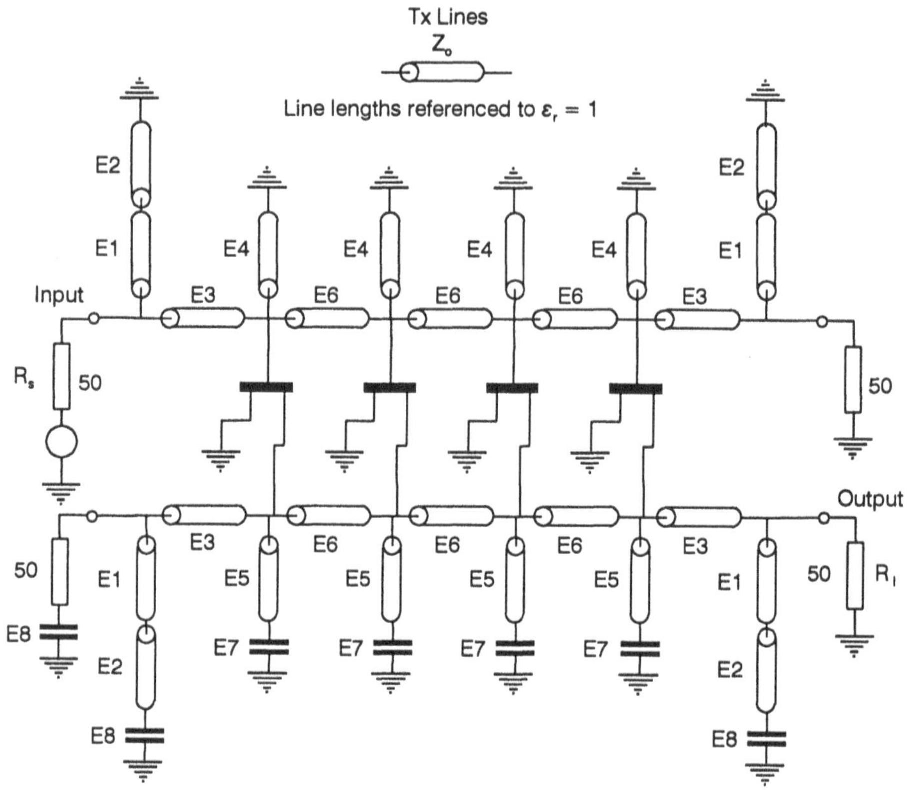

	E1	E2	E3	E4	E5	E6	E7	E8
$Z_o(\Omega)$ or C(pF)	29.86	57.72	51.20	50.00	80.00	56.00	0.2	50
l_{mm}	1.667	1.667	1.667	1.535	2.684	1.520	-	-

Figure 5.26 Distributed amplifier circuit with band-pass filter networks.

it proved impossible to reach the target value of a 30-GHz upper operating frequency. One reason for this is that the FETs used in the amplifier were discrete versions of the FETs used in the monolithic 6- to 18-GHz amplifier, and the inclusion of extra parasitic inductances for bond wire connections increased the effective gate capacitances to more than 0.4 pF. Another reason is associated with the propagation loss along the gate artificial transmission line, which was increased by the use of the band-pass filters. Nevertheless, according to the simulation, the band-pass filters *have* produced an increase in the maximum operating frequency of the amplifier from an original 20 GHz to the new predicted value of 29 GHz.

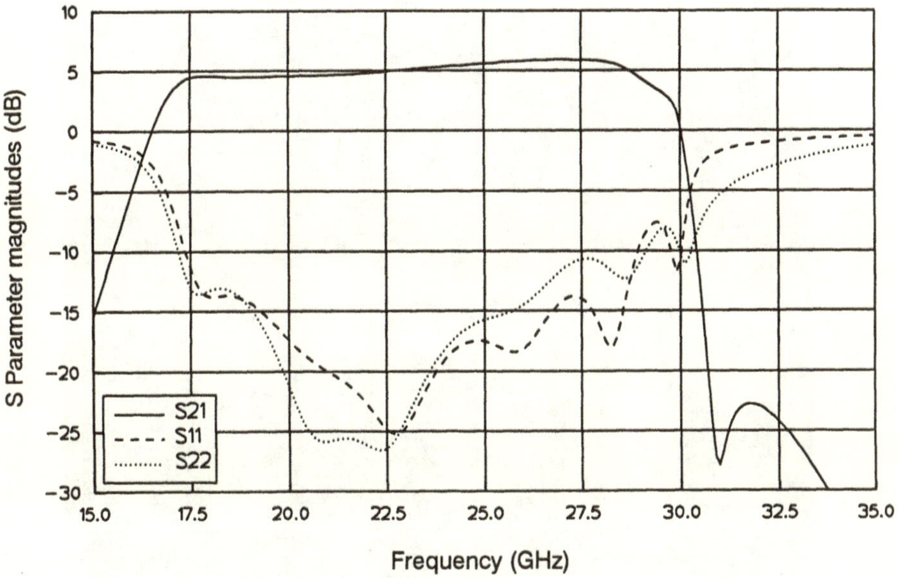

Figure 5.27 Predicted frequency responses.

Figure 5.28 shows a photograph of the 17- to 30-GHz amplifier circuit taken shortly after fabrication in hybrid MIC technology. The circuit measures approximately 10 × 10 mm and was fabricated on 0.25-mm-thick RT Duroid 5880. The four discrete FETs in the middle of the circuit are recessed in 0.5-mm square holes cut in the substrate by a laser and are soldered to the circuit using strips of fine gold ribbon. A row of gold-plated copper blocks provide the short circuits for the stubs along the gate transmission line above the FETs, while a row of chip capacitors provide the necessary decoupling for the stubs along the drain line. The two 50Ω resistors at the ends of the gate and drain transmission lines are of a ceramic, thin-film variety, flip-chip mounted on top of the copper track of the circuit at one end and on top of a 50-pF chip capacitor at the other. Drain dc bias is applied by making a connection to the top of the decoupling capacitor at the end of one of the double-length stubs.

Measured frequency responses of the hybrid amplifier are plotted in Figure 5.29. They were obtained at a time when it was necessary to use two measurements systems, an HP8510 system for frequencies up to 26 GHz and a scalar network analyzer system for those above 26 GHz. For S_{11} and S_{22}, there is generally good agreement with the predicted responses. However, gain (S_{21}) is a little lower than predicted (4.5 ± 0.5 dB over 17.5 to 27.5 GHz), especially toward the band edges. This is caused largely by losses along the gate transmission line that were higher than those predicted by the simulations. The losses are also the cause of the gradual roll-off in the gain response seen toward the edges of the passband.

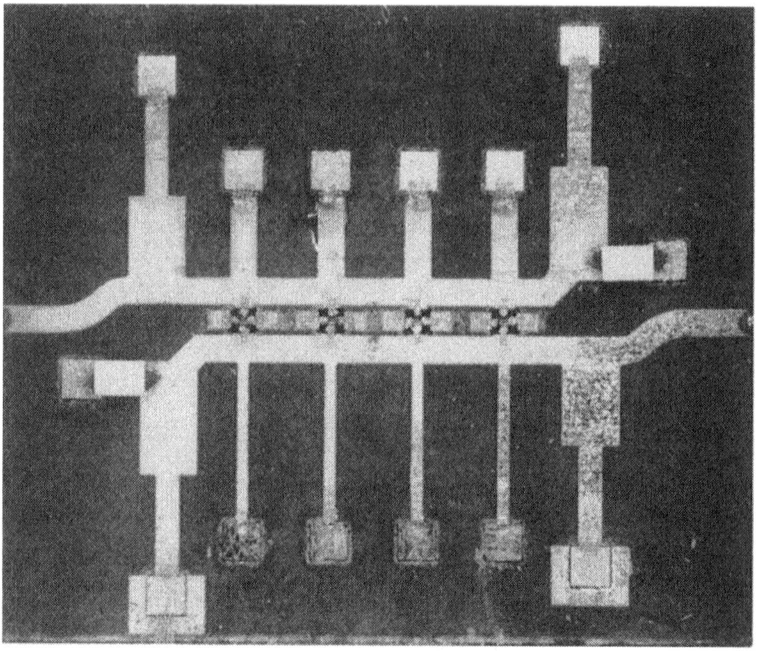

Figure 5.28 The 17- to 30-GHz distributed amplifier.

The upper skirt of the passband in the measured responses is higher and less steep than predicted by the simulations, and the circuit achieves an unexpected gain of 2.5 dB at 30 GHz. This is probably due to an overestimation of the size of the gate and drain parasitic inductances for the simulation. Using the gold ribbon connections to the FETs instead of bond wires appears to have given a marked reduction in the parasitics, which in turn has led to the increased cutoff frequency.

In comparing the measured and predicted results, it is important to remember that the prototype amplifier was assembled by hand in a hybrid technology for which very tight dimensional tolerances are difficult to maintain. For example, as a result of fabricating several of the amplifiers, it was found that errors on the order of only 5 μm in the positioning of the FETs were enough to cause significant changes in amplifier performance. A better performance and a closer agreement between measurement and simulation could have been expected from a similar amplifier circuit fabricated in monolithic technology on GaAs. While a better measured performance would have been desirable, this design exercise has verified the principle of using band-pass networks to extend the frequency of operation of a distributed amplifier and has also served as another interesting example of the application of exact synthesis techniques.

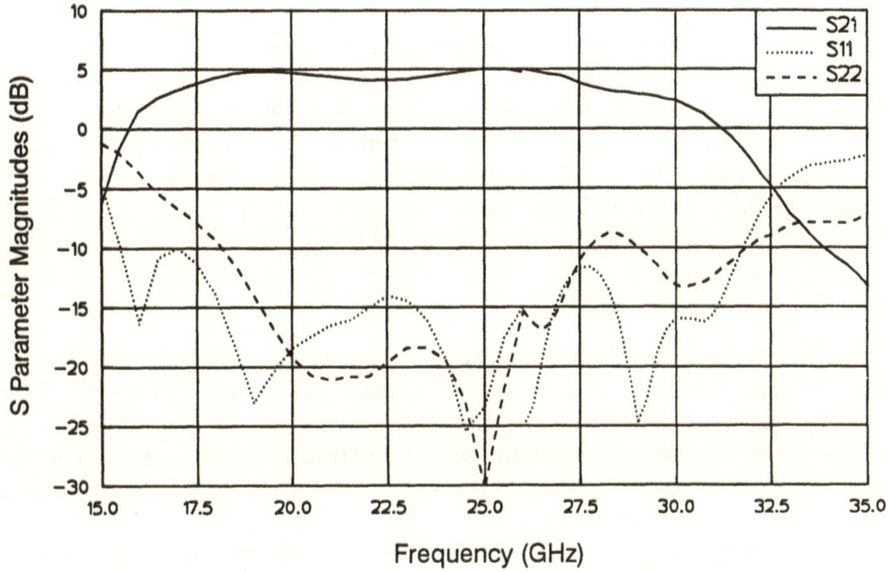

Figure 5.29 Measured frequency responses of 17- to 30-GHz amplifier.

5.3 REACTIVELY MATCHED WIDEBAND POWER AMPLIFIERS

In dealing with the design of a reactively matched amplifier circuit, this section will be the first to give examples of the use of exact synthesis for two-port matching purposes. The section will describe the design and measured performance of a 2.5W MMIC power amplifier for an X-band radar application. Designed at Philips Research Laboratories and fabricated at the PML foundry in France, the amplifier has a gain of approximately 6 dB over the frequency band of 7 to 11 GHz. At the time it was first reported [10], its performance set a new state of the art for a wideband, single-stage, MMIC power amplifier fabricated in Europe. Key aspects of its success were the use of relatively large numbers of relatively small FET devices and the systematic design approach based on exact network synthesis.

As a potential consumer of large quantities of MMICs, military radar systems have been targeted as one of the more commercially viable applications for the technology. GaAs MMIC technology is seen as the only practical solution to reducing the production costs of solid-state transmit/receive (T/R) modules to acceptable levels, and each module needs a power amplifier. For an application relevant to Philips Research Laboratories, two MMIC power amplifiers were required to produce a total of 4W of output power. To overcome losses in power combiners, each amplifier had to have a power output of at least 2.5W. Gain had to be more than

6 dB and efficiency better than 25% over the frequency band of 7.5 to 10.5 GHz. Partly because of the commercial benefit and partly because of a well-established working relationship, use of the D07M process at PML for the fabrication of the amplifiers was also an important objective. Since D07M is a general-purpose 0.7-μm gate-length process, which is not optimized especially for power amplifiers, a significant part of the technical challenge was to find a circuit that exploited its power capability to the absolute maximum.

5.3.1 General Design Approach

After the decision that this will be a single-stage, Class A amplifier, the total FET gate width requirement must be determined for the 2.5W output power. This is determined by considering the location of a dynamic load line plotted on the I/V (current/voltage) characteristics of a FET of unit gate width. Figure 5.30 shows such a load line extending between the knee of the uppermost I/V curve and the value of 12V on the voltage axis, which is the maximum permitted drain-to-source voltage for this particular fabrication process. As indicated, under RF conditions, drain voltage will swing between 1 and 12V, while gate voltage swings between +0.5V

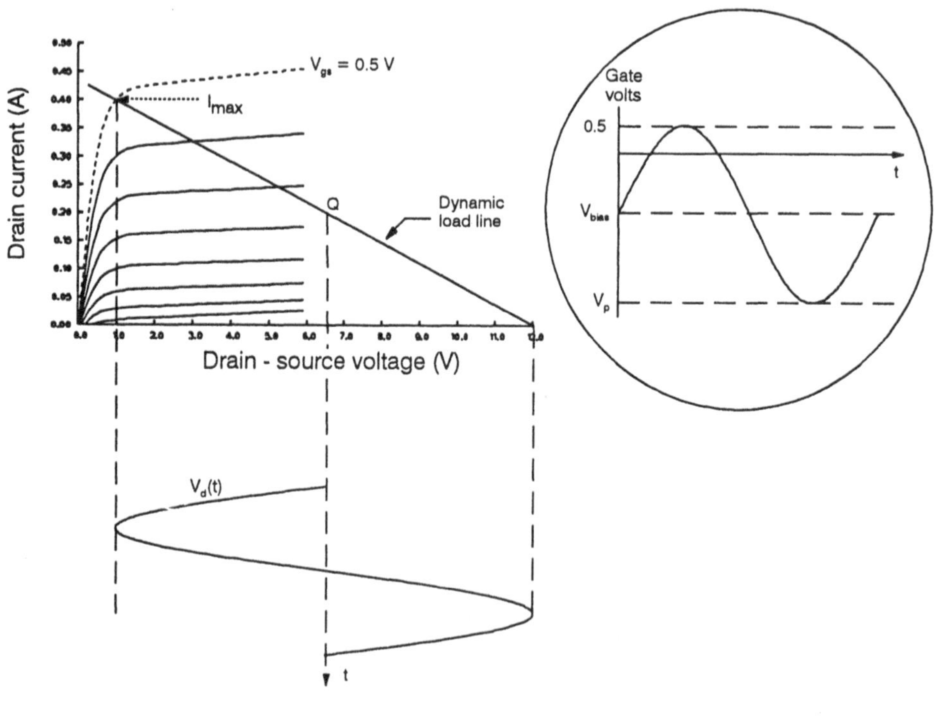

Figure 5.30 FET I/V characteristics and dynamic load line.

and the pinch-off voltage of –3V. The peak-to-peak drain current swing will be 400 mA. The slope of the load line represents the parallel combination of internal resistive losses in the FET and the output load resistor itself. With a value of 75Ω for the effective internal resistance of the FET, the optimum load resistance for maximum output power is 43Ω. These parameters give a maximum power output per unit gate width of 350 mW/mm for Class A operation in the PML D07M process.

A power output per unit gate width of 350 mW/mm makes the total gate width requirement for the 2.5W amplifier in excess of 7 mm. Rather than design the amplifier around one or two large FET devices, the final total of 7.2 mm was split into nine devices of 800-μm gate width each. This approach avoids the performance degradation of large FETs due to excessive channel temperatures, it allows the driving point impedances of the FETs to be raised to more manageable levels, and also allows the use of the more accurate models generally available for FETs with small gate widths. As shown by the photograph of the 2.5W amplifier MMIC in Figure 5.31, the nine FETs are arranged in three groups of three connected between a pair of passive power-dividing structures. It must be emphasized that these structures are not just simple power dividers, but are also the matching networks that transform the resistive impedances of the source and load into the complex impedances that must be presented to the FETs. They behave like a pair

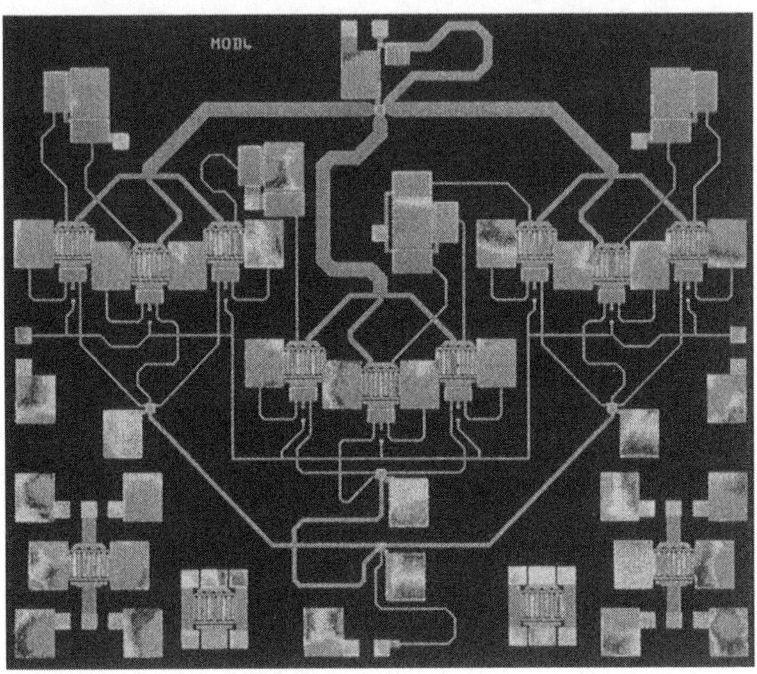

Figure 5.31 The 2.5W amplifier MMIC (4 × 3.4 mm).

of lenses, one spreading the input power across the array of FETs, the other focusing the output power from the FETs toward the single output port. Overall physical dimensions of the circuit are 3.4 × 4 mm.

To overcome the natural –6 dB per octave gain slope of a matched FET, there are several techniques that can be employed, but the one used in this amplifier was to allow the input network to match the device at the top of the frequency band while progressively introducing a mismatch at lower frequencies. The technique has the advantage that at the top of the band it does not waste any power, achieving the maximum possible gain. The disadvantage is that the input reflection coefficient of the amplifier is poor at low frequencies, but since the MMICs will be used in pairs in a balanced configuration, a good input reflection coefficient can easily be restored. As far as output matching is concerned, the main function of the output network is to transform the 50Ω load resistance into an impedance at the FET drains that extracts the maximum available output power over the whole of the required frequency range (7.5 to 10.5 GHz).

5.3.2 Detailed Circuit Design

Despite the apparent complexity of the multiple-branch matching networks of the amplifier, they can be designed relatively simply as a pair of two-port passive networks. Because, in theory at least, the nine signal paths through the amplifier from input to output are electrically identical, the FETs behave as if they are wired in parallel and the multiple branches can be collapsed into a more conventional ladder of two-port elements.

A fundamental premise on which the detailed design of the matching networks is based is that the maximum rated output power will be available at or below the 1-dB gain compression point. In these circumstances, the FETs can be assumed to be pseudolinear devices. While this sacrifices a little output power and efficiency, it vastly simplifies the broadband matching problem and allows the use of a small-signal equivalent circuit to represent the FET. Only in the final stages of the procedure, when the matching networks have been defined, need a large-signal model be used to verify that the desired output power can be achieved.

To begin the design of the matching networks, the small-signal equivalent circuit of an 800-μm FET is analyzed at a frequency near the top of the required 7.5- to 10.5-GHz frequency range. The relevant model, shown in Figure 5.32, is valid for the FET working in a microstrip circuit with via hole source grounding on two sides. Drain and gate bias voltages were 6.5V and –1V, respectively, to give a quiescent drain current of half the saturated value, *Idss*/2. By analyzing the model at 10 GHz, it is possible to derive two very simple equivalent circuits to approximate the input and output impedances of the FET. These give a good first indication of the parasitic capacitances that must be absorbed by the two matching networks and the values of the internal resistive terminations.

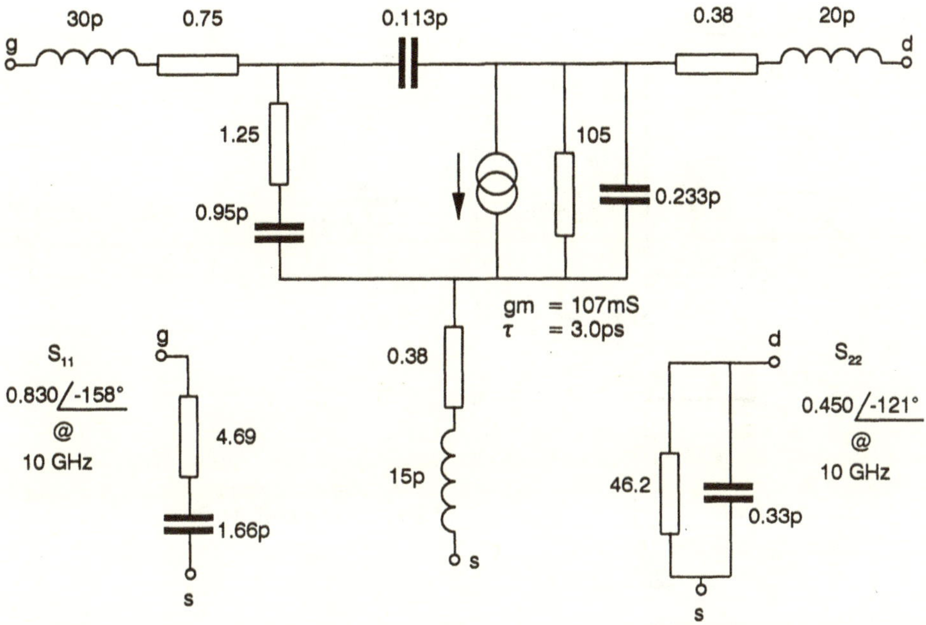

Figure 5.32 FET small-signal equivalent circuit.

5.3.2.1 The Input Matching Network

With nine of the 800-μm FETs wired in parallel, an artificial output matching network comprising a negative capacitor, a transformer, and a 50Ω load is applied to the drains so that a more representative one-port equivalent circuit can be derived for the input impedance of all nine FETs. As indicated in Figure 5.33, the −2.97-pF capacitor is of a magnitude equal to nine times the 0.33-pF equivalent drain capacitance calculated in Figure 5.32. It should therefore cancel the output capacitance of the nine FETs, reducing the output impedance to a pure resistance. The turns ratio of the transformer is then chosen to lower the resistance of the 50Ω load down to a value of 6.03Ω, which from the load line analysis is the value required for extracting maximum power from the FETs (i.e., $6.03 \times 7.2 \approx 43\Omega$). Analyzing this circuit shows that at 10 GHz the real input matching network must be capable of absorbing 16 pF of series capacitance and work into a load resistance of only 0.324Ω.

As a brief aside, it is interesting to apply (5.13) to the simple input equivalent circuit of the FETs and derive a figure for the maximum frequency bandwidth over which a constant match could be achieved. For a 10-dB constant return loss and a fixed upper cutoff frequency of 12 GHz, (5.13) gives a maximum achievable bandwidth of 5.7 to 12 GHz, which comfortably spans the required 7.5- to 10.5-GHz

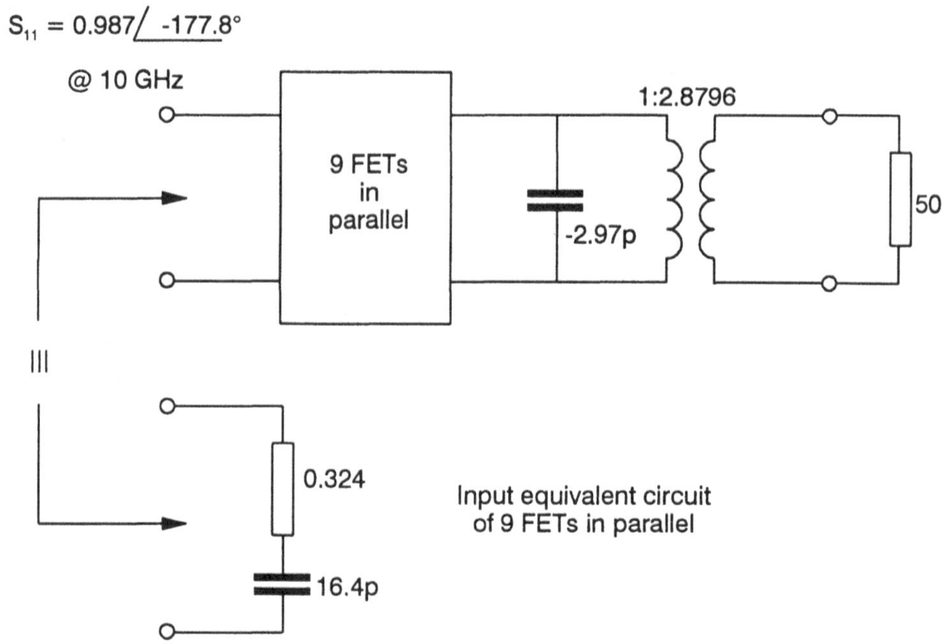

Figure 5.33 Application of an artificial output termination.

working bandwidth of the amplifier. However, the calculation assumes the use of a pure lumped matching network with an infinite number of elements. When an *S*-plane prototype is synthesized for a constant match, the limiting bandwidth is reduced by an amount dependent on the commensurate frequency. For a 20-GHz commensurate frequency, the limiting bandwidth for a 10-dB return loss would be 6.8 to 12 GHz. This is still wider than the 7.5- to 10.5-GHz band, but previous experience shows that the theoretical margin is now small enough to make achieving a constant 10-dB return loss difficult for a real matching network containing a finite number of elements.

Returning to the practical example, the synthesis of the *S*-plane prototype of the input matching network involved making a new, systematic choice of element types and transmission zero distribution. In this particular case, the network topology was not based on any existing circuit configuration, but was created by making a selection from the basic, directly coupled element types available for microstrip circuits, as previously discussed in Chapter 2. The prototype was chosen with a band-pass response to permit the use of a mixture of lumped and distributed elements in the *f*-plane network, thereby minimizing the circuit area. A band-pass response was also chosen because the network needs to contain at least one high-pass element (i.e., a series capacitor) for the absorption of the FET input capacitance and must be able to perform an impedance level transformation between the external 50Ω source resistance and the 0.324Ω of the input resistance to the nine FETs.

To overcome the −6-dB/octave gain slope of the FETs, the *S*-plane prototype must be synthesized for a +6-dB/octave slope over its passband. The result of the synthesis using NETSYN is shown in Figure 5.34. Here, all the elements of the network have been extracted from the right-hand side, the 1Ω resistor on the right having been treated as the source resistance. Consequently, when the whole network is later scaled in impedance by a factor of 0.324, the source resistance will represent the resistive part of the input impedance of the FETs.

Underneath the matching network in Figure 5.34, the input equivalent circuit of the nine FETs has been transformed into the *S*-plane and referenced to the same 1Ω impedance level as the matching network. As indicated, the 16.4-pF *f*-plane capacitance of the FETs has become an *S*-plane capacitor with a value of 0.3518 referenced to 1Ω, at a frequency of 9 GHz and for a commensurate frequency

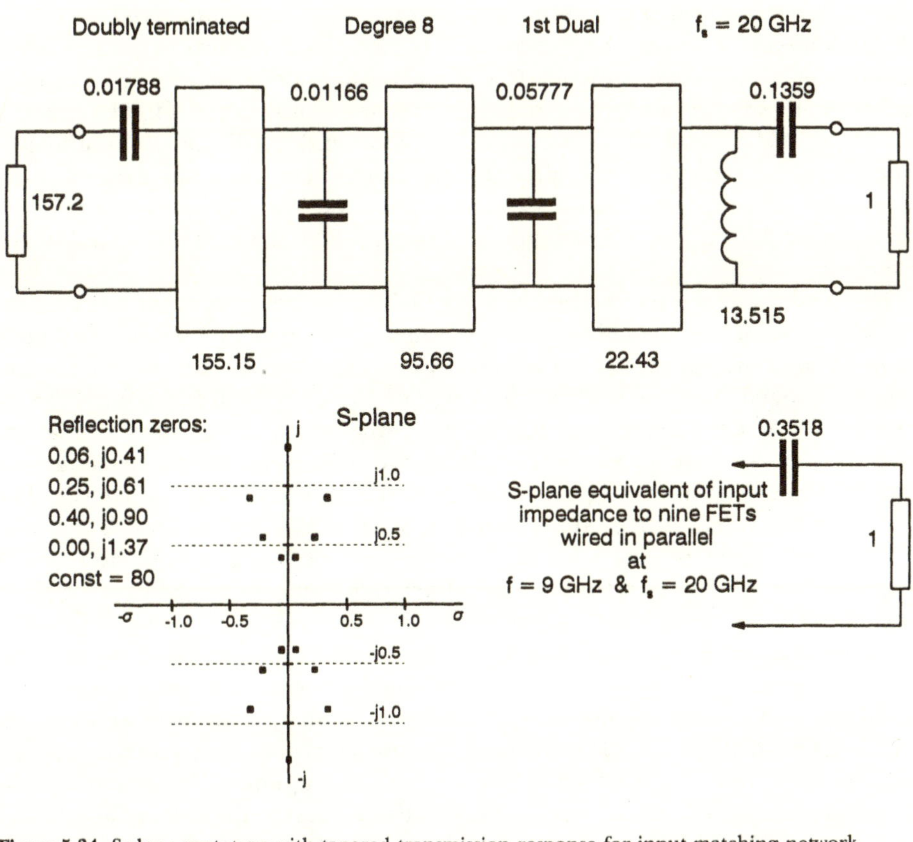

Figure 5.34 *S*-plane prototype with tapered transmission response for input matching network.

(f_s) of 20 GHz. In order to be able to absorb the series capacitance of the FETs, the first element to be extracted in the matching network must be a series capacitor. Its value in Figure 5.34 is 0.1359, which is substantially smaller than the value of the equivalent FET capacitance (0.3518). Hence, the matching network will not only absorb the input capacitance of the FETs, but there will also be a residual capacitance left over that will remain adjacent to the FETs.

The shunt inductor of the S-plane network is present primarily as an impedance-transforming element which is used to adjust the ratio of the two terminating resistors. In some amplifier designs, the inductor would also be used to facilitate access for the gate dc bias supply, but this is not possible in this case because of the presence of the series capacitor. Gate bias will actually be fed to each FET using a resistor whose value is sufficiently high not to affect the RF behavior of the amplifier. At its position on the right side of the network, the shunt inductor gives the network the maximum possible ratio of terminating impedances and ensures that the smaller termination will also be on the right. The unit elements act as element separators and the shunt capacitors make a general contribution to the impedance-transforming properties of the network. The last element of the network to the left-hand side could be either a shunt or series capacitor, but it is more convenient to use a series capacitor and receive the added benefit of an integral dc break. This network topology requires three transmission zeros to be located at $S = (0, j0)$, one for each of the two series capacitors and one for the shunt inductor. There must be three at $S = (1, j0)$, corresponding to the three unit elements, and two at $S = (0, j\infty)$, corresponding to the shunt capacitors. The network is of the first canonical form (i.e., a first dual), it is of degree 8, and it contains no redundant elements.

The simplest way to create the required +6-dB/octave slope in the transmission characteristic of the matching network would have been to specify a band-pass response with the appropriate upper and lower cutoff frequencies and then set the slope parameter T_{sp} equal to −6 dB. The sign of T_{sp} would be negative for NETSYN, because T_{sp} relates directly to insertion loss rather than transmission coefficient. However, to illustrate an interesting alternative approach, the synthesis was performed instead by specifying the location of the passband reflection zeros. This technique was briefly discussed in Chapter 2, Section 2.2.3. For a network of degree 8, the information required by the software when working in this mode is the location of four reflection zeros in the upper right-hand quadrant of the S-plane and the value of a constant multiplier. There is no need to specify the type of frequency response or any of the other more conventional passband parameters.

To begin the process, a doubly terminated network was synthesized in the usual way with a flat transmission characteristic over the f-plane bandwidth of 6 to 12 GHz. A band of 6 to 12 GHz was chosen because it provides a good working margin around the 7.5- to 10.5-GHz band needed for the amplifier. The commensurate frequency was set to 20 GHz, making the passband edges in the S-plane $\omega_1 = 0.5095$ and $\omega_2 = 1.3764$. This preliminary step gives some guidance as to the number of elements required to achieve the required 50:0.324 impedance ratio. It also gives starting values for the location of the reflection zeros. By repeating the synthesis several times, but each time adjusting the displacement of the reflection zeros from the $j\omega$ axis, it was possible to produce the network of Figure 5.34 with the required tapered transmission characteristic. The corresponding frequency responses after transformation into the f-plane have been plotted in Figure 5.35. The transmission coefficient rises from roughly −7 dB at 6 GHz to 0 dB at 12 GHz, which is acceptably close to the required response. The reflection coefficient can be seen to fall from approximately −1 dB at 6 GHz toward a deep null at 12 GHz where one of the reflection zeros was located. The final locations of all the S-plane reflection zeros are given in Figure 5.34, together with the value of the constant multiplier. (Note: The constant multiplier is used internally by the software to scale the value of the characteristic function $|K(j\omega)|^2$.)

A network very similar to the one shown in Figure 5.34 can be synthesized by E-Syn using the more conventional approach to passband specification. A distributed, Tchebycheff, band-pass response should be selected with a passband of 6 to 12 GHz and the electrical length parameter set to 54 deg at 12 GHz. Input and output terminations should both be set to 1Ω. For consistency with the required network topology, it is necessary to select three transmission lines, two low-pass stubs, and three high-pass stubs. Then, after the passband ripple is set to 0.075 dB and the slope parameter T_{sp} to +6.0 dB, the network is extracted in a manual mode. A summary of the parameter set for E-Syn is given below in Table 5.3.

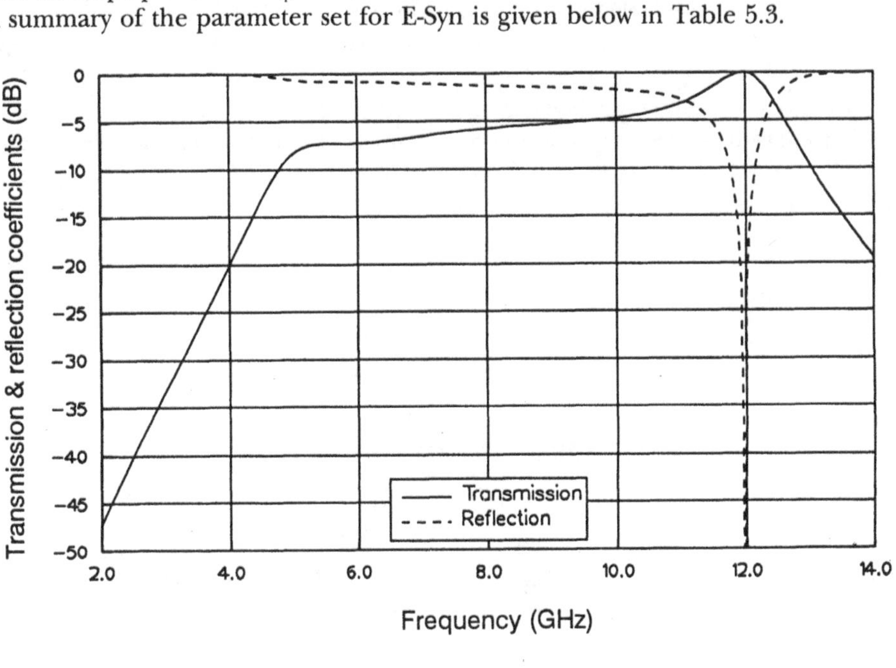

Figure 5.35 Tapered frequency responses of input matching network.

Table 5.3
Prototype Synthesis by E-Syn for the Input Matching Network of the 2.5W Amplifier

Network type	Distributed	Tchebycheff	Band-pass	
Electrical length	54 deg at 12 GHz			
Passband	f_1 = 6 GHz	f_2 = 12 GHz	T_p = 0.075 dB	T_{sp} = +6 dB
Terminations	R_s = 1	R_l = 1		
Elements	Three transmission lines	Two low-pass stubs	Three high-pass stubs	
Mode (F0)	Manual mode of extraction			
NxtEL (F6)	Use successively to extract next element from choice given			
Analyze (F0)	Analysis after extraction			

It may be helpful to make one further point concerning the frequency responses of the input matching network. From Figure 5.35, it will be apparent that the lower passband skirt has been moved a little below the initial setting of 6 GHz to a value nearer 5 GHz. This adjustment was made to alter the ratio to the terminating resistors of the network to the required value (50:0.324) without having to modify the position or split any of the circuit elements. As in the synthesis of more conventional matching networks with flat frequency characteristics, the passband width is a useful parameter for adjusting network element values as long as the electrical specification is not compromised.

With regard to the values of the reactive elements in the *S*-plane prototype, these may seem to span a rather large dynamic range. However, this is not of any major concern at this stage, because the length/impedance trade-offs about to be applied will reduce the range and overcome any problems with physical realization. As indicated in Figure 5.36, the *S*-plane prototype can be transformed into an *f*-plane network containing a mixture of lumped and distributed elements. All the

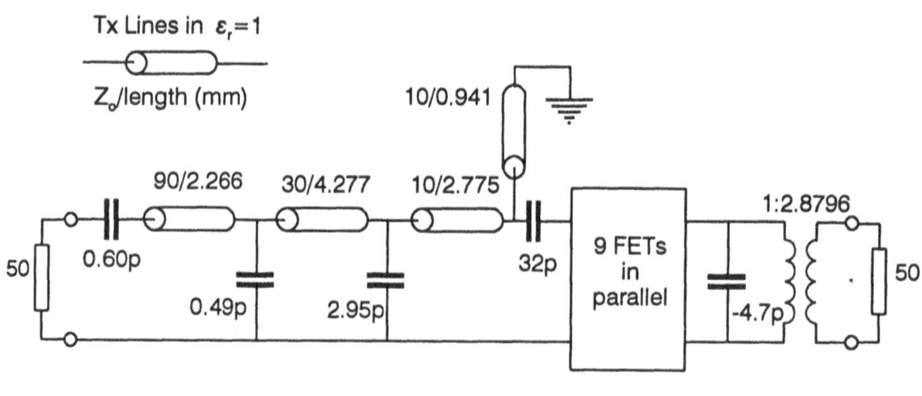

Figure 5.36 *f*-plane input matching network.

S-plane capacitors have become lumped capacitors in the *f*-plane by equating the appropriate reactances at a frequency of 9 GHz. All the unit elements have become transmission lines whose impedances have been adjusted such that in the multiple-branch matching network of the finished amplifier they will all have a 90Ω characteristic impedance. Hence, in the collapsed circuit, the first line has an impedance of 90Ω because it does not need to be split. The second, which splits into three, has an impedance of 30Ω, and the third, which splits into nine, has an impedance of 10Ω. Similarly, the shunt stub, which splits into nine parts, has an impedance of 10Ω.

After the initial application of length/impedance trade-offs and the introduction of lumped elements, the *f*-plane matching network was connected to the parallel combination of the nine FETs and the artificial output matching network, as shown in Figure 5.36. The whole circuit was then subjected to some superficial optimization to correct for the slight disturbances in performance caused by the length/impedance trade-offs and the use of lumped capacitors. The target performance parameters were gain (S_{21}), which was set to a flat value of 9 dB over 6 to 12 GHz, and the output reflection coefficient (S_{22}), which was set for a constant, linear value of zero. With the characteristic impedances of the lines kept fixed, all the other circuit element values were allowed to change, except for the turns ratio of the ideal transformer. It is important not to change the transformer turns ratio so as to preserve the optimum load value previously determined by the load line analysis to give maximum power output. However, the value of the negative capacitor of the artificial output matching network should be allowed to change, and as shown in Figure 5.36, its value has changed to −4.7 pF. The change is due to the loading effect of the input matching network and the nonunilateral nature of the FETs. All the element values shown in the figure are those obtained from the optimization.

5.3.2.2 The Output Matching Network

The multiple-branch output matching network of the amplifier is also derived by applying exact synthesis to a collapsed two-port equivalent circuit. In this case, the matching problem is defined as indicated in Figure 5.37, wherein the output network must absorb the higher drain capacitance of 4.7 pF as a result of inserting the input matching network and must also transform the 50Ω load resistance down to a constant value of 6.03Ω at its input. In order to force the real part of the input impedance to be a constant value of 6.03Ω over the required frequency range, the output network must be synthesized as a singly rather than a doubly terminated network. As in the case of the input matching network, its response should be band-pass to allow subsequent physical realization as a mixture of lumped and distributed elements, but this time its passband is a conventional, flat, equal-ripple characteristic.

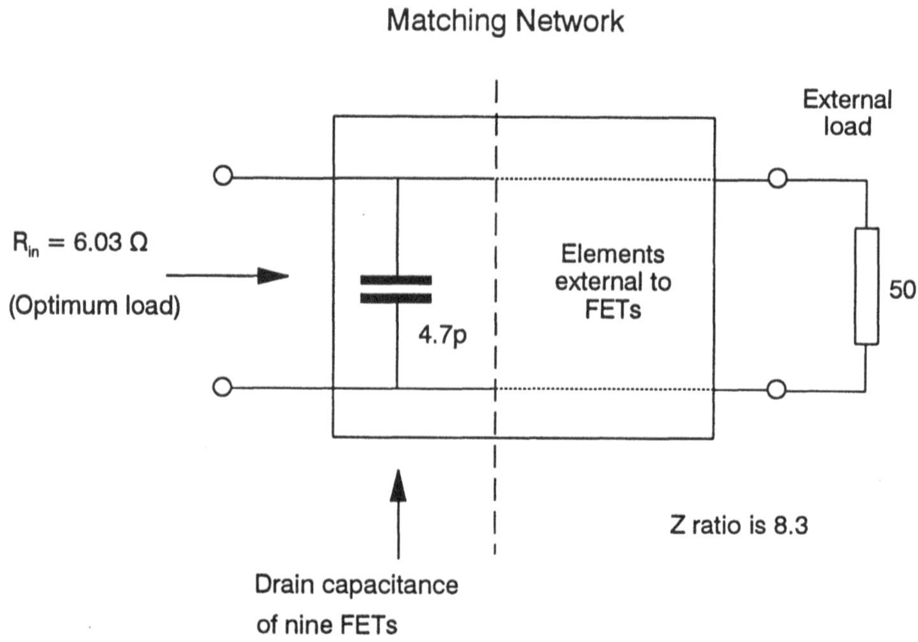

Figure 5.37 Output matching network.

The result of synthesizing the output network by NETSYN is shown in Figure 5.38. As indicated, the network was synthesized for a 6- to 12-GHz *f*-plane bandwidth, which, for a commensurate frequency of 35 GHz, translates into passband edges in the *S*-plane at $\omega_1 = 0.2760$ and $\omega_2 = 0.5975$. NETSYN treats the left-hand end of the network as the input, and since this is a singly terminated network, the real part of the input impedance is automatically normalized to a value of 1Ω. To absorb the drain capacitance of the FETs, it is essential for the network to contain a shunt capacitor adjacent to the input whose value is at least the *S*-plane equivalent of the 4.7-pF drain capacitance. As shown in the figure, an *f*-plane capacitor of 4.7 pF scaled for a 6.03Ω input resistance is equivalent to an *S*-plane capacitor of 3.750 scaled for a 1Ω input resistance. By careful choice of topology and passband specifications, this value has been precisely achieved by the synthesis.

In choosing topology, it is important that the network contain at least one shunt inductor to give access for dc bias and for the purposes of impedance level transformation. This must be at the left-hand end of the network to ensure that there is a rise in impedance level from left to right. However, perhaps the most important aspect of the topology is the deliberate creation of a redundant shunt capacitor in the middle. By choosing three nonredundant unit elements and two nonredundant capacitors, it has been possible to split the capacitor nearest the input into two parts, one of which has been moved through the adjacent pair of

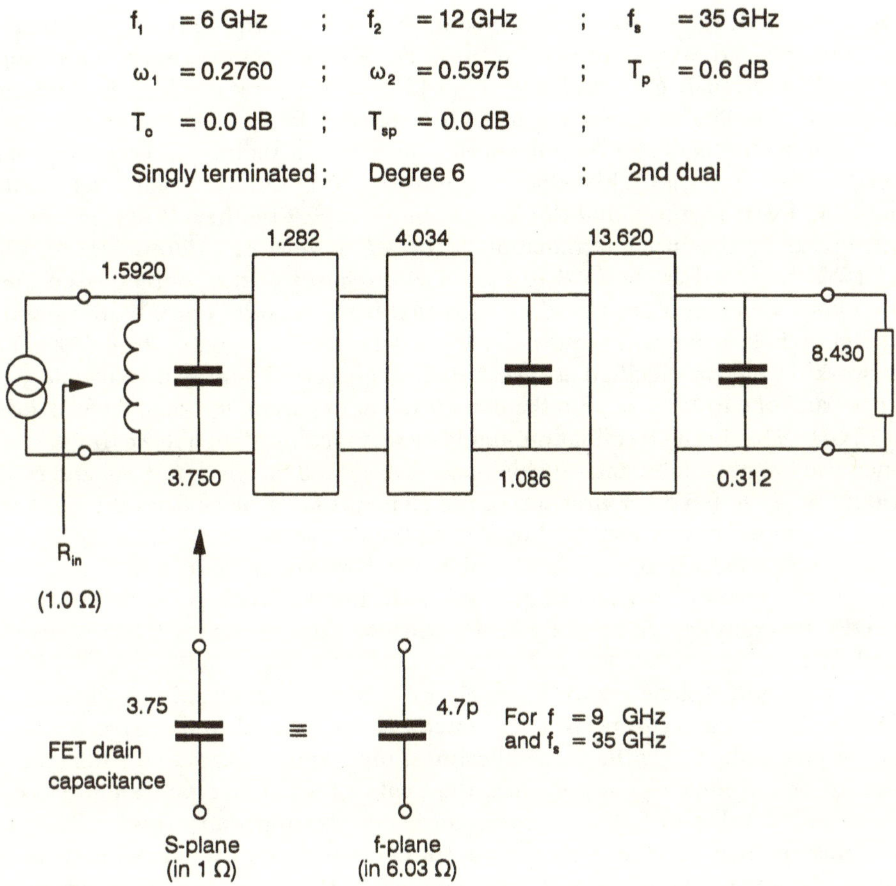

Figure 5.38 Singly terminated *S*-plane prototype of output network.

unit elements using the appropriate Kuroda Identity (Chapter 2, Section 2.4.1). This has provided an extra degree of freedom, allowing the value of the capacitor at the input to be fixed to exactly the required value. Furthermore, the passband width or passband ripple values are now comparatively free to be used for adjusting the value of the load without undue concern over the effect on the value of the capacitor. Hence, to summarize the distribution of transmission zeros for the network shown in Figure 5.38, there is a single transmission zero at $S = (0, j0)$ corresponding to the shunt inductor, three at $S = (1, j0)$ for the unit elements, and two at $S = (0, j\infty)$ for the two nonredundant shunt capacitors. The network is of the second canonical form with a degree of 6.

Once the network topology has been established, the actual synthesis procedure involves choosing a passband width (i.e., 6 to 12 GHz) that both exceeds the operating frequency range of the amplifier and produces a sufficiently large input capacitor. With the passband width fixed, the passband ripple is then adjusted to give the required value of the load resistor. For Figure 5.38, an insertion loss ripple of 0.6 dB eventually gave the correct 1:8.43 impedance ratio, which is equivalent to a ratio of 6.03:50 after scaling up by a factor of 6.03.

For synthesis of the output network by E-Syn, a slightly different approach needs to be employed. This is because in order to synthesize a singly terminated network, E-Syn requires that the load resistor be set to either ZERO or INF and will not allow the input termination to be set to zero or infinity. Hence, what NETSYN regarded as the load resistor, E-Syn regards as the source resistor, and the network elements are therefore extracted from opposite ends. With reference to Figure 5.38, E-Syn will require the input termination to be set to 8.430Ω. The network should be specified as distributed, Tchebycheff, and band-pass and have a passband of 6 to 12 GHz, with the electrical length parameter set to 30.86 degrees at 12 GHz. The output termination should be set to INF, and then three transmission lines, two low-pass stubs, and one high-pass stub should be selected. Passband ripple should be set to 0.6 dB. Extraction of the elements takes place from the right and must be carried out in a manual mode. All the elements except the middle shunt capacitor are extracted completely using the NextEL function key. The middle capacitor is created by specifying its value during a partial extraction using the NextEV function key. A summary of the synthesis parameters for E-Syn is given in Table 5.4.

When the *S*-plane prototype of Figure 5.38 is transformed directly into the *f*-plane, the frequency responses of the network are those shown in Figure 5.39. In accordance with the synthesis specifications, the input resistance function can be seen to be an equal-ripple response, the peaks of which are positioned close to the required value of 6.03Ω. Application of length/impedance trade-offs to the transmission lines and the conversion of shunt open-circuit stubs into shunt capacitors cause some perturbation of these responses. However, this can be corrected and a constant input resistance function can be restored by some minor optimization. The final form of the output network is then that shown to the right of the FET block in Figure 5.40. As with the input matching network, the transmission

Table 5.4
Prototype Synthesis by E-Syn for the Output Matching Network of the 2.5W Amplifier

Network type	Distributed	Tchebycheff	Band-pass	
Electrical length	30.86 deg at 12 GHz			
Passband	f_1 = 6 GHz	f_2 = 12 GHz	T_p = 0.6 dB	T_{sp} = 0 dB
Terminations	R_s = 8.430	R_l = INF		
Elements	Three transmission lines	Two low-pass stubs	One high-pass stubs	
Mode (F0)	Manual mode of extraction			
NxtEL (F6)	Use successively to extract next element from choice given			
NxtEV shift(F6)	Use for partial extraction of middle low-pass stub			
Analyze (F0)	Analysis after extraction			

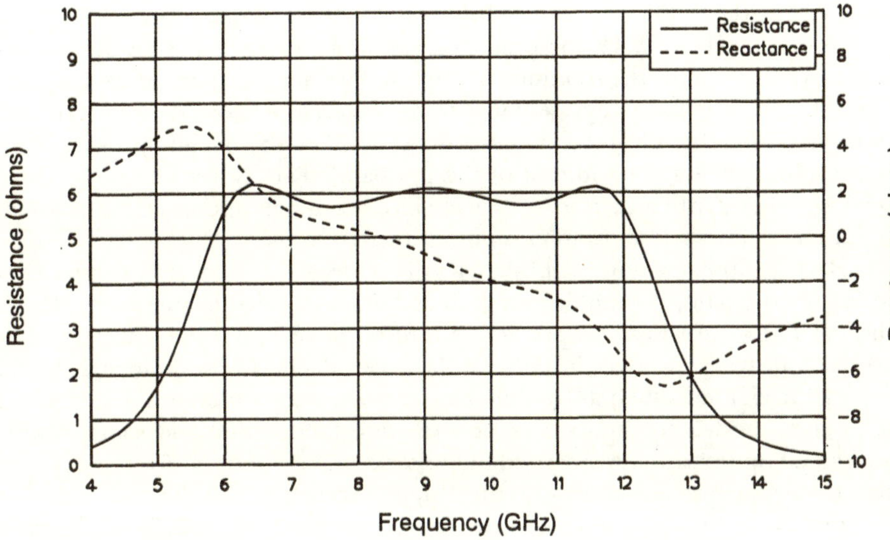

Figure 5.39 Output network responses—impedance vs. frequency.

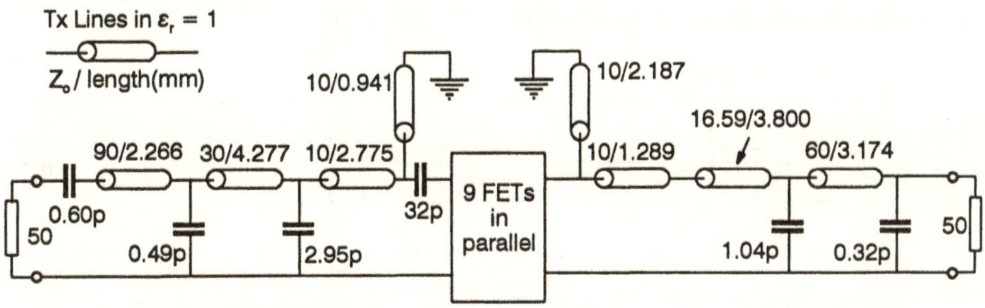

Figure 5.40 Collapsed equivalent circuit of whole amplifier.

line impedances have been chosen with a view to splitting the lines into the shape of the multiple-branch circuit of the complete amplifier. Hence, the first stub and line on the left, which are to be subdivided into nine parts, have a 10Ω impedance; the second line, which will split into three lines of 50Ω each, has an impedance of 16.59Ω; and the last line, which will not be split, has an impedance of 60Ω. The lower impedance lines near the output of the amplifier are needed to carry the relatively high RF current that provides the 2.5W output power.

5.3.2.3 The Complete Amplifier

As a result of the systematic approach to the amplifier design and the use of network synthesis, no further optimization of the collapsed equivalent circuit of the complete amplifier (Figure 5.40) was required in order to give a satisfactory performance. As indicated by Figure 5.41, the predicted gain is an almost flat 8 dB over most of the 6- to 12-GHz range, the rounding of the response near the passband edges being the result of adding some representative transmission line losses. As anticipated, the input match is relatively good at a frequency of 12 GHz, declining progressively toward a low value at the bottom of the passband. The S_{22} responses are also in broad agreement with what is to be expected from a design optimized for maximum output power rather than optimum output match.

Having simulated and established a satisfactory set of frequency responses for the collapsed circuit, the multiple branch circuit was created by successive element splitting. Those nearest the input and output ports were split into three branches and those nearest the FETs in the middle were split into nine branches. Models for parasitic elements such as transmission line bends and junctions, together with the physical models for lumped capacitors, were then added and some carefully constrained optimization was carried out to make some minor performance corrections. Particular care was taken with the output network during the optimization to avoid disturbing the optimum load impedance presented to the FETs for maximum power output. It is important, for example, not to optimize for an improved S_{22} response, as this would substantially degrade the maximum available output power from the amplifier. In the final stages of the design, some refinement of the circuit layout was also made with the help of a 3-D field simulator [11].

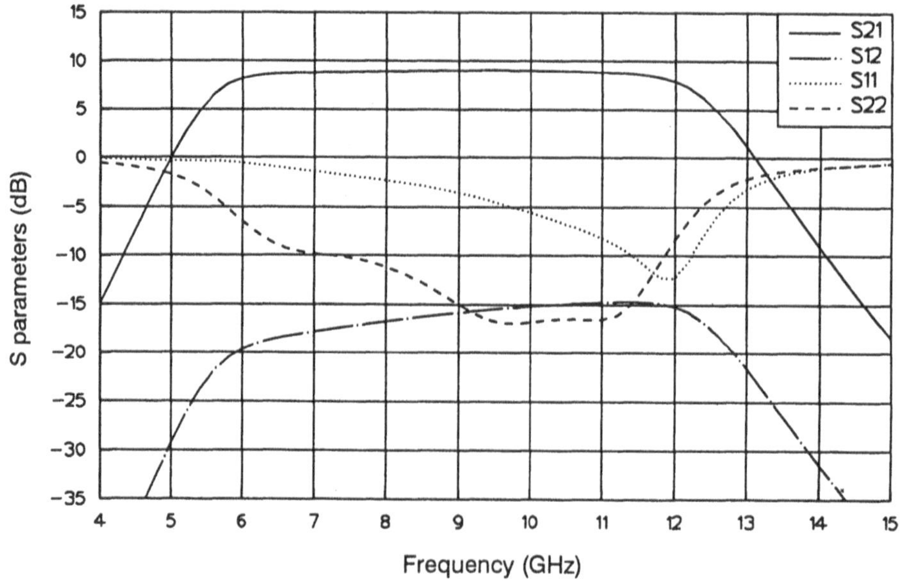

Figure 5.41 Simulated frequency responses without circuit parasitics.

5.3.3 Measured Performance

Two 2-in-diameter wafers of the power amplifiers were fabricated by the D07M process at PML. A large number of the individual amplifiers were first probed to check for basic functionality and then some working examples were soldered onto 1-mm-thick subcarriers. These subcarriers were made of a multilayer copper/molybdenum/copper material giving excellent thermal conductivity and a coefficient of thermal expansion equal to that of GaAs. Each amplifier was surrounded by four chip capacitors mounted on the same subcarrier, providing additional dc supply filtering and convenient posts on which to fix bond wires. Four bond wires were needed to carry the maximum drain supply current of 1.5A, whereas only a single bond wire was required for the gate bias. When biased for maximum output power, the drain voltage must be set to 6.5V and the gate bias to approximately −1V. Small- and large-signal measurements were carried out on the amplifiers after mounting the subcarriers in special housings of the type shown in Figure 5.42.

Network analyzer measurements of the small-signal frequency responses of one of the boxed 2.5W amplifiers have been plotted in Figure 5.43. Compared with the earlier simulations shown in Figure 5.40 for the collapsed equivalent circuit of the amplifier, there has been some loss in gain toward the edges of the 6- to 12-GHz band. However, an excellent gain level (S_{21}) of 6.7 ± 0.5 dB has been

Figure 5.42 The 2.5W amplifier in a coaxial housing.

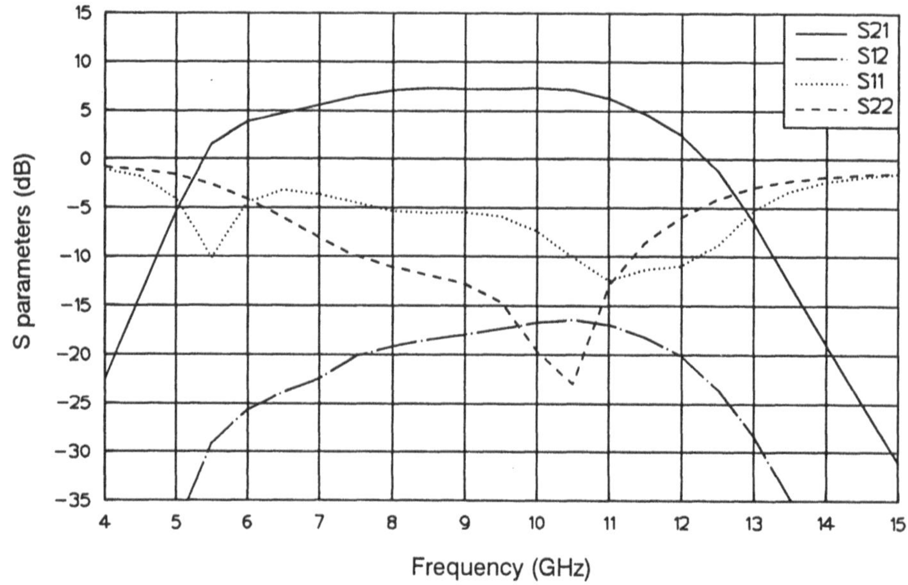

Figure 5.43 Measured small-signal frequency responses.

achieved over the 7.5- to 11.0-GHz band, which comfortably exceeds the original performance specification. Output reflection coefficient (S_{22}) is better than −10 dB over the same band, while the input reflection coefficient (S_{11}) embodies the negative slope seen in the simulations, which serves to flatten the gain response.

The performance of the amplifiers under large-signal conditions is illustrated by the results plotted in Figure 5.44. At 1-GHz intervals between 7 and 11 GHz, input power was swept from a relatively low level up to a level high enough to drive the amplifiers into saturation. As shown in the figure, the output power from the boxed amplifiers reaches 33 dBm (2W) at 1-dB gain compression and at a frequency of 8 GHz. Allowing for box and adapter losses, the output power at the chip itself is calculated to be 33.9 dBm or 2.45W. At 1-dB gain compression, drain efficiency is 27% and power added efficiency is approximately 20%.

5.4 TRAVELING-WAVE MATCHING IN CASCADABLE AMPLIFIER GAIN STAGES

The role of synthesis in creating a new classification of the MMIC amplifier will now be described. These amplifiers, which employ what has become known as the *traveling-wave matching technique* [12], combine the best features of distributed amplifiers with the best features of reactively matched amplifiers to produce circuits that are wideband, process-tolerant, extremely compact, and ideally suited to being

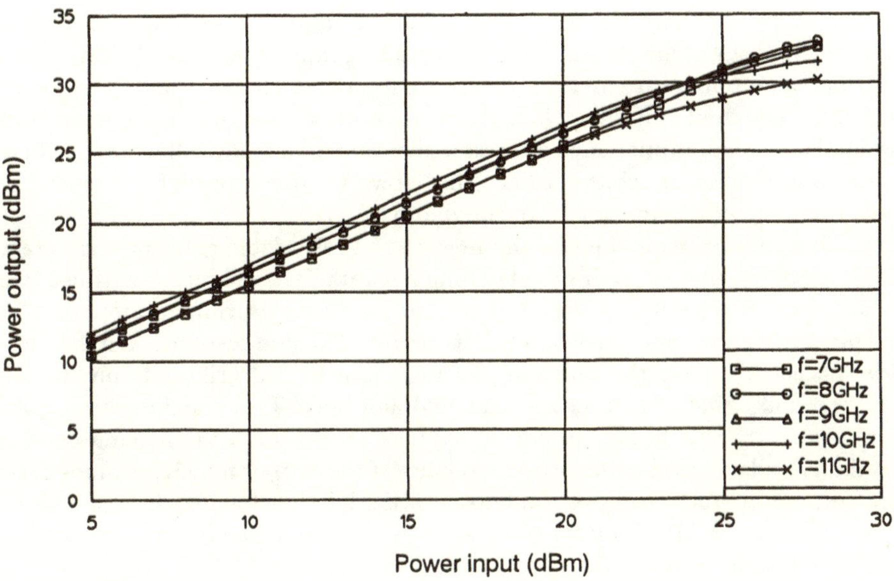

Figure 5.44 Measured output power vs. input power.

cascaded in multistage designs. They have approximately the same gain per stage as distributed amplifiers, with almost the same quality of terminal match, but have twice the efficiency, twice the power output per unit gate width, and occupy as little as a quarter of the surface area of GaAs. The section describes a practical design for a single-stage amplifier with 6-dB gain over the 7- to 14-GHz range, occupying a total surface area of only 0.8 mm^2.

The original incentive behind the invention of the new amplifier circuit was the design of high-gain, broadband, medium-power amplifiers for fabrication in MMIC technology. Since the cost of GaAs MMICs is, to a first approximation, proportional to surface area, it was important to find a circuit that occupied the minimum amount of space. Of the various circuit design concepts that have previously been used for broadband amplifier design, the distributed amplifier and the reactively matched amplifier were the most deserving of further consideration. They have performance characteristics that, to a large extent, are complementary, and any new circuit that could combine the best of both sets of characteristics was considered to be an attractive prospect.

To achieve gain levels of more than 10 dB in a practical MMIC amplifier, it is necessary to use several gain stages connected in cascade. Therefore, the match at the terminals of the individual stages (i.e., S_{11} and S_{22}) must be sufficiently good to prevent undesirable interactions. In this respect, distributed amplifiers are ideal, since they have excellent terminal match characteristics over wide frequency bandwidths. They also have the advantage of being highly process-tolerant and can achieve a flat gain characteristic without degrading input match. To their detriment, distributed amplifiers are inefficient and occupy a comparatively large surface area. Reactively matched amplifiers, in contrast, cannot be cascaded reliably, since they usually have a poor input match as a result of having to degrade reflection coefficient at low frequencies to achieve a flat gain. However, they are highly compact and have twice the efficiency of a distributed amplifier.

The basic principle behind the new type of cascadable gain stage is to take a single MESFET and use an input matching network that, except for having a bandpass frequency response, is the same as that used in a distributed amplifier. This will provide a good input match as well as create a flat gain response. On the output side, instead of using the matching arrangement of a distributed amplifier, the network is like that of a reactively matched amplifier. This matches the amplifier output load directly to the output impedance of the MESFET, making good use of all the available drain current. The quality of the output match becomes dependent on the characteristics of the chosen MESFET, but for the devices typical of general-purpose MMIC foundry processes, a good output match can still be achieved over bandwidths of more than an octave.

5.4.1 Evolution From a Distributed Amplifier

Figure 5.45 is a schematic of a distributed amplifier circuit comprising four FETs connected between two artificial transmission lines. The amplifier's basic mode of operation has been fully described in Sections 5.2.1 and 5.2.2.

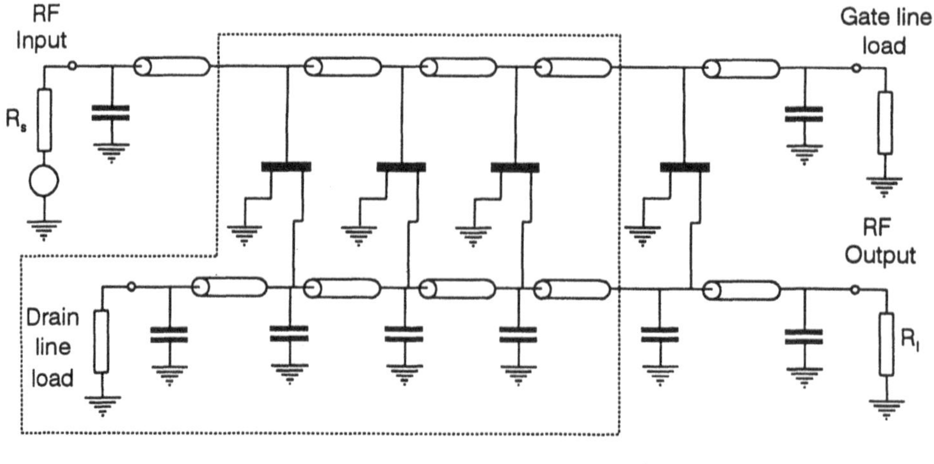

Figure 5.45 A four-FET conventional distributed amplifier circuit.

Distributed amplifiers waste half the available output current from the FETs by allowing current to flow in the wrong direction on the drain artificial transmission line. The wasted current (i.e., that flowing away from the output load) halves efficiency and halves the output power per unit gate width that the amplifier could otherwise achieve. To recover this lost current, several approaches have been considered, including tapering the impedance levels along the length of the drain line [13]. However, the most effective approach is to cut the drain line at a point just to the input side of the last FET so that it becomes impossible for current to flow toward the internal load. As shown in Figure 5.45, this means removing all the circuitry contained within the dotted line, leaving behind a single FET in a configuration that when reshaped is of the form shown in Figure 5.46. Input matching characteristics will have been largely unaffected by the changes. Output match will have been degraded by the removal of the internal load of the distributed amplifier, but match can be restored to acceptable levels by adjustment of the element values of the output network. The circuit is now effectively a distributed amplifier on the input side, but a reactively matched amplifier on the output side.

As previously indicated by (5.22), as long as the characteristic impedance of the artificial transmission lines is free to change, the maximum achievable gain for a loss-free distributed amplifier is not a function of FET size. The same is true for the modified, single-FET distributed amplifier with the reactively matched output network, which, after taking account of the current recovery, has a gain expression of the form

$$G \approx 2\left(\frac{f_t}{f_c}\right)^2 \tag{5.31}$$

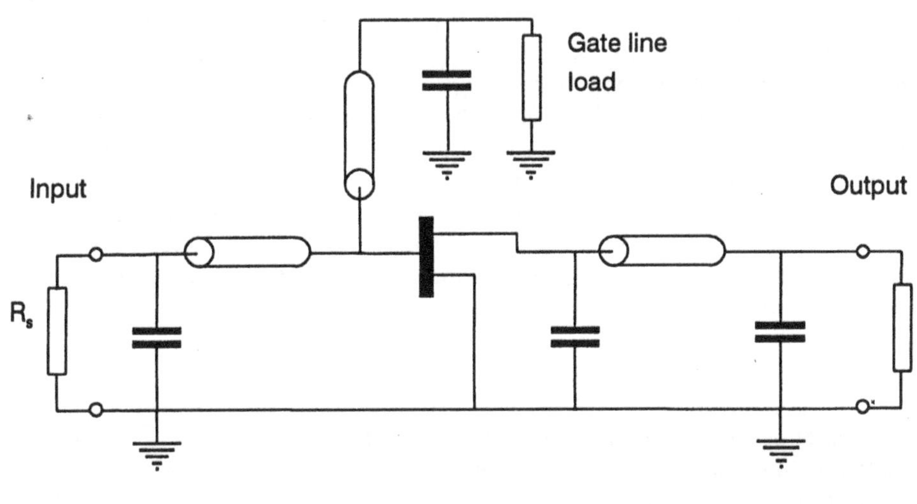

Figure 5.46 Result of eliminating circuit elements from a distributed amplifier.

The current recovery gives a factor of 2 increase in the effective g_m of the FET, which would translate into a factor of 4 increase in power gain if it were not for the adjustment of impedance levels in the output network to restore a good output match, which brings the figure back to a value of approximately 2. Unfortunately, the elimination of all but one of the FETs has reduced the value of n in (5.22) by a factor of 4 or gain by a factor of 16. This means that the net loss in gain in moving from the distributed amplifier to the new circuit of Figure 5.46 is a factor of 8. Typical values of f_t for a MESFET with 0.7-μm gate length are in the region of 16 GHz and, when the effects of circuit losses are taken into consideration, the gain levels that could actually be achieved with the circuit would be unacceptably low except for cutoff frequencies below the order of 1 GHz.

The only way to bring gain back up to acceptable levels without using more or better quality FETs is to increase the size of the single FET while keeping impedance levels the same. An increase on the order of $\sqrt{8}$ would be required, resulting in a FET with a gate width slightly less than the total gate width of the original distributed amplifier. The problem that arises from the bigger FET is that its gate capacitance is much bigger than that of the original FETs. If FET size is increased by a factor of $\sqrt{8}$, cutoff frequency decreases by the same factor for the same value of Z_0 (see (5.18)). It would be possible to restore the cutoff frequency f_c by transforming impedance down at the point of connection of the FET, but as indicated by (5.19), this would reduce gain back to the limit set by (5.31). The solution to the problem, which is one of the key attributes of the new circuit, is to make the gate filter network have a band-pass instead of a low-pass characteristic. For a given impedance level and upper operating frequency, a band-pass network allows a much greater gate capacitance to be accommodated than a low-pass network, which in turn allows the use of the larger FET. The amount by which the size of the FET can be increased depends on the bandwidth requirement of the amplifier. The narrower the bandwidth, the larger the FET and the greater the gain enhancement of the amplifier.

It should not go unnoticed that there is a strong similarity between this application of a band-pass frequency transformation and the application described in Section 5.2.4. In both cases the transformation creates bigger capacitances in the matching networks. Here the bigger capacitance is exploited to give more gain at a particular frequency by using a bigger FET. In Section 5.2.4, the bigger capacitance gave an increase in the maximum operating frequency achievable for the same size of FET.

5.4.2 The Band-Pass Frequency Transformation

A more quantitative explanation of the effect of a band-pass transformation on the input matching network of the cascadable gain stage can be given by transforming the constant K section circuit of Figure 5.7 into its band-pass equivalent of the form shown in Figure 5.47.

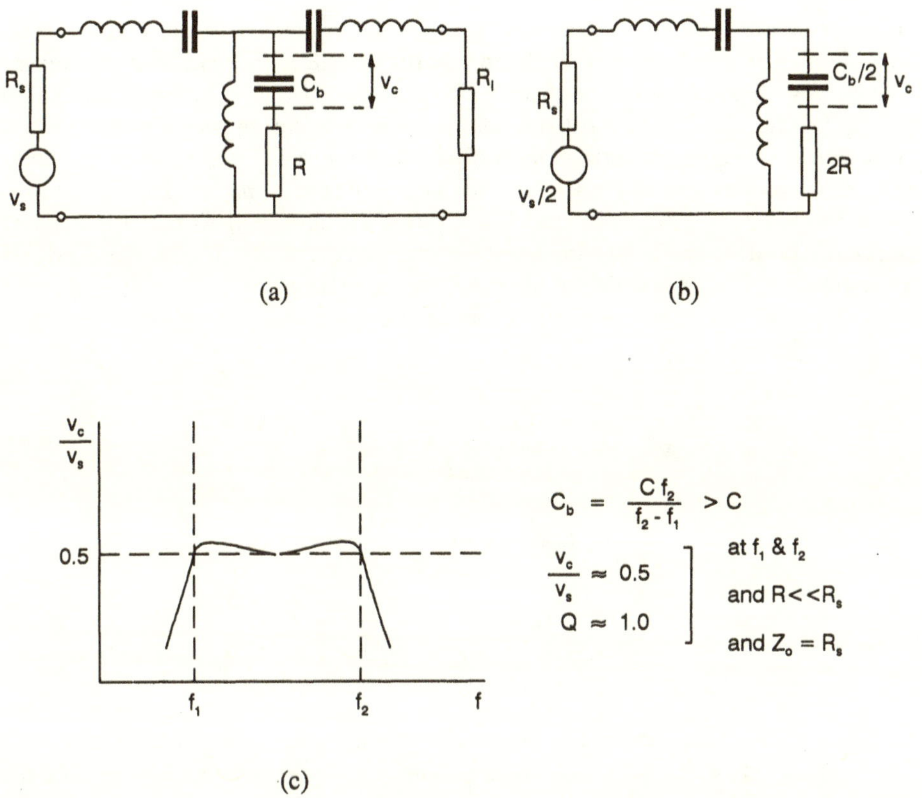

Figure 5.47 Band-pass transformation of a constant K section: (a) band-pass constant K section; (b) half band-pass constant K section; (c) frequency response.

The well-known lumped-element low-pass-to-band-pass frequency transformation is given by

$$\frac{f'}{f_1'} = \frac{f_1}{(f_2 - f_1)}\left[\frac{f}{f_1} - \frac{f_2}{f}\right] \tag{5.32}$$

where the primed variables correspond to the low-pass response and the nonprimed variables correspond to the band-pass response. In this formula, f_1' is the cutoff frequency of the low-pass response, f_1 the lower band edge of the band-pass response, and f_2 the upper band edge of the band-pass response. When applied to the low-pass constant K section, a band-pass version very similar to that shown in Figure 5.47(a) is produced. To be strictly accurate, the series resistive loss of the original shunt capacitor should appear connected at the base of both the inductor and capacitor in the transformed circuit. However, an equivalent series resistor at the base of only the capacitor is a satisfactory approximation and can be more directly associated with the gate losses of the FET. Like the low-pass section, it is possible to simplify the analysis of the band-pass section by creating a half section of the form shown in Figure 5.47(b). On analysis, the frequency response of the transformed network is that of Figure 5.47(c). The response of v_c/v_s is largely flat at or just above a value of 0.5 from the lower cutoff frequency f_1 to the upper cutoff frequency f_2. If $R \ll R_s$, the loaded Q of the circuit at the upper band edge is approximately 1, as was the case for the low-pass section.

After the transformation, the value of the shunt capacitor in the band-pass section is given by

$$C_b = C\left[\frac{f_1'}{f_2 - f_1}\right] \tag{5.33}$$

from which, by setting $f_2 = f_1'$, C_b becomes

$$C_b = C \times \frac{1}{\left(1 - \dfrac{f_1}{f_2}\right)} \tag{5.34}$$

As indicated by (5.34), there has been an increase in the value of the capacitor C by a factor $1/(1 - f_1/f_2)$. Hence, since power gain is proportional to g_m^2, which is proportional to the square of the gate width, which is in turn proportional to the square of the gate capacitance, gain will be increased by a factor of $1/(1 - f_1/f_2)^2$. If we substitute C_b for C in (5.20) and make $f_2 = f_c$, the approximate expression for the gain of the band-pass cascadable amplifier stage in terms of f_t and f_c is now

$$G \approx \left(\frac{f_t}{f_c}\right)^2 \times \frac{2}{\left(1 - \dfrac{f_1}{f_c}\right)^2} \tag{5.35}$$

For f_1 equal to zero, the gain increase is zero, since this must correspond to the original low-pass network. As f_1 rises toward f_c, however, gain progressively increases with the decrease in bandwidth.

Losses have been ignored in (5.35) and the MAG of the FET will always limit the practical gain increase that can be achieved. It should also be noted that the effect of the loss resistor R in the band-pass network will be more pronounced in degrading Q than in the low-pass network. However, taking the example of an octave bandwidth, the increase in gain due to the band-pass transformation will be in the region of a factor of 4 (i.e., 6 dB). This, combined with the current recovery, is almost enough to recover the gain of the original four-FET distributed amplifier. Even for bandwidths as wide as 100%, which is considered to be especially wide in most practical circumstances, the increase in gain is still a worthwhile factor of 1.5^2, or 3.5 dB.

5.4.3 The Basic Cascadable Gain Stage

The steps taken to create the new cascadable amplifier stage from a distributed amplifier can be summarized as follows:

1. Eliminate all but one of the FETs, discarding all the circuitry of the drain line on the side of the internal load;
2. Redesign the remainder of the output network so that it becomes a reactive matching network;
3. Transform the input network to have a band-pass response, choosing bandwidth in accordance with the particular frequency coverage and amplifier gain required;
4. Increase the size of the FET so that its gate capacitance is consistent with the larger value of shunt capacitor in the input network.

A basic schematic of the resulting circuit is given in Figure 5.48. Shown in its present form, the circuit uses the minimum number of elements likely to be needed in practice. For wide-bandwidth applications or when particularly large FET devices must be used, more elements can be added by extending the ladder networks of

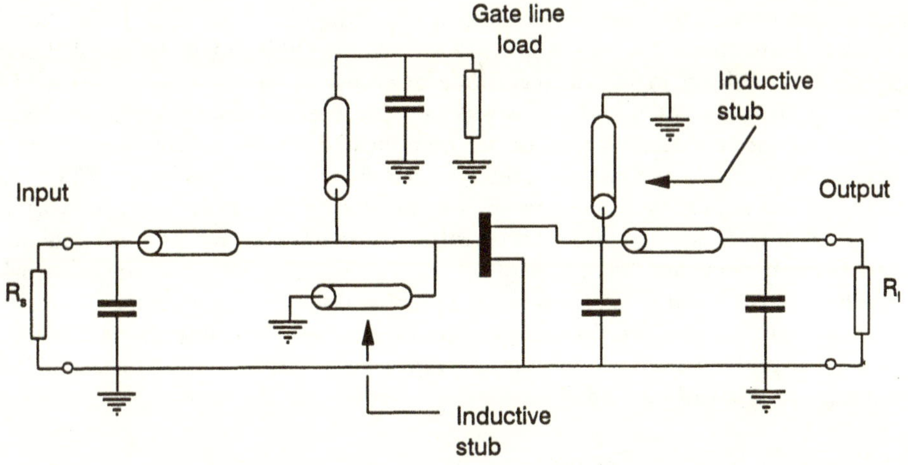

Figure 5.48 Basic schematic of the cascadable gain stage.

lines and shunt capacitors outward on both sides of the FET. It will be apparent that the input network is not a direct transformation of the original low-pass network, but is nevertheless a band-pass filter providing the same level of gain enhancement. The network is generated by exact network synthesis to include only a single shunt inductive stub at the FET. Similarly, the output network is band-pass, allowing an inductive stub to be present at the drain to assist with the matching and provide a point of access for dc supply.

According to computer simulations, the basic cascadable gain stage can achieve gain levels of typically 7 dB over an octave bandwidth centered on 10 GHz when employing a general-purpose, 0.7-μm-gate-length FET of the type produced by the PML foundry. This is only 1 to 2 dB below the sort of gain level that would be achieved by an equivalent reactively matched amplifier design. In marked contrast to the reactively matched design, the new cascadable circuit has a much improved input return loss. For the same octave bandwidth, input return losses can be as good as 20 dB. Output return losses tend to be in the 10- to 15-dB range, depending on the characteristics of the FET.

It is important to recognize that the input matching network is essentially a reactive band-pass filter operating between an external source and an internal load resistor. It does not attempt to match the FET in the sense of injecting power into the FET. For the most part, the FET gate represents a capacitive element within this filter, the voltage on which is induced by a traveling wave (hence the name of the design technique) propagating between the input source resistor and the internal load. As in the case of the conventional distributed amplifier, this causes the gain response of the amplifier to be determined largely by the passband response of the filter network. A flat filter response therefore produces a flat amplifier gain response, unaffected by the natural −6-dB/octave power gain slope of the FET.

At first glance, there is some similarity between the traveling-wave matching technique and the previously reported lossy matching technique [14]. In fact, apart from using an internal load resistor, they are quite different in terms of design approach. The lossy matching technique attempts to match power into a FET loaded by a frequency-dependent lossy element. It treats the matching network and the lossy element separately in a rather complicated, iterative process that leaves unclear the limits that exist on gain and bandwidth for a particular FET. The traveling-wave matching technique, in contrast, treats the whole input network as a single-filter network and gives precise information concerning limits on gain and bandwidth. It provides the values of all the elements of the matching network, including that of the internal load, while ensuring that with the appropriate output matching network present, the resulting gain response of the amplifier will be flat. For a given performance requirement, it also creates a network with the absolute minimum number of elements.

5.4.4 A Practical Design Example

Several cascadable gain stages based on the new circuit concept have been designed around 400-μm and 800-μm FET devices. One of the stages based on an 800-μm device will now be used as a design example. The circuit was required to operate over the 7- to 14-GHz frequency band and be capable of delivering an output power in excess of 200 mW. Consequently it was biased for a Class A mode of operation with a quiescent drain current of *Idss*/2. With the value of 800 μm, the gate width of the FET is approximately the same as the total gate width that might be used in a conventional four-FET distributed amplifier working over the same frequency range.

An equivalent circuit model of the relevant 800-μm FET is shown in Figure 5.49. The FET is a 0.7-μm-gate-length device fabricated by the D07M process at PML, biased at V_{ds} = 3V and V_{gs} = −1V. Quiescent drain current under these bias conditions is *Idss*/2. At a frequency of 15 GHz, which is just above the required operating frequency band of 7 to 14 GHz, values of S_{11} and S_{22} have been calculated and used to derive simple equivalent circuits representing only the input and output impedances of the FET. As the equivalent circuits indicate (Figure 5.49), on the gate side the input matching network will have to absorb a gate capacitance of at least 1.96 pF, whereas on the drain side the output matching network must absorb a drain capacitance of at least 0.228 pF.

5.4.4.1 Input Matching Network

When an artificial output matching network is connected to the drain of the FET as shown in Figure 5.50, S_{11} at 15 GHz is slightly modified and the equivalent series

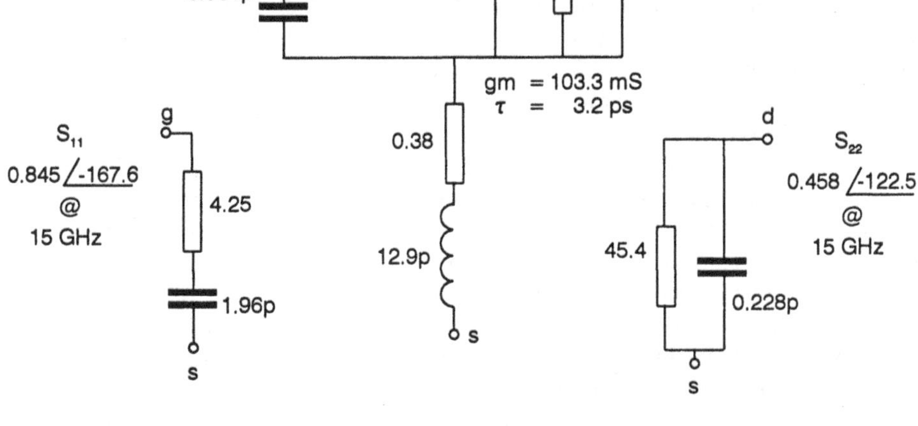

Figure 5.49 Equivalent circuit model of an 800-μm MESFET at V_{ds} = 3V, V_{gs} = −1V (D07M process at PML).

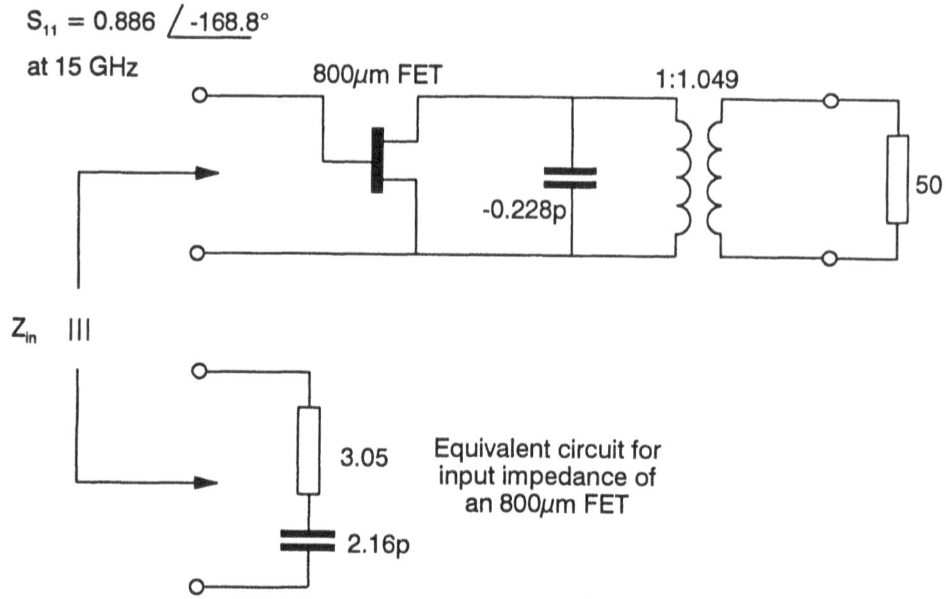

Figure 5.50 Application of an artificial output matching network.

capacitance at the input to the FET is increased to 2.16 pF. This translates into an *S*-plane capacitor with a value 0.158, having chosen an *f*-plane commensurate frequency of 25 GHz. When scaled from a 50Ω to a 1Ω system impedance, the value of the *S*-plane capacitor then becomes 7.88. Hence, the *S*-plane prototype must include a shunt capacitor with a value ≥7.88 at the point where the connection of the FET is intended. After a little experimentation with bandwidth, passband ripple value, and the number of elements, a suitable *S*-plane prototype of the form shown in Figure 5.51 was synthesized using NETSYN. Passband edges were set to $\omega_1 = 0.4706$ and $\omega_2 = 1.2088$, corresponding to *f*-plane band edges of 7 and 14 GHz, respectively. Passband ripple was 0.01 dB.

As indicated by Figure 5.51, the connection point of the FET relative to the *S*-plane prototype is one unit element and one shunt capacitor removed from the terminating load. This is generally the best point of connection because (1) it is the point where the largest capacitor is located, and (2) it is close, but too close to the terminating load. Although it may seem physically desirable to place the FET at the terminating load, the lower loaded *Q* in the vicinity of the load would cause a serious reduction in gain. The shunt inductor is also located at the point of connection of the FET, maximizing the value of the adjacent shunt capacitor. It is possible to move the inductor away from the FET toward the input termination, but this would increase impedances to the right, which would reduce the value of the capacitor and therefore force a reduction in the size of the FET. All the elements

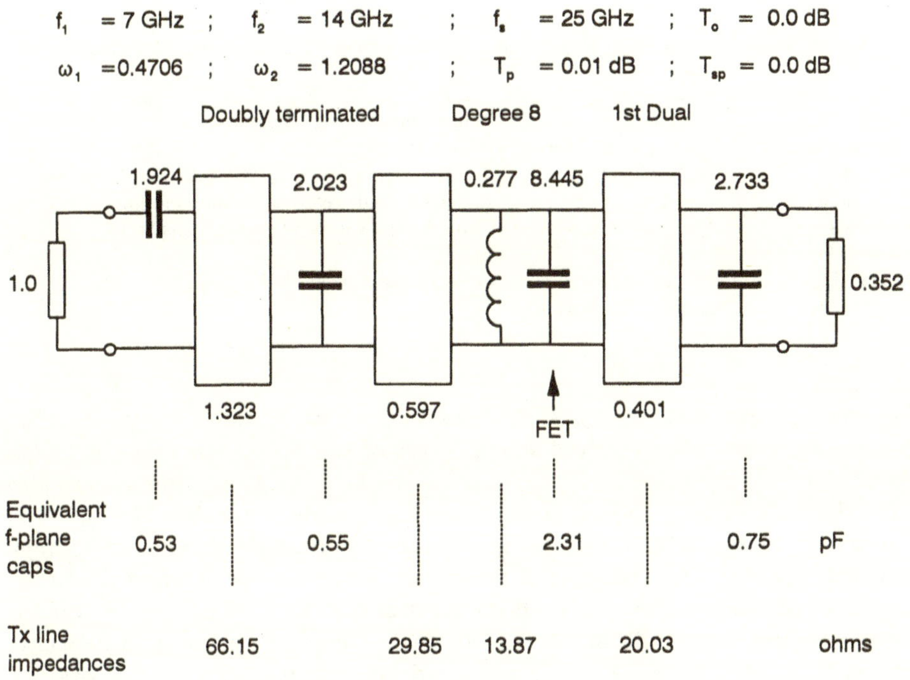

Figure 5.51 *S*-plane prototype of the input matching network.

on the input side of the inductor serve mainly as a means of adjusting the FET capacitor to the most appropriate value. In general, synthesis should begin by specifying the smallest possible number of these elements, and then the number of the unit elements and shunt capacitors should be increased in pairs until the required value of FET capacitor can be achieved. Unless the terminating resistor on the right threatens to become unrealizable after scaling up by 50Ω, its precise value is not important.

The capacitor at the extreme left of the network is a series instead of a shunt element. This has relatively little effect on the value of the rest of the elements, but gives the added bonus of a useful dc break. The degree of the network is 8. It is of the first canonical form with two transmission zeros at $S = (0, j0)$, three transmission zeros at $S = (0, j\infty)$, and three transmission zeros at $S = (1, j0)$.

For synthesis of the network of Figure 5.51 using E-Syn, a summary of the frequency and topological specifications is given in Table 5.5. Since the network contains no redundant elements, the synthesis is straightforward, although it is necessary to extract the elements in a manual mode.

Below the network in Figure 5.51, values of the equivalent *f*-plane capacitors and transmission line impedances have been calculated after scaling to a 50Ω source

Table 5.5
Prototype Synthesis by E-Syn for the Input Matching Network of the Cascadable Gain Stage

Network type	Distributed	Tchebycheff	Band-pass	
Electrical length	50.4 deg at 14 GHz			
Passband	f_1 = 7 GHz	f_2 = 14 GHz	T_p = 0.01 dB	MIL = 0 dB
Terminations	R_s = 1	R_l = 0.352 (only relevant for auto mode)		
Elements	Three transmission lines	Three low-pass stubs	Two high-pass Stubs	
Mode (F0)	Manual			
NxtEL (F6)	Use successively to extract next element from choice given			
Analyze (F0)	Analysis after extraction			

impedance. The capacitor values are based on equating the susceptances of the S-plane elements at ω_2 with the susceptances of the f-plane elements at 14 GHz. As can be seen, the value of the capacitor at the FET is 2.31 pF, which is just above the value calculated in Figure 5.50. When the f-plane network is assembled and connected to the FET, these element values produce an acceptable simulated performance. However, because of the use of lumped capacitors and the need to apply length/impedance trade-offs to the transmissions lines, a little computer optimization is worthwhile. Figure 5.52 shows the complete f-plane equivalent circuit comprising input network, FET, and artificial output matching network, after length/impedance trade-offs have been applied and after optimization.

The input network of Figure 5.52 is similar to that of the basic circuit in Figure 5.48 except for the addition of the extra transmission line and series capacitor. All the transmission line elements now have different lengths but have been assigned the same 90Ω characteristic impedance to simplify physical realization. As shown

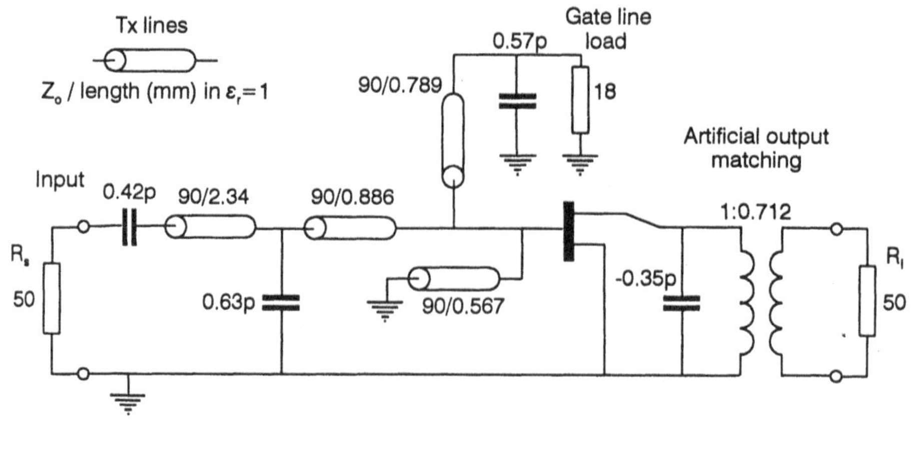

Figure 5.52 The f-plane input matching network, 800-μm FET, and artificial output matching network.

in the figure, the optimization of the circuit included the elements of the artificial output matching network, whose values changed as a result of the presence of the new input network. Plots of the simulated frequency responses of the whole network are given in Figure 5.53. As confirmation of the success of the input matching technique, gain is 6.05 ± 0.05 dB and the input match (S_{11}) is better than −20 dB over practically the whole of the 7- to 14-GHz band.

5.4.4.2 *Output Matching Network*

With further reference to Figure 5.52, the real output matching network of the amplifier must be capable of absorbing a 0.35-pF capacitor at the end connected to the FET and of transforming the effective shunt drain resistance of 98.6Ω down to the 50Ω of the output load. According to (5.7), a lumped-element matching network should be able to achieve an acceptable 12 dB of return loss over an absolute bandwidth of 10.48 GHz. A distributed-element matching network should be able to achieve the same 12-dB return loss over the 4.7- to 14-GHz band for a commensurate frequency of 25 GHz. In both cases the theory suggests a 12-dB return loss should be easy to achieve over the required 7- to 14-GHz bandwidth. However, when the synthesis was attempted with a finite, practical number of elements, this proved only just possible. The resultant *S*-plane prototype produced by NETSYN is shown in Figure 5.54. Passband ripple was set to 0.1 dB and a passband

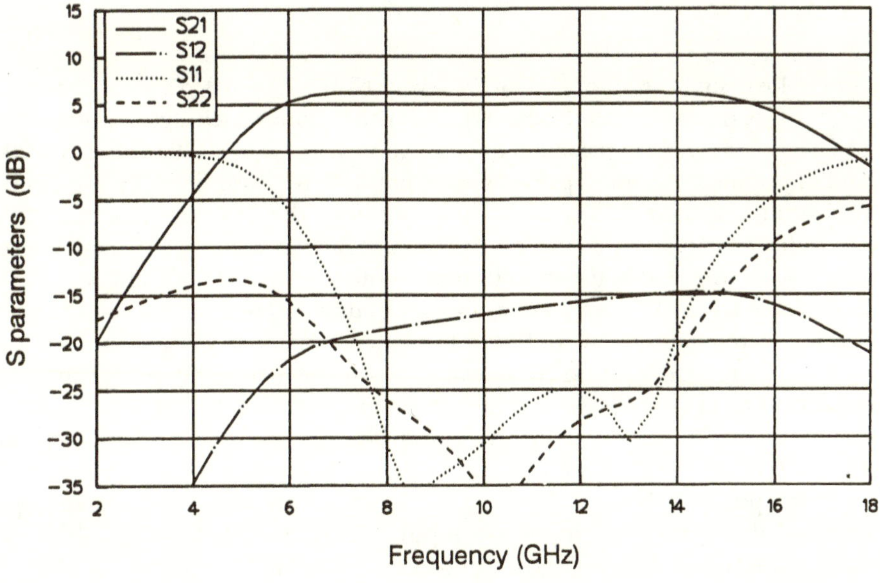

Figure 5.53 Frequency responses of input network, 800-μm FET, and artificial output matching network.

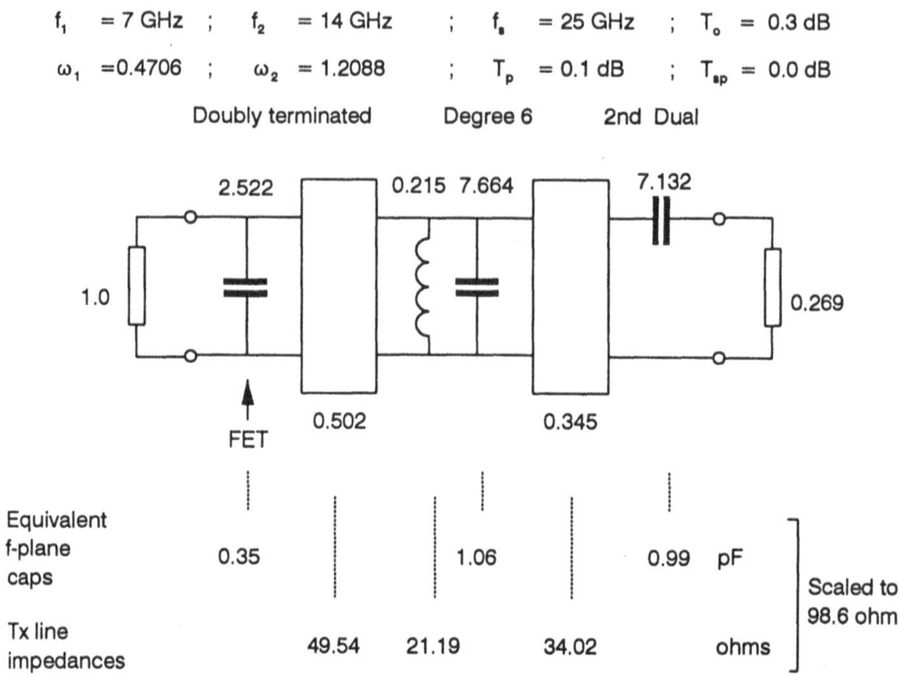

Figure 5.54 *S*-plane prototype of the output matching network.

insertion loss offset was applied of 0.3 dB. (Note: T_o = 0.3 dB is equivalent to $\rho = -12$ dB.) Commensurate frequency, as for the input matching network, was set to 25 GHz, giving *S*-plane passband edge frequencies of $\omega_1 = 0.4706$ and $\omega_2 = 1.2088$.

The topology of the *S*-plane output network is constrained by the need to have a shunt capacitor at the input end to absorb the FET drain capacitance, a series capacitor at or near the output end to act as a dc break, and a shunt inductor somewhere in between to facilitate both access by dc bias and adjustment of impedance levels. The rest can be a ladder of unit elements and shunt capacitors added in whatever numbers are found to be necessary. It is a degree 6 network of the second canonical form, with two transmission zeros at $S = (0, j\infty)$ for the shunt capacitors, two at $S = (1, j0)$ for the unit elements, and two at $S = (0, j0)$ for the inductor and series capacitor. As shown, the network contains no redundant elements. It can be synthesized by E-Syn using the parameter set given in Table 5.6.

In the position shown in Figure 5.54, the shunt inductor produces an impedance ratio of 1:0.269 from input to output, which exceeds the 98.6Ω:50Ω ratio ultimately required. To reduce the impedance ratio it is necessary to move the shunt inductance toward the left, but moving the entire inductance through the adjacent unit element with a Kuroda Identity would cause too great an adjustment.

Table 5.6
Prototype Synthesis by E-Syn for the Output Matching Network of the Cascadable Gain Stage

Network type	Distributed	Tchebycheff	Band-pass	
Electrical length	50.4 deg at 14 GHz			
Passband	f_1 = 7 GHz	f_2 = 14 GHz	T_p = 0.1 dB	MIL = 0 dB
Terminations	R_s = 1	R_l = 0.269 (only relevant for auto mode)		
Elements	Two transmission lines	Two low-pass stubs	Two high-pass stubs	
Mode (F0)	Manual			
NxtEL (F6)	Use successively to extract next element from choice given			
Analyze (F0)	Analysis after extraction			

It is not topologically attractive to split the inductor and move one part through the adjacent unit element, so instead the whole inductor is moved part way through the unit element using the technique discussed in Chapter 2, Section 2.4.3.

As in the case of the input network, values of the equivalent *f*-plane capacitors and transmission line impedances are calculated from the *S*-plane prototype. For the output network, impedances must be scaled up to 98.6Ω to match the effective drain resistance of the FET. Capacitor values are calculated using the same method of equating susceptances, and as shown in Figure 5.54, the value of the first shunt capacitance is now equal to the effective FET drain capacitance of 0.35 pF. After applying length/impedance trade-offs to make all the transmission lines the same 90Ω impedance, the *f*-plane circuit of Figure 5.55 was established. The figure clearly shows the new position of the shunt stub after being moved to the left as mentioned in the previous paragraph. Element values were subjected to some superficial optimization to make minor corrections for the effects of length/impedance trade-offs, and as shown in Figure 5.56, the resultant network then achieves an almost constant 12-dB return loss and 0.3-dB insertion loss over 7 to 14 GHz.

5.4.4.3 The Complete Amplifier

On replacing the artificial output matching network of Figure 5.52 with the real matching network of Figure 5.55, computer simulations predicted an acceptable performance. However, even though there was only slight interaction between the input and output networks, some superficial optimization was worthwhile to achieve the best possible gain and match. The resulting complete amplifier circuit is shown in Figure 5.57 and its frequency responses are plotted in Figure 5.58. Without the addition of circuit parasitics, gain (S_{21}) is seen to be 6.05 ± 0.05 dB, input match (S_{11}) ripples below −25 dB, and output match (S_{22}) is a flat −12 dB over the operating frequency band of 7 to 14 GHz. It may be interesting to note that the MAG of the 800-μm FET was 8.7 dB at 14 GHz. The excellent input match will ensure that the gain stage can be cascaded without difficulty, despite the slightly inferior output match.

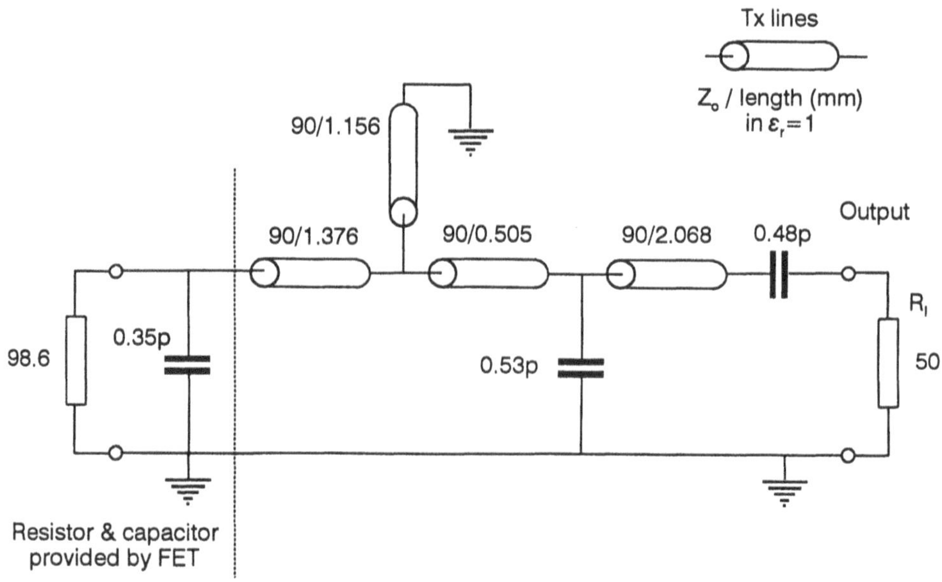

Figure 5.55 The *f*-plane output matching network.

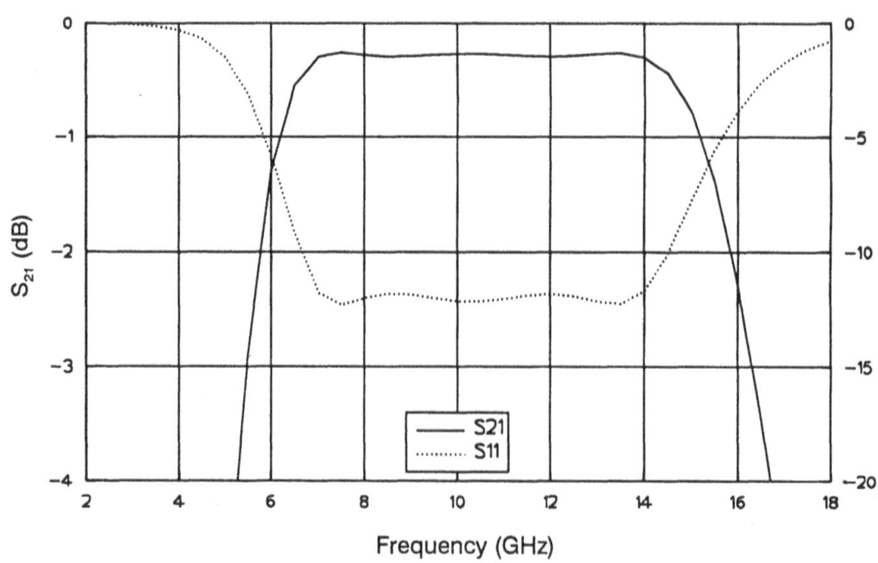

Figure 5.56 Frequency responses of the output matching network.

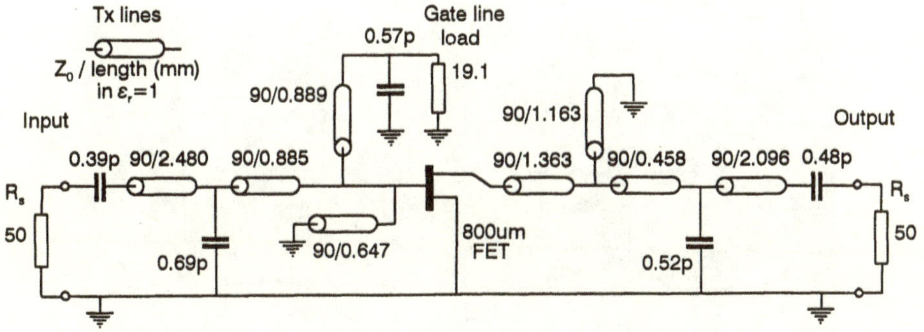

Figure 5.57 Schematic of the complete 800-μm cascadable gain stage.

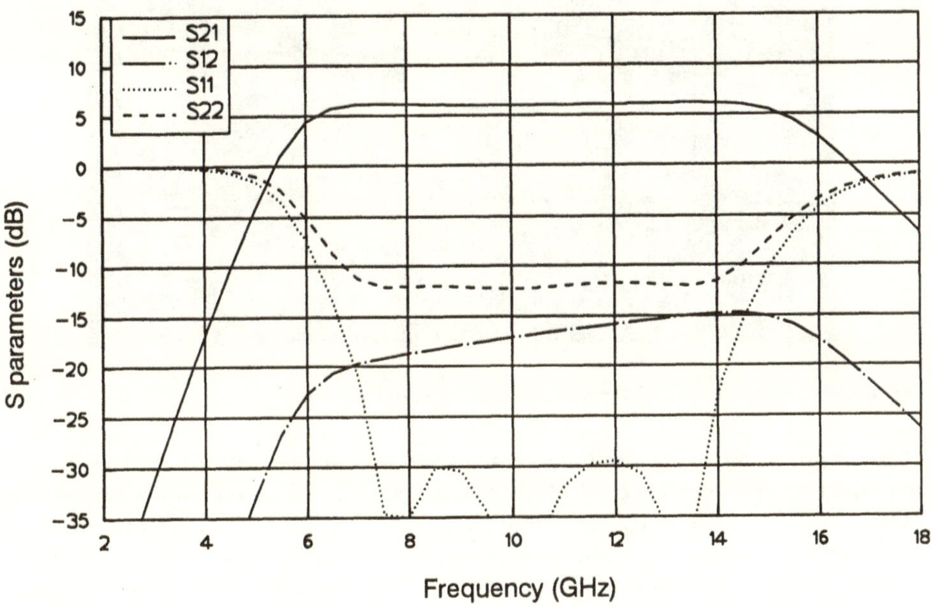

Figure 5.58 Simulated frequency responses without parasitics.

A photograph illustrating the layout of the corresponding amplifier MMIC, measuring only 0.8 × 0.9 mm, is given in Figure 5.59. The input matching network occupies the left half of the picture and the RF input is at the bottom. Input signal travels past the FET from bottom to top, suffering relatively little loss until it is dissipated in the internal load. The shunt stub of the input network is seen connected to the center of the gate electrode, while the other end is split into a pair of tracks grounded by two decoupling capacitors. Gate bias is applied via the internal load resistance (top left), which is also decoupled by a large capacitor. All the tracks

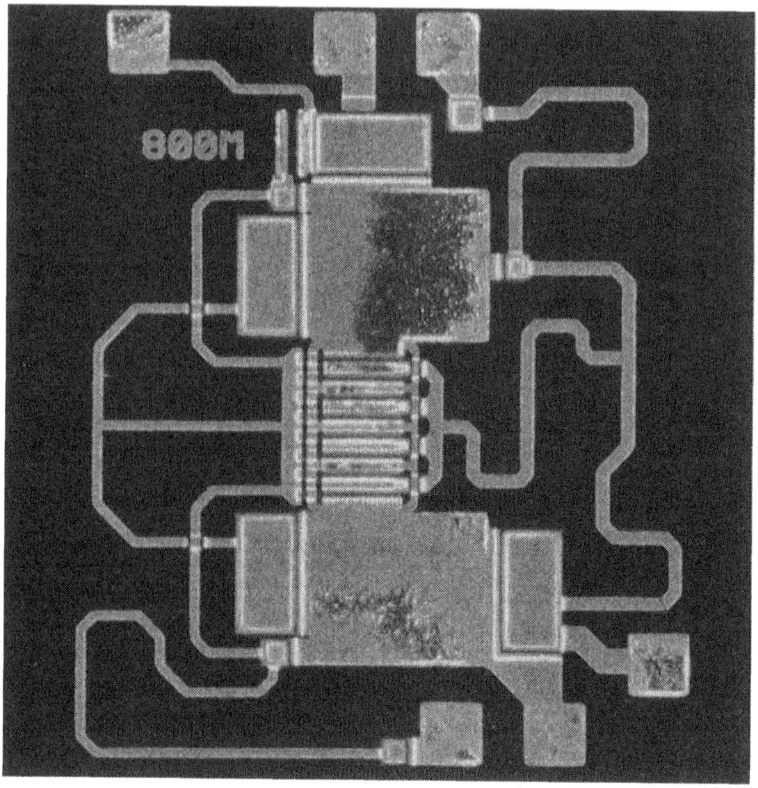

Figure 5.59 Photograph of MMIC cascadable gain stage (0.9 × 0.9 mm).

of the input network are of 90Ω characteristic impedance. On the drain side, line impedances have been lowered a little compared to the circuit of Figure 5.57, giving wider and longer tracks that are better able to carry the higher current levels. As in the input network, the shunt stub is decoupled to ground (bottom right of the picture) to allow application of drain dc bias. RF output is at the top of the layout. Normal dc bias conditions are V_{ds} = 3V and V_{gs} = –1V, corresponding to a drain current of half the saturated value, Idss.

The circuit of Figure 5.59 is the result of adding parasitic elements to the pseudoideal circuit of Figure 5.57 and adjusting transmission line lengths and capacitor values in a final phase of computer optimization. The results are plotted in Figure 5.60. Since losses had already been included in the simulation of the pseudoideal circuit, there is no significant drop in of gain, which is still a flat 6 dB from 7 to 14 GHz. Some minor changes can be seen in S_{22}, and the general level of S_{11} has risen to approximately –17 dB. However, the modest adjustments to the circuit elements that were necessary to compensate for circuit parasitics have almost fully restored the very attractive performance of Figure 5.58.

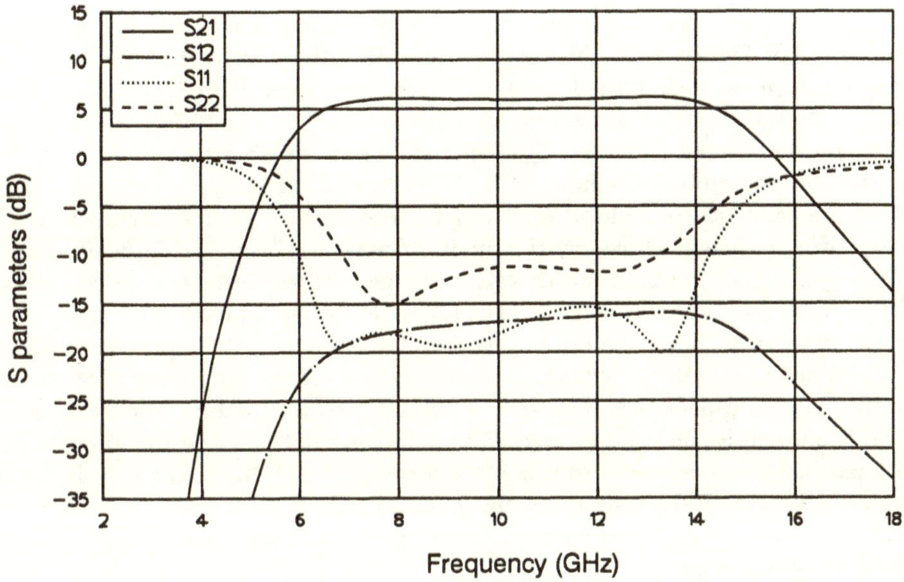

Figure 5.60 Simulated frequency characteristics including parasitics.

Although intended for small-signal operation with an applied drain voltage of 3V, the amplifier will deliver useful levels of output power if drain voltage is increased to 6V. From load line considerations and the basic dc parameters of the 800-μm FET, the output power capability of the amplifier should be in excess of 150 mW. Theory suggests power levels of 250 mW could be reached if the output matching network were optimized for maximum power output. These figures compare very favorably with the power output capability of a conventional distributed amplifier with the same total gate periphery.

5.4.4.4 Measured Performance

Two wafers of MMICs containing numerous examples of these single-stage amplifiers have been processed at the Philips Microwave foundry in France. As previously mentioned, fabrication was by the D07M process for which FET gate length is 0.7 μm. After an initial inspection to check for visual defects, several tens of the circuits were taken from various sites across both wafers and measured under small-signal conditions using RF probes. Relatively little performance variation between circuits was observed, and the results obtained from one of the most typical circuits are plotted in Figure 5.61. Bias conditions were V_{ds} = 3V, V_{gs} = −1V, and I_{ds} = 116 mA.

Comparing the measured responses of Figure 5.61 with the simulated responses of Figure 5.60, the most noticeable difference is that the measured responses appear to be shifted up in frequency by between 1 and 2 GHz. Otherwise, the measured responses agree well with the theory. The value of S_{11} is better than −14 dB from 7 to 14 GHz, and S_{22} is an almost flat −12 dB from 9 to 17 GHz. It is worthwhile remembering that during the design of the output stage, −12 dB was the target value calculated for S_{22}. The gain level at 5.5 dB is fractionally less than the 6 dB predicted. However, this in itself may not be particularly significant, because the g_m of the FET and therefore the gain of the amplifier are very strong functions of the gate bias voltage V_{gs}. The reason for the general shift up in frequency, particularly at the upper end of the passband, can be traced to a problem with the processing of the wafers. According to the process control monitors (PCM) on the wafers, all the capacitor values were at least 12.5% below their intended values. This is enough to alter the tuning of the matching networks. At the lower end of the passband, however, the shift is more associated with the folding of the shunt stubs in the circuit. Folding the stubs tends to reduce their effective inductance, and since they are high-pass elements, their influence is felt most strongly near the lower passband edge.

Finally, the power output capabilities of the 800-μm cascadable gain stage have been measured using a scalar network analyzer system and a power source that could be swept from 0 to +25 dBm. Like the small-signal measurements, these

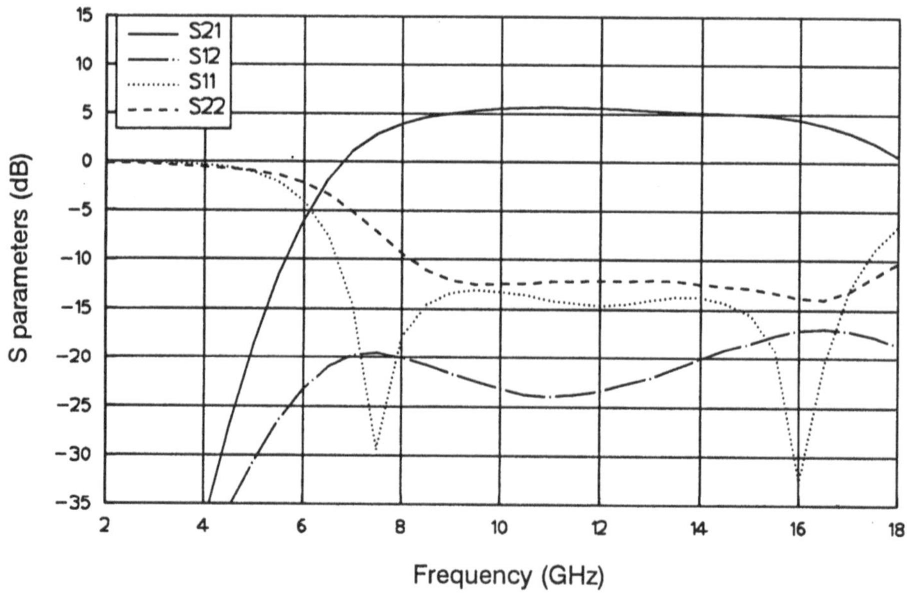

Figure 5.61 Measured frequency responses of the cascadable amplifier MMIC.

power measurements were taken using RF probes with the circuit held flat under vacuum on the table of the probe station. Drain voltage was 6.5V and gate voltage −1V. The resulting plots of power output versus power input are given in Figure 5.62 for several frequencies in the 8- to 14-GHz range. As indicated by the figure, power output reaches +25 dBm (316 mW) at some frequencies before there is any significant sign of gain compression. This is rather more than predicted and a significant achievement from such a small circuit.

5.5 A TWO-STAGE AMPLIFIER WITH INTERSTAGE MATCHING

Highly compact and stable, MMIC amplifiers can now be fabricated by cascading several single-stage modules of the kind shown in Figure 5.59. However, if only two stages are required and if space saving is especially important, it is worthwhile considering the design of the stages as an integrated pair with a dedicated interstage matching network in between. In this case, the traveling-wave matching technique can be applied with equal beneficial effect to the interstage network as it was to the input network. The interstage matching network will ensure that undesirable interactions between the FETs are kept to a minimum. It will incorporate the gain correction needed for the second stage without compromising the excellent tolerance of the whole circuit to modeling inaccuracies and process variations.

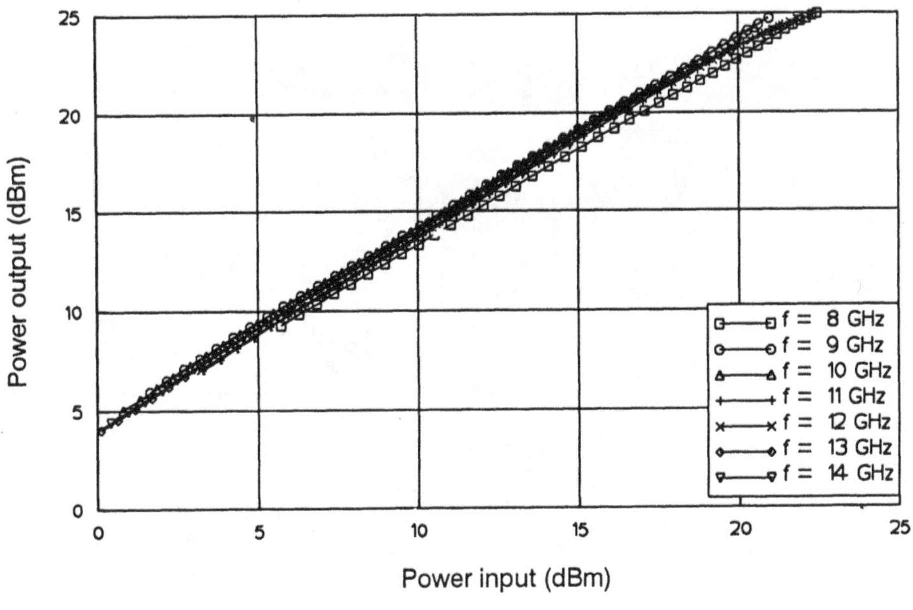

Figure 5.62 Measured output vs. input power for the 800-μm cascadable gain stage.

A two-stage amplifier will be described as a design example. It is an MMIC amplifier based on the use of an 800-μm FET driving a pair of 1,200-μm FETs to give an overall gain of 12 dB from 7 to 14 GHz. A layout of the complete amplifier circuit measuring approximately 2.4 × 1.7 mm is shown in Figure 5.63. Intended for use as a medium-power driver amplifier, the circuit has been designed to produce an output power of approximately 0.8W when biased for Class A operation with V_{ds} = 6.5V and V_{gs} = −1V.

Despite the use of two output FETs and what are clearly multiple-branch matching networks, the circuit of Figure 5.63 is almost perfectly symmetrical about a horizontal center line and can therefore be treated as a cascade of more simple two-ports. Collapsing the circuit about the center line allows its representation by an equivalent circuit of the form shown in Figure 5.64. This comprises an input network almost identical with the type used in the previous single-stage design example, followed by the 800-μm FET, an interstage network, a pair of 1,200-μm FETs wired in parallel, and a reactive output matching network. The topology of the collapsed output network is also exactly the same as that of the single-stage design, but it is synthesized as a singly terminated network to maximize output power, as was the case in the 2.5W power amplifier design of Section 5.3.

Synthesis of the whole two-port network of Figure 5.64 is split into three parts. First the input network is synthesized after applying an artificial output matching

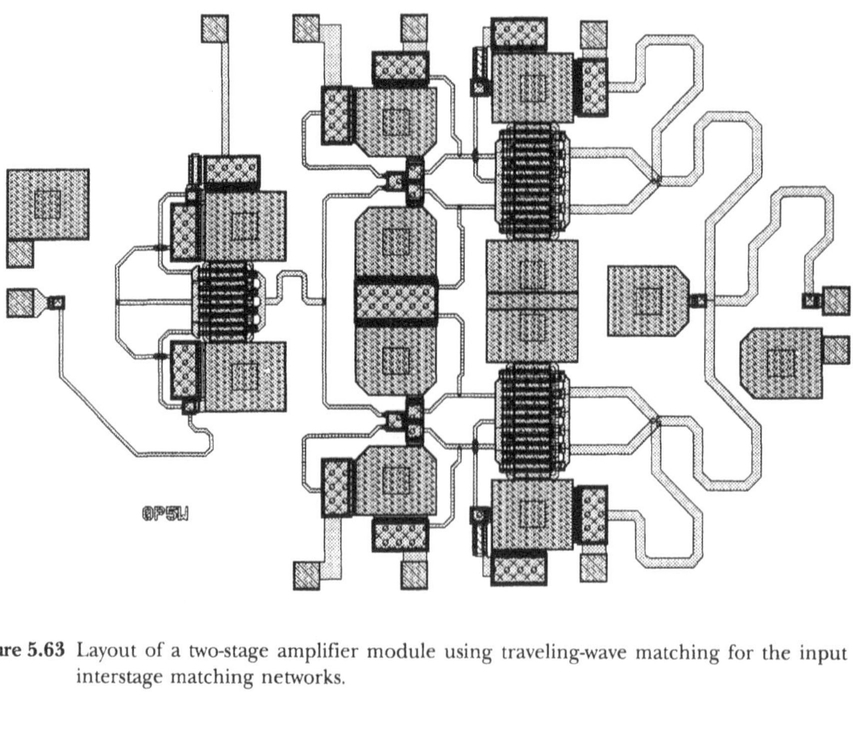

Figure 5.63 Layout of a two-stage amplifier module using traveling-wave matching for the input and interstage matching networks.

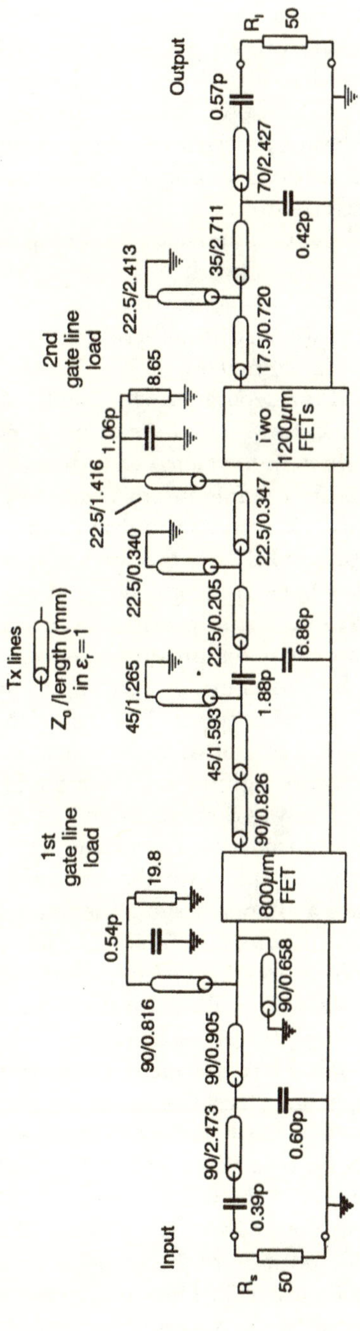

Figure 5.64 Collapsed equivalent circuit of the two-stage amplifier.

network to the 800-μm FET. Then the interstage network is synthesized after applying both artificial input and output matching networks. Finally the output matching network is synthesized to absorb the drain capacitance of the pair of 1,200-μm FETs and to present the optimum load impedance for extraction of maximum power. Since the relevant procedures for the input and output networks have already been covered in previous sections, it is appropriate to describe only the procedure for the interstage network.

As illustrated in Figure 5.64, the interstage matching network resembles the combination of an output matching network and an input matching network after a series capacitor, a shunt capacitor, and an interconnecting transmission line have been eliminated. This is not at all surprising, since the required distribution of elements to perform the impedance matching, the dc bias coupling, and the dc bias decoupling is the same as if two separate stages had been cascaded. In dealing with a single interstage network, however, there is no necessity to match to an intermediate 50Ω resistance, which in turn allows a reduction in component count and further reduces the risk of undesirable interactions between the stages. At the end nearest the input, the network absorbs the drain capacitance of the 800-μm FET. Just to the right of the 800-μm FET the two lines in cascade are derived from a single length of line that has been split into two to help with the subsequent layout of the circuit. The next shunt stub provides access for dc drain supply to the 800-μm FET and the series capacitor provides a necessary dc break between the FETs. Both the stub and the capacitor are also important for establishing the required impedance levels across the network. The rest of the network leading up to the 1,200-μm FETs is the same as that of the input matching network except that the shunt stub has been moved slightly back from the FETs, part way through the adjacent transmission line. This was done to make a final adjustment of impedance levels at the FET. All the transmission lines have been subjected to length/impedance trade-offs, so that when split into the multiple branches of the real matching network, they all have a characteristic impedance of 90Ω.

The *S*-plane prototype of the interstage matching network as synthesized by NETSYN is given in Figure 5.65. Its topology is consistent with the topology of the *f*-plane network except that in this nonredundant form, the first and second unit elements of the network have not yet been split into two. In order to absorb the drain capacitance of the 800-μm FET, the element on the extreme left of the network must be a shunt capacitor. Its value must be greater than 1.4 in the *S*-plane, which corresponds to the 0.35 pF of the FET drain capacitance (see Figure 5.52) after scaling impedances up by a factor of 54. The scaling by a factor of 54 corresponds to a load resistance of 54Ω, which by load line considerations is the load that will extract maximum power from the 800-μm FET. Connection of the two 1,200-μm FETs will be at the position of the second capacitor nearest the right of the network, as indicated in Figure 5.65. This time, to be able to absorb the gate capacitance of the FETs, the value of the *S*-plane capacitance must be at least 25.5, which is equivalent to a total gate capacitance of 6.48 pF when the two 1,200-μm

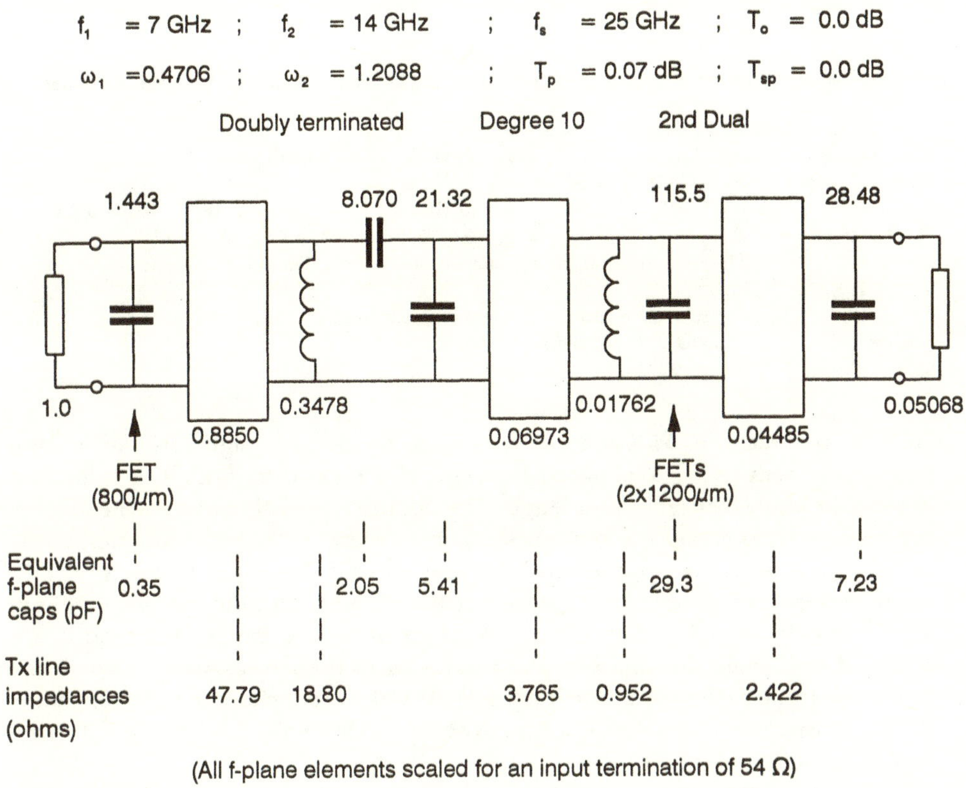

Figure 5.65 *S*-plane prototype of interstage matching network.

FETs are loaded at the drains by an artificial output matching network. The capacitor in the prototype (value 115.5 in the *S*-plane) currently far exceeds this minimum value, but is easily reduced by some simple network transformations.

As with the previous single-stage amplifier, passband edges of 7 and 14 GHz in the *f*-plane give *S*-plane passband edges of ω_1 = 0.4706 and ω_2 = 1.2088 for a commensurate frequency of 25 GHz. By experimenting with different passband ripple values, it was found possible to achieve almost precisely the required capacitor value on the left of the network if T_p was set to 0.07 dB. The network has three transmission zeros at $S = (0, j0)$ with respect to the two shunt inductors and the series capacitor, it has four at $S = (0, j\infty)$ with respect to the four shunt capacitors, and has three at $S = (1, j0)$ with respect to the three unit elements. The network is doubly terminated, of degree 10, and is of the second canonical form. It can be synthesized by E-Syn using the parameter set given in Table 5.7, extracting all the elements one by one in a manual mode.

Values of equivalent *f*-plane capacitors and transmission line impedances are given beneath the network in Figure 5.65. These are all the result of scaling imped-

Table 5.7
Prototype Synthesis by E-Syn for the Interstage Matching Network of the Two-Stage Amplifier

Network type	Distributed	Tchebycheff	Band-pass	
Electrical length	50.4 deg at 14 GHz			
Passband	f_1 = 7 GHz	f_2 = 14 GHz	T_p = 0.07 dB	MIL = 0 dB
Terminations	R_s = 1	R_l = 0.0507 (only relevant for auto mode)		
Elements	Three transmission lines	Four low-pass stubs	Three high-pass stubs	
Mode (F0)	Manual			
NxtEL (F6)	Use successively to extract next element from choice given			
Analyze (F0)	Analysis after extraction			

ances up by a factor of 54. Capacitor values may seem rather high and transmission line impedances rather low toward the right of the network, but all the elements prove to be easily realizable after length/impedance trade-offs and after the subdivision necessary to create the multiple-branch network of the real amplifier. Some of the elements will be split into pairs, while others will be split into sets of four. As previously mentioned, the capacitor at the connection point of the 1,200-μm FETs needs to be reduced in value by a factor of approximately 0.22, and this is achieved by moving the adjacent shunt inductor to the left, part way through the unit element (see Chapter 2, Section 2.4.3). As well as reducing the capacitor value, the transformation raises the impedances of the other elements to the right, further easing the realizability problem.

As a final comment on the *S*-plane prototype, it is important to use a doubly and not a singly terminated network for interstage matching. The doubly terminated network will not be as effective for extracting maximum power from the driving FET, but the capacitor values in the doubly terminated network are larger than those in a singly terminated network for the same set of frequency parameters. The net result is that the output stage will have significantly higher gain.

When the *S*-plane network is first transformed into the *f*-plane, the shunt capacitor on the left must remain temporarily in place to represent the drain capacitance of the 800-μm FET. For the purposes of simulation, an ideal transformer must also be added temporarily at the left to transform the 54Ω terminal resistance down to a more convenient 50Ω. At the connection point of the 1,200-μm FETs, the shunt capacitor must be replaced by a pair of 1,200-μm FET models wired in parallel, the drains of which are connected to a suitable artificial output matching network. In this situation, the network is then optimized to establish a flat gain of approximately 6.5 dB and good values of S_{11} and S_{22} over the 7- to 14-GHz band.

Once good interstage frequency responses have been established, the 800-μm FET complete with its input matching network can be substituted for the first shunt capacitor of Figure 5.65. A check is made of the frequency responses of the pair of stages with the 1,200-μm FETs still connected to an artificial output matching network. In this configuration, the gain of the amplifier is found to be a flat 12 dB over 7 to 14 GHz. Some slight optimization of the elements of the input network gives an improvement in S_{11}, but it is important not to change any of the values of the elements in the interstage for fear of degrading the large signal performance of the 800-μm FET. Finally, the output network is added without any further optimization.

A simulation of the complete collapsed circuit of the amplifier produces the frequency responses shown in Figure 5.66(a). As indicated, gain (S_{21}) is a flat 12 dB over the 7- to 14-GHz frequency range and input match (S_{11}) is better than –25 dB. Output match (S_{22}) is generally less than –10 dB, but this is to be expected as a result of designing the output matching network for maximum output power. All the circuit element values relevant to these predicted responses are those given in Figure 5.64.

The final form of the two-stage amplifier shown in Figure 5.63 was created by expanding the collapsed two-port equivalent circuit of Figure 5.64 into a multiple-branch network. More realistic physical models were then added for the various lumped and distributed elements, allowing a final phase of optimization to be undertaken to correct for the presence of the circuit parasitics. After the optimization, the frequency responses obtained were very similar to those of the collapsed circuit shown in Figure 5.66, and therefore, for the sake of brevity, the final set of predicted frequency responses will not be shown. The measured frequency responses of the finished circuit, however, are plotted in Figure 5.66(b), where it can be seen that a 12-dB gain has been achieved over most of the central region of the passband. The measurement was made using RF probes connected to an HP8510 network analyzer. As indicated, there has been a rounding of the shape of the gain response as a consequence of extra circuit losses and a general shift of all the responses up in frequency by about 1 GHz, as was previously the case for the single-FET cascadable gain stage. The cause can once again be attributed to the extra parasitics associated with the folded transmission lines that were not fully accounted for in the simulations.

5.6 A FIVE-STAGE, 1.5W AMPLIFIER MMIC WITH >25 dB GAIN

As a definitive demonstration of how effectively the new type of gain stage can be cascaded, a high-gain, five-stage amplifier has been designed for fabrication as a single MMIC. Figure 5.67 shows the geometrical layout of the whole amplifier measuring only 4.2 × 3.4 mm. Starting from the input of the amplifier, three individual stages have been cascaded, which are of a type similar to that described in Section 5.4.4. The first is like the cascadable gain stage described in Section 5.4.4 but based on a 400-μm FET operating at the saturated drain current (Idss). The second and third, using 800-μm FETs, are identical to the design of Section 5.4.4 except that the third has an output matching network optimized for maximum power output. These three stages together produce a gain of approximately 18 dB

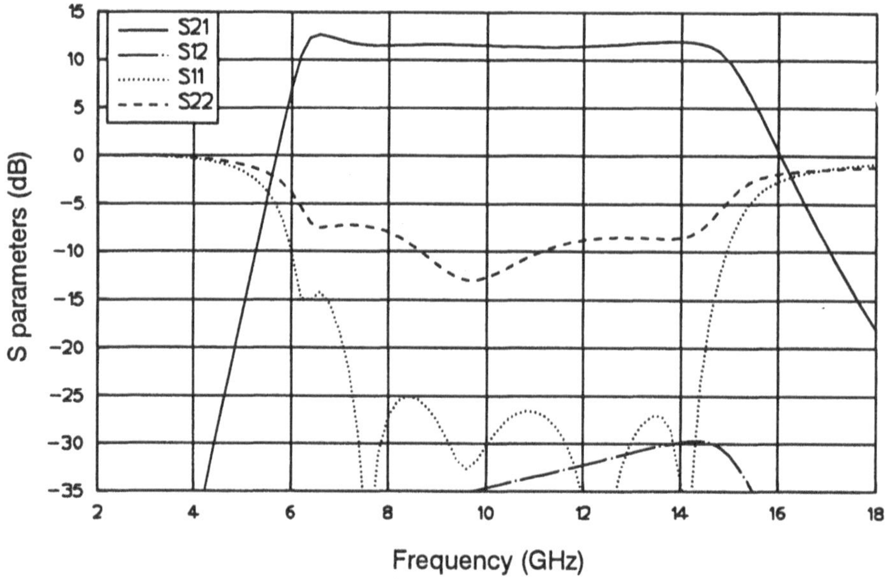

(a)

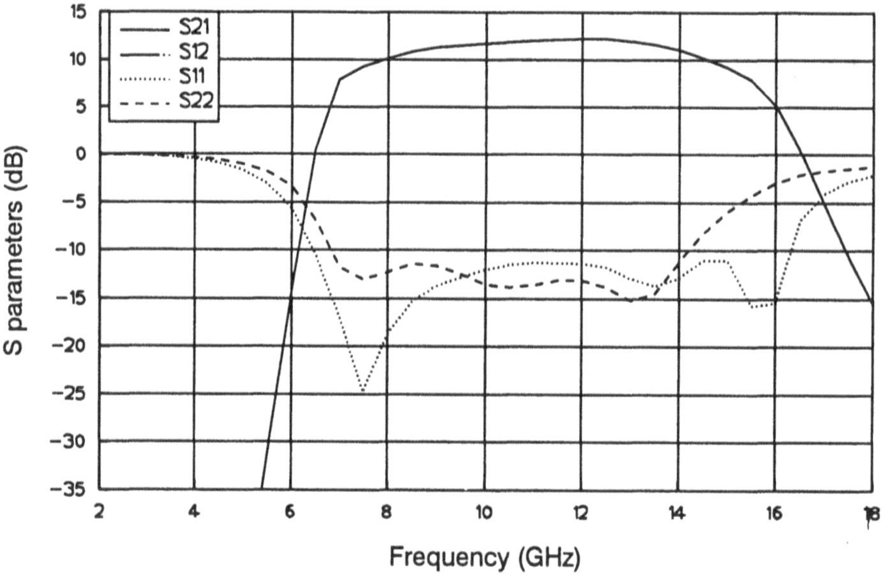

(b)

Figure 5.66 Simulated and measured frequency responses of the two-stage amplifier: (a) simulated responses; (b) measured responses.

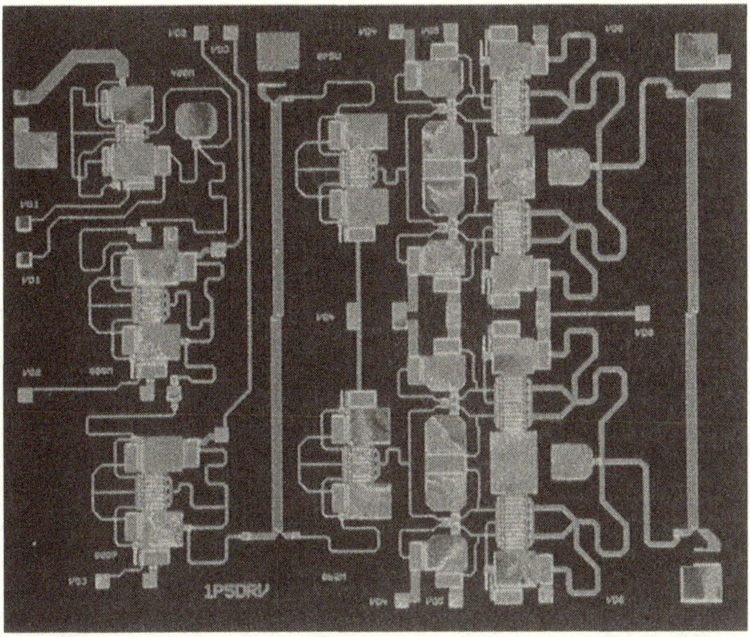

Figure 5.67 Layout of five-stage, 1.5W amplifier MMIC.

over the 7- to 14-GHz range as indicated by the simulated frequency responses plotted in Figure 5.68. An excellent input match has been achieved with relatively little gain ripple as a consequence of their interconnection. As indicated by Figure 5.67, the three stages are contained within a 1-mm-wide strip stretching across the full width of the MMIC, which represents approximately a quarter of the total surface area of GaAs. The three-stage chain is capable of producing output power of 250 mW at the 1-dB gain compression point, which should be more than adequate for driving the following pair of output stages.

The output stages of the amplifier comprise a pair of the two-stage modules of the type illustrated in Figure 5.63 connected in a balanced arrangement between two 2.5-dB Lange [15] couplers. Together their combined gate periphery should be sufficient to deliver output power approaching 1.5W before losses in the Lange couplers are taken into account. Employing the Lange couplers is a convenient means of combining the output power from the two modules, but also ensures that the amplifier will have a good output match. A plot of the simulated frequency responses of the balanced pair of output stages is given in Figure 5.69, demonstrating a gain of 11 dB over 7 to 14 GHz and confirming the beneficial effect of the couplers on the input and output reflection characteristics.

A computer simulation of the entire five-stage amplifier has been carried out for which all circuit parasitics were included. The results are plotted in Figure 5.70.

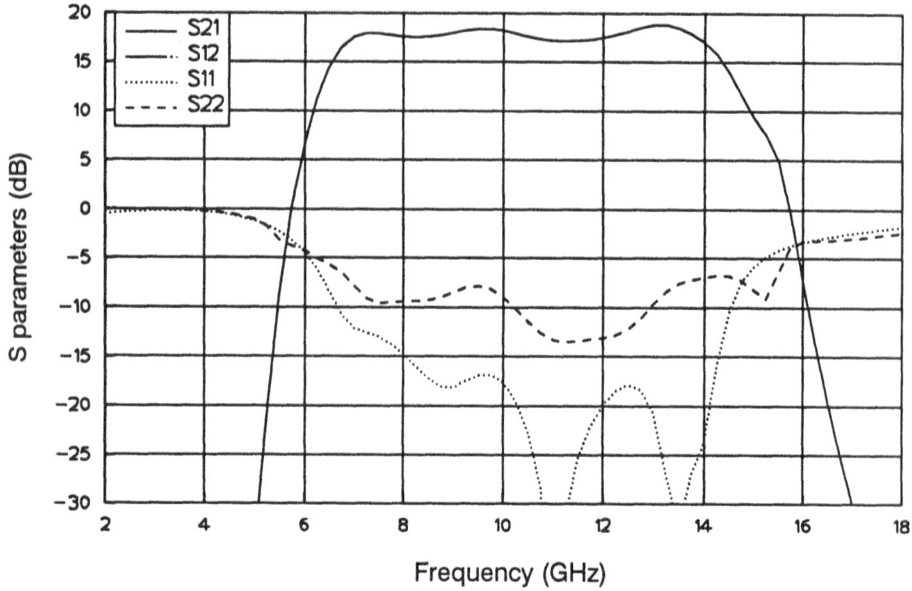

Figure 5.68 Simulated frequency responses of the first three stages.

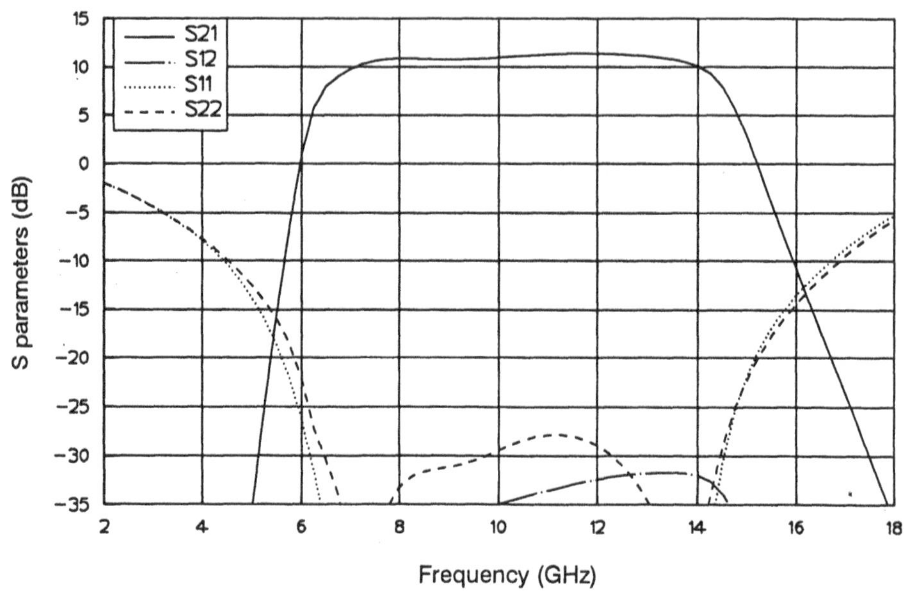

Figure 5.69 Simulated frequency responses of the balanced output stage.

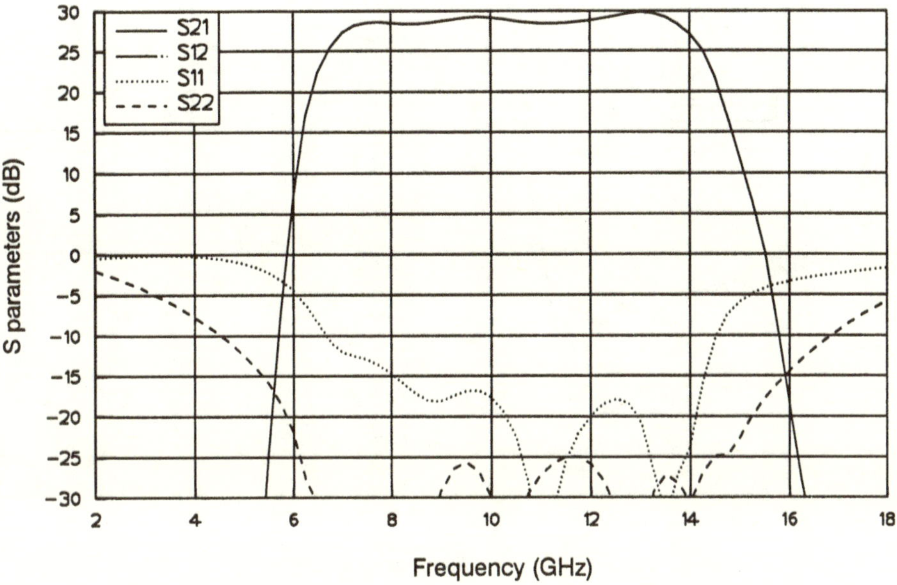

Figure 5.70 Simulated frequency responses of the whole five-stage, 1.5W amplifier.

Gain is approximately 29 ± 1 dB from 7 to 14 GHz. The input reflection coefficient, as determined by the traveling-wave matching of the input stage, is better than −12 dB, and the output reflection coefficient as determined by the output Lange coupler is better than −25 dB over the same frequency band.

After fabrication and initial visual inspection, several examples of the five-stage amplifier MMICs were soldered onto carriers and all the necessary bond wires were attached for the dc supplies. Bond wires were attached to external 50-pF chip capacitors, which provided additional bias supply decoupling as well as acting as convenient bonding posts. Small-signal RF measurements were performed using direct RF probing, while large-signal measurements were performed with the circuits installed in a microstrip housing. An HP 8510 network analyzer was used for the small-signal measurements and a scalar network analyzer for the large-signal measurements. For the small-signal measurements, all the stages except the first stage were operated with a gate voltage of −1V and a drain voltage of 3V (note: for first stage V_{gs} = 0V). For the large-signal measurements, the drain voltage was increased to 6V.

The results of the small-signal measurements on one of the circuits have been plotted in Figure 5.71. As indicated, the gain (S_{21}) level in the passband is slightly lower than predicted (Figure 5.70) and the passband has been shifted up in frequency by approximately 1 GHz. This was to be expected in view of the results obtained for the previous single- and two-stage amplifier designs on which this five-

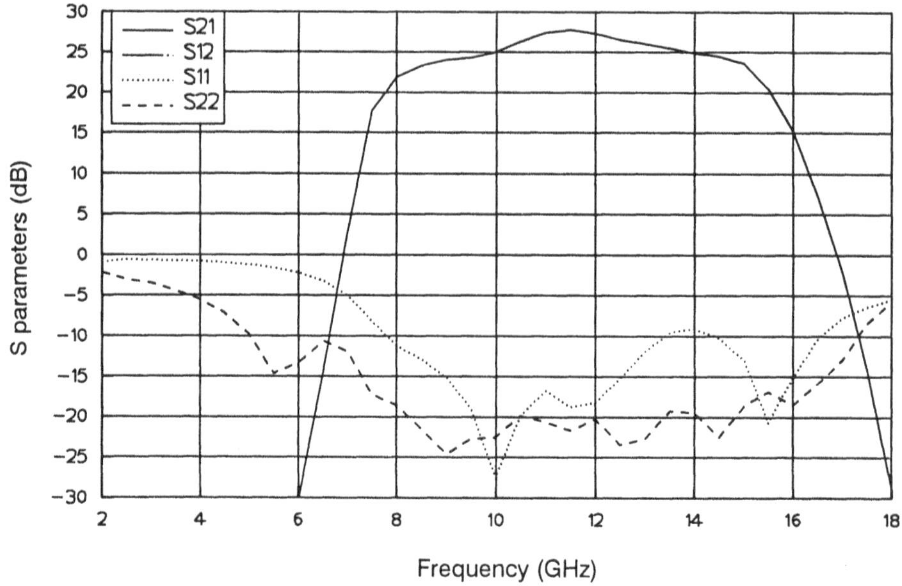

Figure 5.71 Measured small-signal frequency responses of the five-stage MMIC amplifier.

stage circuit was based (i.e., the causes were extra circuit losses and the uncompensated folding of transmission lines). However, despite the drop in gain, an extremely useful gain of 25 ± 3 dB from 8 to 15 GHz has been achieved. S_{11} is substantially better than −10 dB except at 8 and 14 GHz, and S_{22} is generally better than −20 dB over 8 to 14 GHz.

Results of the large-signal measurements are plotted in Figure 5.72. Input power to the amplifier was swept from a low level to up to a maximum of +18 dBm for frequencies in the range 8 to 14 GHz. Except for the 8- and 8.5-GHz frequencies, the saturated output power of the amplifier is over +31 dBm, or 1.26W. This is a commendable output power from an amplifier with a gain of 25 dB. The Lange coupler accounted for more than 0.5 dB loss of output power. Depending on the frequency, Figure 5.72 shows the power output at 1-dB gain compression to be approximately +29 dBm (800 mW). However, the 1-dB compression point of a multistage amplifier like this is a rather pessimistic measure of output power capability, because the 1-dB compression figure is the summation of the gain compressions of all the individual gain stages, each one of which will be operating well inside its own 1-dB compression point.

It is hoped that this example and the other practical examples described in the chapter will have consolidated the theoretical treatment of Chapter 2 and given further insight into the methodology of exact synthesis as applied to active circuit design.

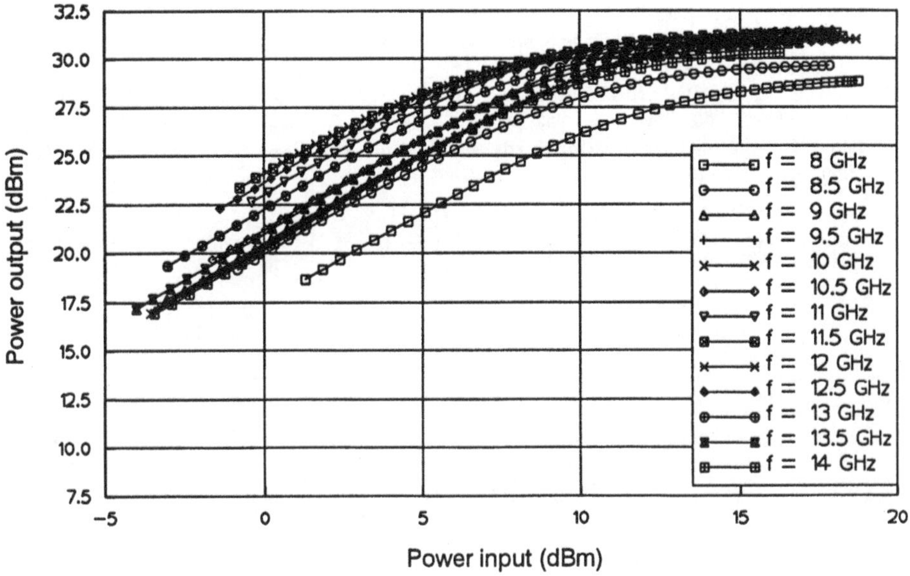

Figure 5.72 Measured large-signal performance of the five-stage MMIC amplifier.

References

[1] Fano, R. M., "Theoretical Limitations on the Broadband Matching of Arbitrary Impedances," *J. Franklin Institute*, January 1950, pp. 57–83.

[2] Bode, H. W., *Network Analysis and Feedback Amplifier Design*, New York: D. Van Nostrand Co., 1945.

[3] Percival, W. S,. "Thermionic Valve Circuits," British Patent 460562, January 1937.

[4] Ayasli, Y., L. D. Reynolds, J. L. Vorhaus, and L. Hanes, "2-20 GHz GaAs Travelling-Wave Amplifier," *IEEE Trans. MTT*, Vol. MTT-30, July 1982, pp. 976–981.

[5] Ayasli, Y., "Decade Bandwidth Amplification at Microwave Frequencies," *Microwave J.*, April 1984, pp. 71–78.

[6] Minnis, B. J., "A 2-18 GHz Multistage Distributed Amplifier With 40 dB Gain," IEE colloquium, *Multi-Octave Active and Passive Components and Antennas*, Digest no. 1987/35, March 1987, pp. 3/1–3/4.

[7] Gamand, P., M. Fairburn, C. Varin, and J. C. Meunier, "1 to 20 GHz Monolithic Distributed Amplifier Using GaAs MESFETs or HEMTs," *Proc. 17th European Microwave Conf.*, September 1987, pp. 249–254.

[8] Beyer, J. B., S. N. Prasad, R. C. Becker, J. E. Nordman, and G. K. Hohenwarter, "MESFET Distributed Amplifier Design Guidelines," *IEEE Trans. MTT*, Vol. MTT-32, March 1984, pp. 268–275.

[9] Storer, J. E., *Passive Network Synthesis*, New York: McGraw-Hill, 1957, pp. 85–91.

[10] Minnis, B. J., and M. Healy, "2.5 W MMIC Power Amplifiers for X Band Phased Arrays," *JFIT 93 Conf. Digest*, March 1993, pp. 71–76.

[11] Milsom, R. F., "FACET—A CAE System for RF Analogue Simulation Including Layout," *26th ACM/IEEE Design Automation Conf.*, 1989, pp. 622–625.

[12] Minnis, B. J., "The Travelling Wave Matching Technique for Cascadable MMIC Amplifiers," *IEEE Trans. MTT*, Vol. 42, No. 4, April 1994.

[13] Jones, K. E., G. S. Barta, and G. C. Herrick, "A 1 to 10 GHz Tapered Distributed Amplifier in a Hermetic Surface Mount Package," *1985 IEEE Gallium Arsenide Integrated Circuit Symposium*, Technical Digest, pp. 137–140.
[14] Niclas, K. B., "On Design and Performance of Lossy Match GaAs MESFET Amplifiers," *IEEE Trans. MTT*, Vol. MTT-30, November 1982, pp. 1900–1907.
[15] Lange, J., "Interdigitated Stripline Quadrature Hybrid," *IEEE Trans. MTT*, Vol. MTT-17, December 1969, pp. 1150–1151.

Summary and Conclusions

This book has presented a universal design procedure for microwave circuits and components and has demonstrated its application with numerous practical examples. The examples have included a wide range of both passive and active circuits. Exact synthesis has the overwhelming advantage over the "classical" circuit design procedure of guaranteeing to find a network that satisfies a particular set of frequency specifications. As long as the number and types of elements have been specified, it will calculate their values in whatever position they are located in the network. The skill of the circuit designer comes into play in choosing the element types, their location, and the fine detail of the target frequency responses so that the network will be physically realizable. Whether or not it is physically realizable, a valid prototype network is always guaranteed. All the practical examples described were either linear or pseudolinear circuits, but the universal design procedure is equally effective in designing highly nonlinear circuits such as detectors, mixers, switches, and very-high-power amplifiers. This is because the major contribution of the universal design procedure to the design of a microwave circuit is in the creation of the passive matching networks. The matching networks for large-signal circuits must simply satisfy a wider range of boundary conditions pertaining to both fundamental and harmonic frequency components.

The universal design procedure as presented concerns itself mainly with the amplitude frequency responses of a circuit and makes no reference to, for example, noise performance or absolute power output. However, this does not mean that the procedure will not cope with designing low-noise amplifiers or power amplifiers. It simply means that other, complementary theorems must be used to establish the required matching conditions at the active devices. Low-noise and power amplifiers, in common with all other microwave circuits, need matching networks and it is usually the design of these that represents the most difficult part of the procedure.

It is in this difficult part where the benefit of the universal design procedure is to be felt most.

No other design technique is believed to be quite so versatile. No other technique can give such a high probability of finding optimum solutions for the general case. No other technique is quite so direct and no other technique relies so little on optimization. Some narrowband circuits with well-defined physical structures may be designed more efficiently by alternative, dedicated design procedures. However, there is no effective alternative for wideband circuits or for circuits with unconventional or previously undefined physical structures.

Exact network synthesis does not deserve its reputation for being complicated and only relevant to filter design. Once some simple rules and concepts have been properly understood, it is an extremely simple technique to use. It is hoped that through the work described herein, more engineers will be encouraged to adopt this procedure and exploit the yet untapped potential of exact synthesis techniques.

APPENDIX A

Some Useful Network Transformations

A.1 TRANSFORMING A SECOND-ORDER REACTANCE BRANCH INTO UNIT ELEMENTS AND VICE VERSA

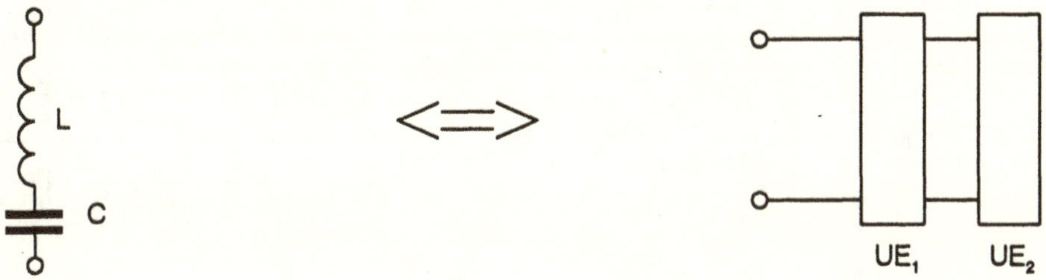

$$L = \frac{\mathrm{UE}_1^2}{\mathrm{UE}_1 + \mathrm{UE}_2} \qquad \mathrm{UE}_1 = L + \frac{1}{C}$$

$$C = \frac{\mathrm{UE}_1 + \mathrm{UE}_2}{\mathrm{UE}_1\mathrm{UE}_2} \qquad \mathrm{UE}_2 = \left(L + \frac{1}{C}\right)\frac{1}{LC}$$

A.2 TRANSFORMING A FOURTH-ORDER REACTANCE BRANCH INTO UNIT ELEMENTS AND VICE VERSA

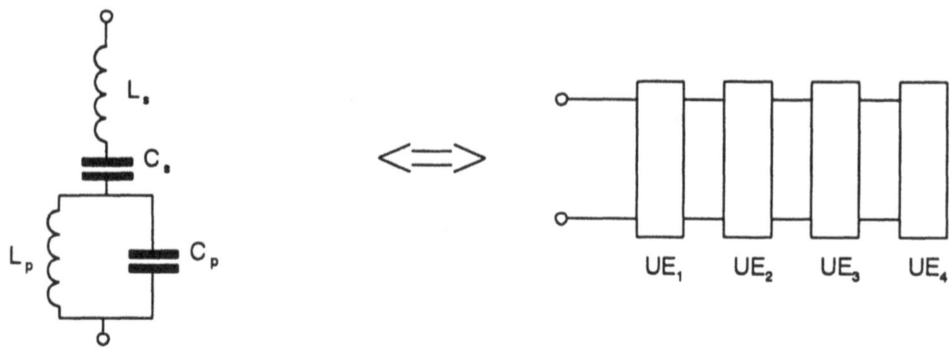

$$C_s = YE_1 + YE_2 + YE_3 + YE_4 \qquad YE_1 = \frac{C_s(L_pC_p + 1)}{(L_pC_p + 1)(1 + C_sL_s) + C_sL_p}$$

$$\frac{1}{L_s} = \frac{YE_1}{YE_2}\left(YE_1 + YE_2 + \frac{YE_1\,YE_3}{YE_2\,YE_4}(YE_2 + YE_4)\right) \qquad YE_2 = \frac{YE_1^2\left(C_pL_s - \dfrac{1}{L_pC_s}\right) + \dfrac{YE_1}{L_p}}{YE_1\left(1 + \dfrac{L_s}{L_p} + \dfrac{C_p}{C_s} + \dfrac{2}{L_pC_s}\right) - \dfrac{1}{L_p}}$$

$$n^2 = \frac{YE_2}{YE_1} + \frac{YE_4}{YE_3} + \left(\frac{YE_1 + YE_2}{YE_1\,YE_2}\right)(YE_3 + YE_4) - C_sL_s - \frac{YE_2\,YE_4}{C_sL_sYE_1\,YE_3} \qquad YE_i = \frac{1}{\mathrm{UE}_i}$$

$$C_p = \frac{YE_2\,YE_4}{n^2L_sYE_1\,YE_3} \qquad YE_4 = \frac{C_s - YE_1 - YE_2}{\left(\dfrac{YE_2}{L_sC_sL_pC_pYE_1} + 1\right)}$$

$$L_p = \frac{n^2}{C_s} \qquad YE_3 = C_s - YE_1 - YE_2 - YE_4$$

A.3 TRANSFORMING A FOURTH-ORDER REACTANCE BRANCH INTO TWO SECOND-ORDER BRANCHES AND VICE VERSA

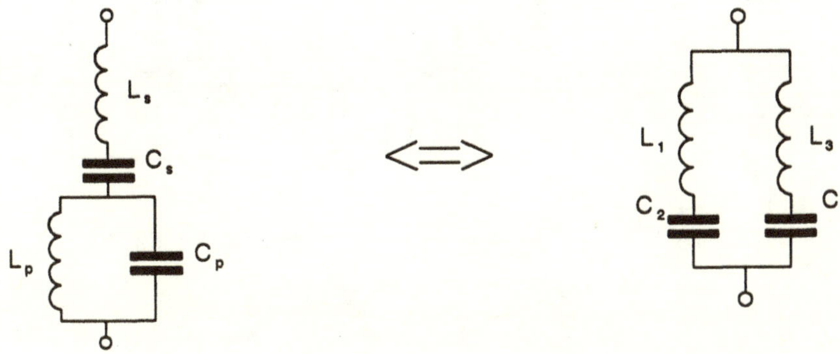

$$\omega_{01,2}^2 = \frac{(C_pL_p + C_sL_s + L_pC_s) \pm \sqrt{(C_pL_p + C_sL_s + L_pC_s)^2 - 4C_pL_pC_sL_s}}{2\ C_p L_p C_s L_s}$$

$$L_s = \frac{L_1L_3}{L_1 + L_3}$$

$$C_s = C_2 + C_4$$

$$L_p = \frac{L_1C_2 + L_3C_4}{C_2 + C_4} - \frac{C_2C_4(L_1 + L_3)}{(C_2 + C_4)^2} - \frac{L_1L_3}{L_1 + L_3}$$

$$C_p = \frac{C_2C_4(L_1 + L_3)}{L_p(C_2 + C_4)}$$

$$L_1 = \frac{L_s\left[1 - \left(\frac{\omega_{02}}{\omega_{01}}\right)^2\right]}{1 - L_sC_s\omega_{02}^2}$$

$$C_2 = \frac{1}{L_1\omega_{01}^2}$$

$$L_3 = \frac{L_s\left[1 - \left(\frac{\omega_{02}}{\omega_{01}}\right)^2\right]}{1 - L_sC_s\omega_{01}^2}$$

$$C_4 = \frac{1}{L_3\omega_{02}^2}$$

A.4 TRANSFORMING A CAPACITOR L SECTION INTO A T SECTION

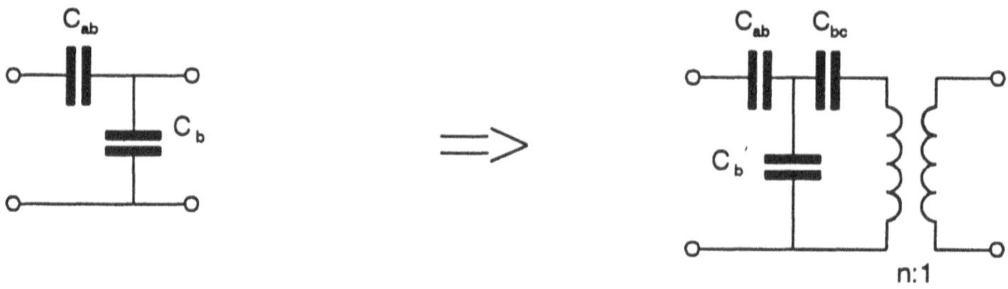

$$C'_{ab} = \frac{C_b C_{ab}}{C_b + C_{ab}(1-n)} \quad C'_b = \frac{C_b}{n} \quad C'_{ab} = \frac{C_b}{n(n-1)}$$

$$\text{For } C'_{ab} = C'_{bc} \quad \text{make} \quad n = \sqrt{1 + \frac{C_b}{C_{ab}}}$$

A.5 TRANSFORMING A CAPACITOR L SECTION INTO A π SECTION

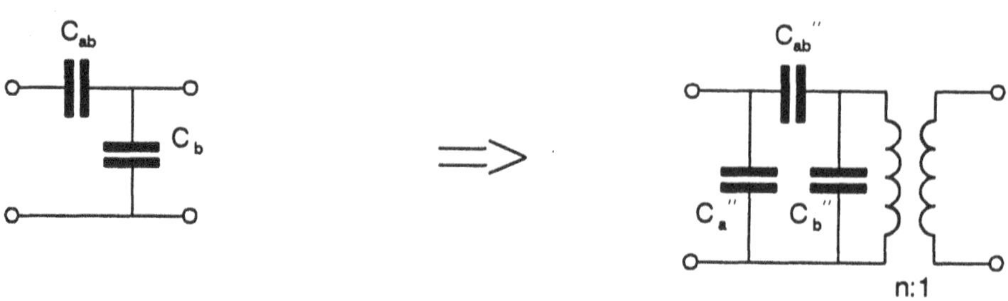

$$C''_a = C_{ab}\left(1 - \frac{1}{n}\right) \quad C''_{ab} = \frac{C_{ab}}{n} \quad C''_b = \frac{1}{n}\left(\frac{C_b}{n} + C_{ab}\left(\frac{1}{n} - 1\right)\right)$$

$$\text{For} \quad C''_a = C''_b \quad \text{make} \quad n = \sqrt{1 + \frac{C_b}{C_{ab}}}$$

A.6 TRANSFORMING A CAPACITOR T SECTION INTO A π SECTION AND VICE VERSA

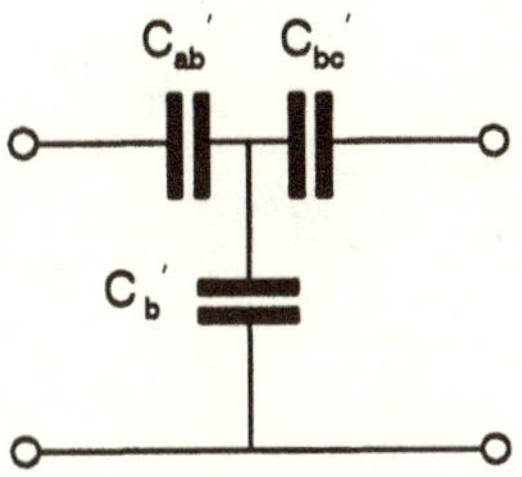

<=>

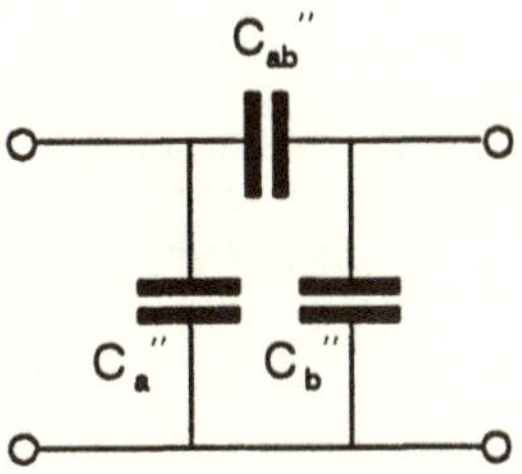

$$C'_{ab} = \frac{C_T}{C''_b}$$

$$C'_b = \frac{C_T}{C''_{ab}}$$

$$C'_{bc} = \frac{C_T}{C''_a}$$

where $C_T = C''_a C''_{ab} + C''_a C''_b + C''_{ab} C''_b$

$$C''_a = \frac{C'_{ab} C'_b}{C_T}$$

$$C''_{ab} = \frac{C'_{ab} C'_{bc}}{C_T}$$

$$C''_b = \frac{C'_{bc} C'_b}{C_T}$$

where $C_T = C'_{ab} + C'_b + C'_{bc}$

APPENDIX B

Library of Coupled-Line Sections and Equivalent Circuits

B.1 GENERAL RELATIONS

Normalized distributed capacitances are related to odd and even mode characteristic admittances by the formulas

$$C_a = aY_{oea} \qquad Y_{oea} = \frac{C_a}{a}$$

$$C_b = aY_{oeb} \qquad Y_{oeb} = \frac{C_b}{a}$$

$$C_{ab} = \frac{a}{2}(Y_{ooa} - Y_{oea}) \qquad Y_{ooa} = \frac{1}{a}(C_a + 2C_{ab})$$

$$C_{ab} = \frac{a}{2}(Y_{oob} - Y_{oeb}) \qquad Y_{oob} = \frac{1}{a}(C_b + 2C_{ab})$$

where $a = 377/\sqrt{\epsilon_{reff}}$ and ϵ_{reff} is the effective dielectric constant of the transmission line medium. If the coupled line section is symmetrical, $C_a = C_b$, $Z_{oe} = 1/Y_{oe}$, and $Z_{oo} = 1/Y_{oo}$, where Z_{oe} and Z_{oo} are even and odd mode characteristic impedances, respectively. The voltage coupling ratio C_v and the section impedance Z_{sec} are then given by

$$C_v = \frac{Z_{oe} - Z_{oo}}{Z_{oe} + Z_{oo}} \quad \text{and} \quad Z_{sec} = \sqrt{Z_{oe} \times Z_{oo}}$$

B.2 COUPLED LINES WITH OPEN CIRCUITS AT OPPOSITE ENDS

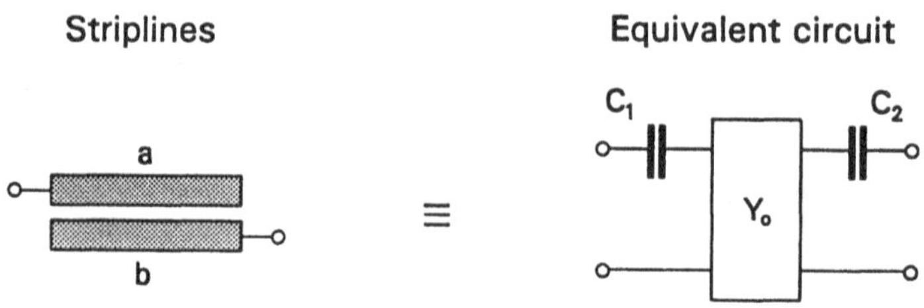

$$C_a = a\left(\frac{C_1 Y_0}{Y_0 + C_1 + C_2}\right) \qquad C_1 = \frac{C_aC_b + C_aC_{ab} + C_bC_{ab}}{aC_b}$$

$$C_b = a\left(\frac{C_2 Y_0}{Y_0 + C_1 + C_2}\right) \qquad C_2 = \frac{C_aC_b + C_aC_{ab} + C_bC_{ab}}{aC_a}$$

$$C_{ab} = a\left(\frac{C_1 C_2}{Y_0 + C_1 + C_2}\right) \qquad Y_0 = \frac{C_aC_b + C_aC_{ab} + C_bC_{ab}}{aC_{ab}}$$

If $C_1 = C_2$, $Z_1 = 1/C_1$, $Z_2 = 1/C_2$, and $Z_0 = 1/Y_0$, then

$$Z_{oe} = Z_1 + 2Z_0 \quad \text{and} \quad Z_{oo} = Z_1 = Z_2$$

B.3 COUPLED LINES WITH OPEN CIRCUITS AT THE SAME END

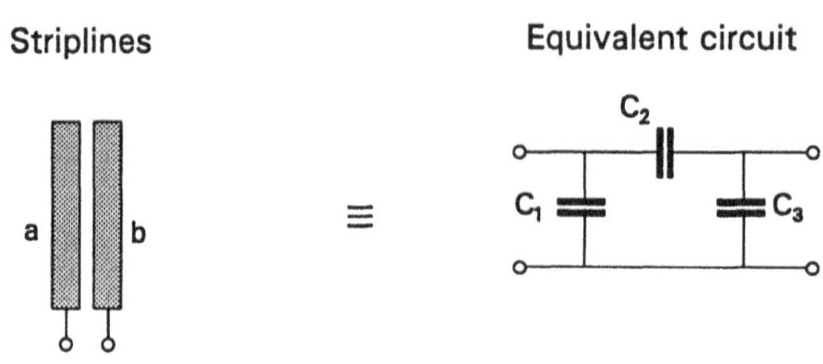

$$C_a = aC_1 \qquad C_1 = \frac{C_a}{a}$$

$$C_b = aC_3 \qquad C_3 = \frac{C_b}{a}$$

$$C_{ab} = aC_2 \qquad C_2 = \frac{C_{ab}}{a}$$

If $C_1 = C_3$, then

$$Z_{oe} = \frac{1}{C_1} = Z_1 \quad \text{and} \quad Z_{oo} = \frac{1}{C_1 + 2C_2} = \frac{Z_1 Z_2}{Z_2 + 2Z_1}$$

B.4 COUPLED LINES WITH SHORT CIRCUITS AT OPPOSITE ENDS

Striplines

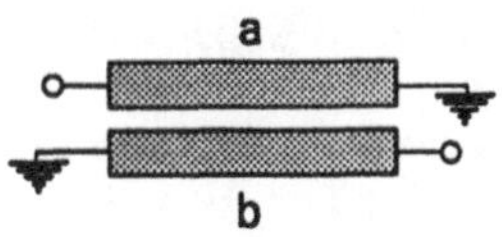

≡

Equivalent circuit

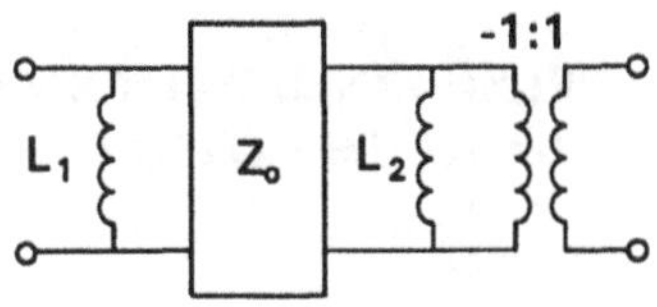

$$C_a = \frac{a}{L_1} \qquad L_1 = \frac{a}{C_a}$$

$$C_b = \frac{a}{L_2} \qquad L_2 = \frac{a}{C_b}$$

$$C_{ab} = \frac{a}{Z_0} \qquad Z_0 = \frac{a}{C_{ab}}$$

If $L_1 = L_2$, then

$$Z_{oe} = L_1 = L_2 \quad \text{and} \quad Z_{oo} = \frac{Z_0 L_1}{Z_0 + 2L_1}$$

B.5 COUPLED LINES WITH A SHORT CIRCUIT AT ONE END AND A BRIDGE AT THE OTHER

Striplines

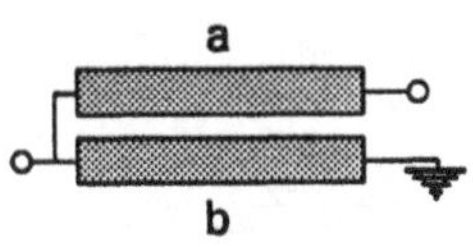

≡

Equivalent circuit

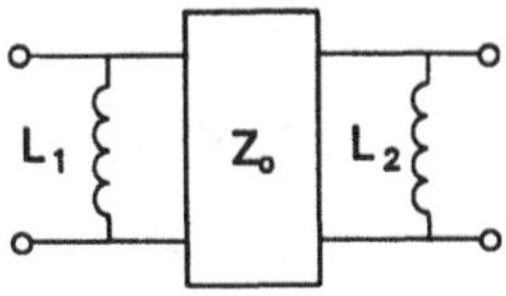

$$C_a = \frac{a}{Z_0} \qquad Z_0 = \frac{a}{C_a}$$

$$C_b = \frac{a}{L_1} \qquad L_1 = \frac{a}{C_b}$$

$$C_{ab} = \frac{a}{L_2} \qquad L_2 = \frac{a}{C_{ab}}$$

For symmetrical coupled lines, $C_a = C_b$ and $L_1 = Z_0$.

B.6 COUPLED LINES WITH AN OPEN CIRCUIT AT ONE END AND A BRIDGE AT THE OTHER

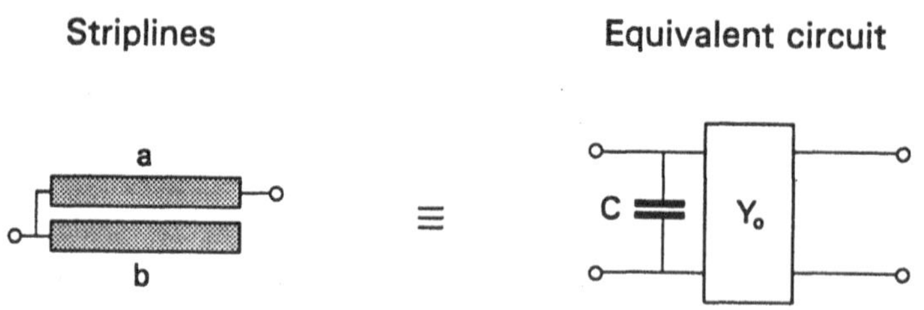

$$C_a = \frac{a}{2}(Y_0 + C) \qquad Y_0 = \frac{1}{a}\left(\frac{C_aC_b + C_aC_{ab} + C_bC_{ab}}{C_b + C_{ab}}\right)$$

$$C_{ab} = \frac{a}{4}\left(\frac{Y_0^2 - C^2}{C}\right) \qquad C = \frac{1}{a}\left(\frac{C_b^2}{C_b + C_{ab}}\right)$$

(when $C_a = C_b$)

For symmetrical coupled lines,

$$Z_{oe} = \frac{2}{Y_0 + C} \qquad Z_0 = \frac{Z_{oe} + Z_{oo}}{2}$$

$$Z_{oo} = CZ_0Z_{oe} \qquad C = \frac{Y_0Z_{oo}}{Z_{oe}}$$

B.7 COUPLED LINES WITH AN OPEN CIRCUIT AT ONE END AND A SHUNT OPEN-CIRCUIT STUB AT THE OTHER

Striplines

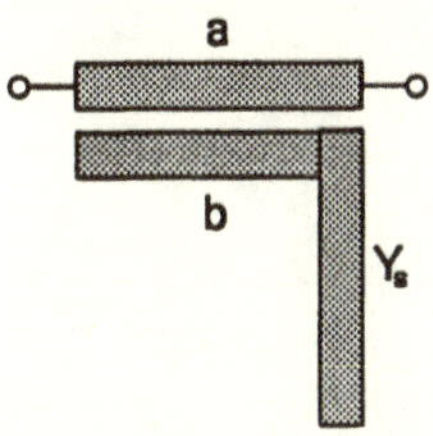

≡

Equivalent circuit

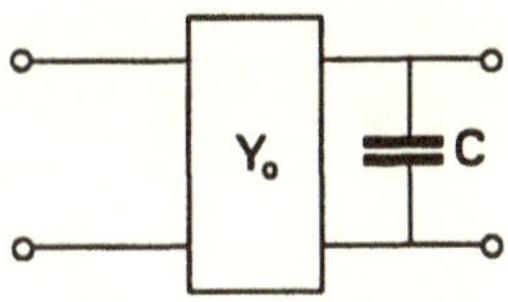

For $C_a = C_b$, choose Y_s and then

$$C_a = a\left(\frac{Y_s(C_1 + Y_0)}{Y_s - C_1} - \frac{\sqrt{Y_s C_1 (Y_s + Y_0)(C_1 + Y_0)}}{Y_s - C_1}\right) \qquad C_1 = \frac{C_{ab}^2 Y_s}{(C_a + C_{ab})(C_a + C_{ab} + aY_s)}$$

$$C_{ab} = a\left(\frac{\sqrt{Y_s C_1 (Y_s + Y_0)(C_1 + Y_0)}}{Y_s - C_1}\right) \qquad Y_0 = \frac{1}{a}\left(C_b + \frac{C_a C_{ab}}{C_a + C_{ab}}\right)$$

List of Technical Publications (December 1994)

1. B. J. Minnis, "Printed Circuit Coupled-Line Filters for Bandwidths up to and Greater Than an Octave," *IEEE Trans. MTT,* Vol. MTT-29, No. 3, March 1981, pp. 215–222.
2. ———,"Classes of Sub-Miniature Microwave Printed Circuit Filters With Arbitrary Passband and Stopband Widths," *IEEE Trans. MTT,* Vol. MTT-30, No. 11, November 1982, pp. 1893–1900.
3. ———,"Sub-Miniature, Microwave Printed Circuit Filters With Arbitrary Passband and Stopband Widths," *1982 IEEE MTT-S International Microwave Symposium Digest,* pp. 535–537.
4. ———, "Printed Circuit Filters Covering 10%–100% Bandwidths," IEE colloquium, *Microwave Filters,* Digest No. 1982/4, January 1982, pp. 3/1–3/4.
5. R. N. Bates and B. J. Minnis, "Some Experimental Distributed Amplifiers in Hybrid Form," IEE colloquium, *Distributed and Travelling Wave Solid-State Circuits and Devices,* Digest No. 1985/45, April 1985, pp. 3/1–3/5.
6. B. J. Minnis, "Direct Frequency Synthesisers for MTI Radar," IEE colloquium, *Low Noise Oscillators and Synthesisers,* Digest No. 1986/13, January 1986, pp. 11/1–11/4.
7. ———, "A Printed Circuit Stub Tuner for Microwave Integrated Circuits," *IEEE Trans. MTT,* Vol. MTT-35, No. 3, March 1987, pp. 346–349.
8. ———, "Decade Bandwidth Bias Ts for MIC Applications up to 50 GHz," *IEEE Trans. MTT,* Vol. MTT-35, No. 6, June 1987, pp. 597–600.
9. B. J. Minnis and D. M. Nugent, "Control and De-embedding Software for HP8510 Network Analyzer," IEE colloquium, *The Technology of Measuring Microwave and Millimetric Components,* Digest No. 1986/121, November 1986, pp. 2/1–2/6.
10. B. J. Minnis, "A 2–18 GHz, Multistage, Distributed Amplifier With 40 dB Gain," IEE colloquium, *Multi-Octave Active and Passive Components and Antennas,*" Digest No. 1987/35, March 1987, pp. 3/1–3/4.
11. ———, "Novel Variation of Distributed Amplifier for Millimetre-Wave Operation," *Electronics Letters,* Vol. 24, No. 9, April 1988, pp. 513–514.
12. B. J. Minnis et al., "The State of the Art for Broadband MMIC Amplifier Design in Philips UK," IEE colloquium, *Microwave and Millimetre Wave Monolithic Integrated Circuits,* Digest No. 1988/117, November 1988, pp. 4/1–4/5.
13. B. J. Minnis et al., "A Novel Monolithic Distributed Mixer Design," IEE colloquium, *Microwave and Millimetre Wave Monolithic Integrated Circuits,* Digest No. 1988/117, November 1988, pp. 13/1–13/6.

14. B. J. Minnis, "Microwave Circuit Design by Synthesis," invited lecture given as part of a postgraduate course at Essex University, January 1989.
15. ———, "Designing Distributed Amplifiers for Prescribed Gain Slope," IEE colloquium, Multi-Octave Active and Passive Components and Antennas," Digest No. 1989/75, May 1989, pp. 4/1–4/5.
16. ———, "Extending the Frequency Range of Distributed Amplifiers Using Bandpass Filter Structures," *Microwave J.*, August 1989, pp. 109–122.
17. L. M. Devlin and B. J. Minnis, "A Versatile Vector Modulator Design for MMIC," *IEEE MTT-S International Microwave Symposium Digest*, June 1990, paper L-7, pp. 519–522.
18. P. Gamand, B. J. Minnis, et al., "Faisabilite D'oscillateurs HEMT Monolithiques en Bande Millimetrique," *7emes Journees Nationales Microondes*, Grenoble, March 1991.
19. B. J. Minnis and M. Healy, "New Broadband Balun Structures for Microwave Monolithic Integrated Circuits," *1991 IEEE MTT-S International Microwave Symposium*, Paper M-2, pp. 425–428.
20. P. Gamand, B. J. Minnis, et al., "Millimetre Wave Monolithic HEMT Oscillators Fabricated With a Standard 0.5μm Process," *1991 European Microwave Conf. Proc.*, pp. 167–171.
21. M. Healy and B. J. Minnis, "Wideband Power Combiners for GaAs MMICs," *1991 European Microwave Conf. Proc.*, pp. 1035–1040.
22. M. Healy, B. J. Minnis, R. F. Milsom, and K. J. Scott, "Application of FACET to MMIC Design," IEE colloquium, *Computer Based Tools for Microwave Engineers*, Digest No. 1991/152, October 1991, pp. 5/1–5/4.
23. M. Healy and B. J. Minnis, "Electromagnetic Layout Analysis Improves MMIC Design," *Microwave Engineering Europe*, February 1993, pp. 23–27.
24. B. J. Minnis and M. Healy, "2.5 W MMIC Power Amplifiers for X Band Phased Arrays," *JFIT 93 Conf. Digest*, March 1993, pp. 71–76.
25. B. J. Minnis and S. Gourrier, "Amplificateur MMIC 7-14 GHz Avec Puissance de Sortie 2.5 W Pour Application de Radar a Antenne Active," CNES conf. digest, *Seminaire Antennes Actives—MMIC*, April 1994, pp. 74–79.
26. B. J. Minnis, "The Travelling Wave Matching Technique for Cascadable MMIC Amplifiers," *IEEE Trans. MTT*, Vol. 42, No. 4, April 1994.
27. ———, "A High Gain, Multi-Stage Power Amplifier for X-Band Applications Exploiting the Travelling Wave Matching Technique," *Microwave Engineering Europe*, September 1994, pp. 53–62.
28. ———, "Compact Cascadable MMIC Amplifiers Using the Travelling Wave Matching Technique," CAE, Modelling and Measurement Verification Workshop, Wembley, London, October 1994, pp. 54–60.
29. ———, "A 2-stage, X-band MMIC Amplifier With 2.5 W Output Power," IEE Tutorial Workshop on MMICs, Digest No. 1994/216, November 1994, pp. 7/1–7/4.

Glossary

3-D	three-dimensional
ac	alternating current
BP	band-pass
CAD	computer-aided design
dc	direct current
FET	field-effect transistor
GaAs	gallium arsenide
HEMT	high-electron mobility transistors
HP	high-pass
LC	inductor/capacitor
LCR	inductor/capacitor/resistor
LO	local oscillator
LP	low-pass
MAG	maximum available gain
MESFET	metal semiconductor FET
MIC	microwave integrated circuit
MIL	minimum insertion loss
MIM	metal-insulator-metal
MMIC	monolithic microwave integrated circuit
nichrome	nickel and chromium
PC	personal computer
PCM	process control monitor
PML	Philips Microwave Limeil
PRL	Philips Research Laboratories
PTFE	polytetrafluoroethylene

QW	quarter wave
RF	radio frequency
TEM	transverse electromagnetic
T/R	transmit/receive
UE	unit elements

About the Author

Brian Minnis graduated in 1973 with a B.Sc. (with honors) in electronics from the University of Kent at Canterbury, which also awarded him a Ph.D. in 1994. He began his career at MEL in the United Kingdom as a microwave engineer working on the design of broadband microwave components for electronic warfare (EW) systems. After leaving MEL and after a short spell in teaching, he joined Marconi Space and Defence Systems to work first on wideband radomes for EW and subsequently on system design for satellite communications.

In 1978, Dr. Minnis joined Philips Research Laboratories, where he took responsibility for research on microwave filters. The practical success of several new filter structures can be attributed to the application of exact synthesis techniques. Since then, the techniques have been refined, generalized, and applied to many different types of microwave components and integrated circuits. Over the last eight years, he has been responsible for most of the microwave component activities at Philips Research Laboratories (U.K.). Recently he has been specializing in the design of several types of MMICs. These have included wideband low-noise amplifiers and power amplifiers for use in the 1- to 20-GHz range. His most recent work has been on multifunction single-chip downconverters for 35-GHz radar applications.

To date, Dr. Minnis has published 30 papers, most of which describe new microwave components or circuits designed by synthesis.

Index

The Artech House Microwave Library

Acoustic Charge Transport: Device Technology and Applications, R. Miller, C. Nothnick, and D. Bailey

Advanced Automated Smith Chart Software and User's Manual, Version 2.0, Leonard M. Schwab

Algorithms for Computer-Aided Design of Linear Microwave Circuits, Stanislaw Rosloniec

Analysis, Design, and Applications of Fin Lines, Bharathi Bhat and Shiban K. Koul

Analysis Methods for Electromagnetic Wave Problems, Eikichi Yamashita, editor

Automated Smith Chart Software and User's Manual, Leonard M. Schwab

C/NL2 for Windows: Linear and Nonlinear Microwave Circuit Analysis and Optimization, Software and User's Manual, Stephen A. Maas and Arthur Nichols

Capacitance, Inductance, and Crosstalk Analysis, Charles S. Walker

Computer-Aided Analysis, Modeling, and Design of Microwave Networks: The Wave Approach, Janusz A. Dobrowolski

Design of Impedance-Matching Networks for RF and Microwave Amplifiers, Pieter L. D. Abrie

Designing Microwave Circuits by Exact Synthesis, Brian J. Minnis

Dielectric Materials and Applications, Arthur von Hippel, editor

Dielectrics and Waves, Arthur von Hippel

Digital Microwave Receivers, James B. Tsui

Electric Filters, Martin Hasler and Jacques Neirynck

Electrical and Thermal Characterization of MESFETs, HEMTs, and HBTs, Robert Anholt

E-Plane Integrated Circuits, P. Bhartia and P. Pramanick, editors

Feedback Maximization, Boris J. Lurie

Filters with Helical and Folded Helical Resonators, Peter Vizmuller

Frequency Synthesizer Design Toolkit Software and User's Manual, Version 1.0, James A. Crawford

Fundamentals of Distributed Amplification, Thomas T. Y. Wong

GaAs MESFET Circuit Design, Robert A. Soares, editor

GASMAP: Gallium Arsenide Model Analysis Program, J. Michael Golio *et al.*

GSPICE for Windows, Sigcad Ltd.

Handbook of Microwave Integrated Circuits, Reinmut K. Hoffmann

Handbook for the Mechanical Tolerancing of Waveguide Components, W. B. W. Alison

HELENA: HEMT Electrical Properties and Noise Analysis Software and User's Manual, Henri Happy and Alain Cappy

HEMTs and HBTs: Devices, Fabrication, and Circuits, Fazal Ali, Aditya Gupta, and Inder Bahl, editors

High-Power Microwave Sources, Victor Granatstein and Igor Alexeff, edtiors

High-Power GaAs FET Amplifiers, John Walker, editor

High-Power Microwaves, James Benford and John Swegle

Introduction to Microwaves, Fred E. Gardiol

Introduction to Computer Methods for Microwave Circuit Analysis and Design, Janusz A. Dobrowolski

Introduction to the Uniform Geometrical Theory of Diffraction, D. A. McNamara, C. W. I. Pistorius and J. A. G. Malherbe

LINPAR for Windows: Matrix Parameters for Multiconductor Transmission Lines, Software and User's Manual, Antonije Djordjevic, Miodrag B. Bazdar, Tapan K. Sarkar, Roger F. Harrington

LOSLIN: Lossy Line Calculation Software and User's Manual, Fred E. Gardiol

Lossy Transmission Lines, Fred E. Gardiol

Low-Angle Microwave Propagation: Physics and Modeling, Adolf Giger

Low Phase Noise Microwave Oscillator Design, Robert G. Rogers

MATCHNET: Microwave Matching Networks Synthesis, Stephen V. Sussman-Fort

Matrix Parameters for Multiconductor Transmission Lines: Software and User's Manual, A. R. Djordjevic *et al.*

Microelectronic Reliability, Volume I: Reliability, Test, and Diagnostics, Edward B. Hakim, editor

Microelectronic Reliability, Volume II: Integrity Assessment and Assurance, Emiliano Pollino, editor

Microwave and Millimeter-Wave Diode Frequency Multipliers, Marek T. Faber, Jerzy Chamiec, Miroslaw E. Adamski

Microwave and RF Circuits: Analysis, Synthesis, and Design, Max Medley

Microwave and RF Component and Subsystem Manufacturing Technology, Heriot-Watt University

Microwave Circulator Design, Douglas K. Linkhart

Microwave Engineers' Handbook, Two Volumes, Theodore Saad, editor

Microwave Materials and Fabrication Techniques, Second Edition, Thomas S. Laverghetta

Microwave MESFETs and HEMTs, J. Michael Golio *et al.*

Microwave and Millimeter Wave Heterostructure Transistors and Applicatons, F. Ali, editor

Microwave and Millimeter Wave Phase Shifters, Volume I: Dielectric and Ferrite Phase Shifters, S. Koul and B. Bhat

Microwave and Millimeter Wave Phase Shifters, Volume II: Semiconductor and Delay Line Phase Shifters, S. Koul and B. Bhat

Microwave Mixers, Second Edition, Stephen Maas

Microwave Transmission Design Data, Theodore Moreno

Microwave Transition Design, Jamal S. Izadian and Shahin M. Izadian

Microwave Transmission Line Couplers, J. A. G. Malherbe

Microwave Tubes, A. S. Gilmour, Jr.

Microwaves: Industrial, Scientific, and Medical Applications, J. Thuery

Microwaves Made Simple: Principles and Applicatons, Stephen W. Cheung, Frederick H. Levien *et al.*

MMIC Design: GaAs FETs and HEMTs, Peter H. Ladbrooke

Modern GaAs Processing Techniques, Ralph Williams

Modern Microwave Measurements and Techniques, Thomas S. Laverghetta

Monolithic Microwave Integrated Circuits: Technology and Design, Ravender Goyal *et al.*

Nonuniform Line Microstrip Directional Couplers, Sener Uysal

PC Filter: Electronic Filter Design Software and User's Guide, Michael G. Ellis, Sr.

PLL: Linear Phase-Locked Loop Control Systems Analysis Software and User's Manual, Eric L. Unruh

RF Design Guide: Systems, Circuits, and Equations, Peter Vizmuller

Scattering Parameters of Microwave Networks with Multiconductor Transmission Lines: Software & User's Manual, A. R. Djordjevic *et al.*

Solid-State Microwave Power Oscillator Design, Eric Holzman and Ralston Robertson

Terrestrial Digital Microwave Communications, Ferdo Ivanek *et al.*

Time-Domain Response of Multiconductor Transmission Lines: Software and User's Manual, A. R. Djordjevic *et al.*

Transmission Line Design Handbook, Brian C. Waddell

TRAVIS Pro: Transmission Line Visualization Software and User's Manual, Professional Version, Robert G. Kaires and Barton T. Hickman

TRAVIS Student: Transmission Line Visualization Software and User's Manual, Student Version, Robert G. Kaires and Barton T. Hickman

Yield and Reliability in Microwave Circuit and System Design, Michael Meehan and John Purviance

For further information on these and other Artech House titles, contact:

Artech House
685 Canton Street
Norwood, MA 02062
617-769-9750
Fax: 617-769-6334
Telex: 951-659
e-mail: artech@world.std.com

Artech House
Portland House, Stag Place
London SW1E 5XA England
+44 (0) 171-973-8077
Fax: +44 (0) 171-630-0166
Telex: 951-659
e-mail: bookco@artech.demon.co.uk

www.ingramcontent.com/pod-product-compliance
Lightning Source LLC
LaVergne TN
LVHW101457200926
848483LV00001B/147
* 9 7 8 0 8 9 0 0 6 7 4 1 3 *